Biochemistry

Biochemistry

Edited by **Artie Weissberg**

SYRAWOOD
PUBLISHING HOUSE
New York

Published by Syrawood Publishing House,
750 Third Avenue, 9th Floor,
New York, NY 10017, USA
www.syrawoodpublishinghouse.com

Biochemistry
Edited by Artie Weissberg

International Standard Book Number: 978-1-68286-069-4 (Hardback)

Printed in the United States of America.

Contents

Preface

This comprehensive book provides an in-depth insight into the concepts and applications of biochemistry. While understanding the long-term perspectives of the topics, the book makes an effort in highlighting their impact as a modern tool for the growth of the discipline. Some of the significant concepts and topics encompassed in this book are bio-molecular recognition, genomic analysis, metabolic engineering and gene therapy. The major sections covered in this extensive book deal with the various applications of biochemistry within different fields like pharmaceuticals, bioengineering, etc. It will serve as a valuable source of reference for graduate and post graduate students.

This book is a comprehensive compilation of works of different researchers from varied parts of the world. It includes valuable experiences of the researchers with the sole objective of providing the readers (learners) with a proper knowledge of the concerned field. This book will be beneficial in evoking inspiration and enhancing the knowledge of the interested readers.

In the end, I would like to extend my heartiest thanks to the authors who worked with great determination on their chapters. I also appreciate the publisher's support in the course of the book. I would also like to deeply acknowledge my family who stood by me as a source of inspiration during the project.

Editor

A survey of avian malaria parasite in Kano State, Northern Nigeria

Karamba K. I*, Kawo A. H., Dabo N. T and Mukhtar M. D

Department of Biological Sciences, Bayero University, Kano, Nigeria.

As an attempt to keep abreast of the variety of avian *Plasmodium* parasite in Kano State, a total of 218 blood films were made from wild (116) and domesticated (102) bird species collected between January and July, 2009 period. The slides were examined for the presence of *Plasmodium* parasite (parasitaemia value). Birds examined were six *Columbidae livia* (pigeons), two *Cisticola cantans* (Singing cisticola), two *Crinifer piscato* (Western grey plantain eaters), two *Lamprotornis superbus* (Buffalo weavers), five *Stigmatopelia senegalensis* (Laughing doves), twenty-six *Ploceus cucullatus* (Black-headed weavers), forty-three *Amadina fasciata* (Cut-throat finches), five *Lamprotoni caudatus* (Long-tailed glossy starlings), twenty-four *Uraeginthus bengalus* (Cordon bleu finches), fifty poultry chickens, forty-six local chickens *(Gallus gallus)*, three *Nigrita Spp* (Negro finches) and four *Streptopchia decipiens* (African mourning doves). Results of the investigation showed that *Plasmodium circumflexicum*, *Haemoproteus columbae* and *Plasmodium gallinaecium* were present. The frequencies of occurrence in the birds' species were: 19.56% for local chickens, 50% for pigeons, 13.95% for Cut-throat finches, 50% for Grey plantain eaters, 33.3% for Negro finches and 0% for other birds. Overall, 6.89% of all the wild birds screened were infected as against 11.7% in domesticated birds. Domesticated birds had higher parasetaemia value (100 to 1000) cells per field than the wild birds (10 to 100) cells per field. However, the distribution of the parasites among the different species of the host birds was not statistically significant (P > 0.05). Chloroquine was found to be potent at 2.2 mg/ml concentration upon infected pigeons. A lineage of *H. columbae* named as type COLIV03 in MalAvi database was identified from the pigeons. This finding has thus called immediate massive screening of pigeons in Kano for this new variant of *Plasmodium* species with a view to elucidating its molecular and virulence nature as agent of avian malaria.

Key words: Kano, *Plasmodium*, pigeon, survey, malaria.

INTRODUCTION

Humans are not the only animals suffering from malaria resulting from *Plasmodium* infection (Jennings et al., 2006). Avian malaria is an arthropod-borne disease where protozoan blood parasites (*Plasmodium* species) are transmitted to birds by mosquitoes (Derraik and Maguire, 2005; Jennings et al., 2006). The causative agent of malaria in birds was discovered in 1885, only five years after it was first recognized in man (Carlton, 1938). Although, malaria parasites have been reported from many species of birds throughout the world, very little is known of their prevalence among the avifauna or the effects on the avian population (Carlton, 1938). In its broadest sense malaria, which originally meant "bad air", may today be interpreted as applying to the disease case used by the blood-inhabiting protozoa of the family *Plasmodium*. At least three genera of *Plasmodium* have been reported from birds: *Lecocytozoon, Heamoproteus* and *Plasmodium*. All these forms live within red blood cells. Carlton (1938) reported that mosquito transmission was first worked out with birds in India by Ross in 1898. Host susceptibility to avian malaria varies, and some widespread species, such as the *Passer domesticus* (exotic sparrow), *Turdus philomelos* (song thrush) and *T. merula* (black bird) may be asymptomatic carriers (Laird, 1950a, b). This research was carried out in Kano State. This is because no such study was ever carried out in the State and the need to explore the possibility of the infection in the area was long overdue. The study was therefore aimed at determining the presence or otherwise of *Plasmodium* parasites in birds, the assessment of the parasitaemia value and evaluating the *in-vivo* efficacy of chloroquine in the examined birds against the *Plasmodium* parasites.

MATERIALS AND METHODS

Description of the study area

The study area was the environs of Kano State. The State lies between latitude $13^{o}N$ in the North and $11^{o}N$ in the South and longitude $8^{o}E$ in the West and $10^{o}E$ in the East. The state capital is located on latitude $12^{o}N$ and longitude $8.3^{o}E$. It is within the semi-arid Sudan savannah zone of West Africa about 840 km from the edge of the Sahara desert. It has a mean height of about 472.45m above sea level and has two seasonal periods categorized on the basis of moisture as dry and rainy seasons. The temperature of Kano usually ranges between a maximum of $33^{o}C$ and a minimum of $15.85^{o}C$ although, sometimes during the harmattan it falls down to as low as $10^{o}C$. The average rainfall is between 63.3 ± 48.2mm in May and 133.4 ± 59 in August the wettest month. The soil in most part of Kano State is light or moderately leached, yellowish brown and sandy just like most other savanna parts of Northern Nigeria. The soil fertility supported agriculture. The natural vegetation as earlier mentioned is semi-arid Sudan savanna, which is sandwiched by the Sahel savanna in the north and the Guinea savanna in the south. The savanna has been described as the zone that provides opportunity for optimal human attainment. This is because it is rich in faunal and floral resources, and as such suitable for both cereal agriculture and livestock rearing (Shekarau, 2010).

Sample collection sites

Eight local government areas (LGAs) in Kano State were chosen for this study in order to diversify the sampling area. They included Nassarawa, Kano Municipal, Wudil, Bebeji, Doguwa and Gwale. Amongst the areas chosen were places where there are water bodies, a lot of trees and animals, which provide conducive breeding spaces for both wild birds and mosquitoes. Such areas included the Botanic garden in Bayero University, Kano (Gwale LGA), Zoological garden (Kano Municipal LGA), Tiga (Bebeji LGA) and Falgore forest reserve (Doguwa LGA). Market places are chosen in order to have bird's representation of locally kept birds such as pigeons and chickens. Poultry farm was taken into consideration in the study so that the poultry birds kept within the State would have a good representation too. No specificity was given to particular bird specie but rather any bird was observed in the study to get a general over view of the infection in all birds' species.

Collection and handling of birds species used for the study

The wild birds were captured using birds trap and the domesticated birds were with hand.

The wild birds were collected from the following areas:

1. Botanical Garden of Biological Science Department, Bayero University, Kano.
2. Falgore Forest along Jos Road, Kano – Doguwa LGA.
3. Zoological Garden, Kano State – Kano Municipal.
4. Getso Town – Gwarzo LGA.
5. Tiga Dam - Bebeji Local Government.

The domesticated birds were collected at:

1. Farida Farms, Hotoro GRA - Poultry Chickens – Nassarawa LGA
2. Wudil Local Market – Local Chickens – Mudil LGA
3. Sharada Local Market – Piegions – Kano Municipal

The birds were handled carefully in cages of appropriate size and were fed adequately. The birds were identified by Suleiman Abubakar Fagge; Director of Wild Life Services, Zoological Garden, Kano and the parasites were confirmed by Veterinary Dr. Yahuza Aliyu of the Epidemiology unit, Kano State Ministry of Agriculture and Natural Resources.

Collection of blood from the study birds

A total of 218 blood smears of different species of birds were collected from different areas in Kano State. They include 102 domestic and 116 wild birds. The domestic birds were divided into 50 hybrid (exotic) chickens, 46 local chickens and 6 pigeons. Blood samples were taken from brachial vein of the avian hosts using the procedure described by Rukhsana (2005) and Chesebrough (2000). The brachial vein area was cleansed with Swab moistened with 70% v/v alcohol, and it was allowed to dry. Using a sterile lancet, the area was pricked and squeezed gently to obtain a large drop of blood. The blood was collected using heparinized capillary tube. Using a microscopic slide, a small drop of blood was dropped at the center of the slide and a larger drop on another microscopic slide. The thin film was spread using a smooth edge slide spreader (slide). Without delay, the large drop was spread, to make a thick smear. The area of about 15 × 15 mm was covered. It was mixed evenly to avoid red cells forming marked rouleaux during staining. Using a grease pencil the slide was labeled with number for identification. The blood was left to air dry with the slide in a horizontal position and it was placed in a separate box covered with a lid to protect it from insects and dust and it was kept in a warm sunny place in order for the film to dry quickly it was then removed from the sun immediately it was dried (Chesebrough, 2000).

Preparing thick and thin blood films

The thick and thin blood films were made on different slides that have frosted ends for easy labeling (Chesebrough, 2000).

Standardizing the amount of blood used and area covered by the thick film

To ensure good staining, standardization and reproducible results, the amount of blood used particularly in thick film was kept as constant as possible and the blood was spread evenly over the specified area of the slide (Chesebrough, 2000).

Staining of slides

Using absolute methanol (methyl alcohol), the blood films were fixed by spreading on the microscopic slide (Chesebrough, 2000).

Table 1. Varieties of birds screened for *Plasmodium* species from some parts of Kano.

Birds species	EXC	LC	PG	WGPE	AMD	LD	LTGS	BFW	CTF	CB	NF	SC	BHW
Location													
Botanical garden BUK				1	3	2	2	2					
Falgore forest				1		1	2			24			
Kano Zoo garden					1	2			19		3		
Getso town									24			2	
Tiga town													26
Wudil LGA		46											
Nassarawa LGA	50												
Kano Municipal			6										
Total	50	46	6	2	4	5	4	2	43	24	3	2	26

EXC – exotic chickens; LC- local chickens; PG – pigeons; WGPE – Western Grey plantain eater; AMD – African mourning dove; LD – laughing dove; LTGS – long tailed glossy starling; BFW – Buffalo weaver; CTF – cut throat finches; CB – Cordon Bleu; NG – Negro Finches; SC – Singing Cisticola; BHW – black headed weaver.

The slides were placed horizontally on a staining rack. A small drop of absolute methanol was added to the thin film. The absolute methanol was not used on the thick film in order to prevent lysis of the red cells and make the thick film unreadable. Malaria parasite in thick and thin blood films were stained at the pH 7.1 - 7.2 using Giemsa stain (Chesebrough, 2000). Immediately before use, Giemsa stain was diluted at 10% solution for 10 minutes. 45 ml of buffered water, pH 7.1 to 7.2 was measured in a 50 ml cylinder and 5 ml of Giemsa stain was added to 50 ml mark then it was mixed gently. The slides were placed on a staining rack. It was immersed in a staining trough. Thick blood films were allowed to dry thoroughly while the thin blood films were fixed using methanol for 2 min. The diluted stains were poured on the staining trough for 10% solution for 10 min. The stain from the staining container and slide were washed with clean water to avoid the films being covered with a fine deposit stain. The back of each slide was cleaned and placed on a draining rack to air dry.

Microscopic estimation of parasitaemia

The blood films were examined microscopically using the 100 oil-immersion objectives. After the thick film was completely air dried, a drop of oil immersion was applied to the area of the film that appears mauve colored. The oil was spread to cover the area. This is to enable the film to be examined first at low magnification. An area that is not too thick is selected and it was examined under the ×100 oil-immersion objective. The parasites and pigments were examined. The parasites were identified and approximate number of parasites; trophozoites of plasmodium parasite, was reported (Cheesbrough, 2000). When the thin film was completely dried, a drop of oil immersion was dropped to the lower third of the film. It was examined under 40 objectives to check the staining, morphology and distribution of the cells and to detect malaria schizonts, gametocytes and trophozoites (Chesebrough, 2000).

Parasite identification using PCR technique

The specimen was placed on a filter paper and was kept in the laboratory to dry. It was covered with wire gauze to avoid tampering by flies. The specimen was forwarded to Professor Staffan Bensch of Lund University Sweden for molecular typing (Wiersch et al., 2005).

Statistical analysis of the data

Data for number and variety of birds recruited for the study were presented in frequency distribution table as percentages and means. Variation in occurrence of infection and its rates between wild birds and domesticated ones are tested by Chi-Square tool at 5% level of probability (Mukhtar, 2005).

RESULTS

Variety of birds collected and screened for Plasmodium species

Table 1 shows the variety of birds screened. It includes the following: Pigeons (2.75%), Western grey plantain eaters (0.91%), African mourning dove (1.83%), Laughing dove (2.29%), Long-tailed glossy starling (2.29%), Buffalo weaver (0.91%), Cordon bleu (11.01%), Cut-throat finches (19.72%), Singing cisticola (0.91%), Black-headed weaver (11.93%), Local and exotic chickens (21.10%) as well as Negro finches (1.38%).

Incidence of malaria parasite in the variety of birds screened

Tables 2 to 5 show the incidence of avian plasmodia in wild varieties of birds caught around Kano State. Cut-throat finches were found at Getso rural area of Gwarzo LGA. Eighty percent of the five catch were infected with *Plasmodium circumflexicum* at a parasitaemia value of 1 to 10 per microscopic field of their blood. The remaining 20% showed parasitic load of 10 to 100 per microscopic field. At Falgore forest reserve in Tudun Wada LGA, Grey plantain eater was identified to possess a *Plasmodium* species whose identification was terminated at the genus level. Around Kano municipal local government area (precisely in Zoological garden), Negro finch and Cut-throat

Table 2. Density and suspected species of *Plasmodium* found in wild birds.

S/No.	Bird species	Site	Slide Code	Suspected *plasmodium* species	Degree of parasitaemia
1.	Cut throat finch	Getso	Getso 13	*Plasmodium circumflexicum*	+
2	Cut throat finch	Getso	Getso 13	*Plasmodium circumflexicum*	+
3	Cut throat Finch	Getso	Getso 10	*Plasmodium circumflexicum*	+
4	Cut throat finch	Getso	Getso 9	*Plasmodium circumflexicum*	+
5	Cut throat finch	Getso	Getso3	*Plasmodium circumflexicum*	++
6	Grey plantain eater	Falgore	Falgore A	*Plasmodium Spp.*	+
7	Negro finch	KMC	Zoo 10	*Plasmodium circumflexicum*	+
8	Cut throat finch	KMC	Zoo 6	*Plasmodium circumflexicum*	+

+ = 1-10 per 100 high power field; ++ = 11-100 per 100 high power field.

Table 3. Density and suspected species of *Plasmodium* in domesticated birds.

S/No.	Bird species	Site	Slide code	Suspected *plasmodium* species	Degree of parasitaemia
1	Pigeon	KMC	P1	*Haemoproteus columbae*	+
2	Pigeon	KMC	P2	*Haemoproteus columbae*	++
3	Pigeon	KMC	P3	*Haemoproteus columbae*	++
4	Local chicken	Wudil	LDP	*Plasmodium gallinaecium*	+++
5	Local chicken	Wudil	LDO	*Plasmodium gallinaecium*	++
6	Local chicken	Wudil	LDN	*Plasmodium gallinaecium*	+++
7	Local chicken	Wudil	LDM	*Plasmodium gallinaecium*	+
8	Local chicken	Wudil	LDS	*Plasmodium gallinaecium*	+++
9	Local chicken	Wudil	LDQ	*Plasmodium gallinaecium*	+++
10	Local chicken	Wudil	LDA	*Plasmodium gallinaecium*	+
11	Local chicken	Wudil	LDC	*Plasmodium gallinaecium*	+++
12	Local chicken	Wudil	LDE	*Plasmodium gallinaecium*	++

+ = 1 to 10 per 100 high power fields; ++ = 11 to 100 per 100 high power fields; +++ = 1 to 10 in very high power field; ++++ = More than ten in very high power field.

Table 4. Occurrence of *Plasmodium* parasites in domesticated birds in Kano.

S/No.	Bird type	Locality	No. examined	No. infected	Frequency (%)
1	Local chickens	Wudil LGA	46	9	19.56
2	Poultry chickens	Nassarawa LGA	50	NPF	0.0
3	Pigeons	KMC	6	3	50.0
	Total		102	12	12.24

NPF = No parasite found.

finch were found to be infected with *P. circumflexicum* at 1 to 10 parasitic densities per microscopic field. The varieties of domesticated birds showing the infection with the parasites were pigeons 25% and local chickens 75% found from Kano municipal and Wudil LGAs respectively (Table 3). Pigeons were heavily infected with *H. columbae* with a density of 10 to 1000 per microscopic field. The infected local chickens showed also 10 to 1000 parasites per microscopic field and the best identified were *P. gallenaecium*. Based on the PCR analysis (Table 6), one new lineage of *H. columbae* namely COLIV03

was identified at Lund University of Sweden by Prof. Staffan Bensch. In terms of frequency of occurrence (Table 4), local chickens had 19.56% as obtained from Wudil. Exotic poultry chickens from Nassarawa LGA had zero percent incidence of infection.

DISCUSSION

Among the 116 wild species of birds observed it was found that only eight wild birds from three different

Table 5. Occurrence of *Plasmodium* parasites found in wild birds in Kano.

S/No.	Bird species	Locality	Number examined	Number infected	Percentage frequency
1	Cut throat finches	Getso, Zoo	43	6	13.95
2	Grey plantain eater	Falgore, BUK	2	1	50.0
3	Negro finches	Zoo	3	1	33.33
4	Laughing dove	BUK, Falgore	5	0	0.0
5	African mourning dove	BUK, Zoo	4	0	0
6	Long tailed glossy starling	BUK, Falgore, Zoo	5	0	0.0
7	Buffalo weaver	BUK	2	0	0.0
8	Singing cisticola	Getso	2	0	0.0
9	Cordon blue finches	Falgore, BUK	24	0	0.0
10	Black headed weaver	Tiga	26	0	0.0
Total			116	8	9.28

Table 6. Results of *Plasmodium* species based on PCR method analysis.

S/No.	Bird speci	Locality	Filter paper code	Result	Plasmodium specie	Remarks
1.	Pigeon	Tarauni market	PE	Positive	*Haemoproteus columbae*	Identified as Existing Lineage in the Mal Avi Data Base HAECOL1
2.	Pigeon	Tarauni market	PF	Positive	*Haemoproteus columbae*	Identified as new Lineage in the Mal Avi Data Base. It is termed COLIV03
3.	Pigeon	Tarauni market	PD	Negative	-	
4.	Local chicken	Tarauni market	CC	Negative	-	
5.	Local chicken	Tarauni market	CD	Negative	-	
6.	Local chicken	Tarauni market	CE	Negative	-	
7.	Cut throat finch	Zoological garden	W3	Negative	-	
8.	Cut throat finch	Zoological garden	W4	Negative	-	
9.	Cut throat finch	Zoological garden	W10	Negative	-	

species were actually infected with the *Plasmodium* species. The dominant bird specie infected was the Cut-throat finches (*Amadina fasciata*). About seven were infected and six out of the seven were sampled from Getso town of Gwarzo LGA. Only one was sampled from Zoological garden in Kumbotso LGA indicating its scarcity in the urban settlement. In the domesticated birds, pigeons and local chickens were the most infected birds' species (Figure 1). This may be as a result of a prolonged evolutionary association of the host with the parasite. Furthermore, the domestication and lack of proper care by the keepers of the birds might have exposed them to the parasite in spite of the fact that the birds are known to have high disease resistance (Wikipaedia, 2008; Valkiunas et al., 2009b). Less care was given to them and this might have made them prone to infection by plasmodiasis of birds. Another factor could be lack of proper education or enlightenment on how to keep the poultry chickens on the part of the local keepers (Baker, 1976).

The goal of this study was the identification of plasmodium parasites in birds using microscopy. The parasites were seen and the level of infection evaluated in the wild and the domesticated birds. Identification of the genus *Plasmodium* was done microscopically by the use of Giemsa staining technique. For more elaborate study to identify the species of the *Plasmodium* found in different types of birds' species a more effective method must be used which is called molecular typing. Molecular Typing using Polymerase Chain Reaction Mechanism, which amplifies DNA and the particular specie is identified as noted by Valkiunas et al. (2005, 2009a). Because of high sensitivity PCR may detect very small numbers of sporozoites in birds. Standard microscopy protocols are insufficiently sensitive to detect such light parasetaemias of both sporozoites and gametophytes; this could explain some discrepancies between levels of infection prevalence recorded using these two methods in parallel (Valkiunas et al., 2005).

For the exotic chickens, no single one was found to be infected with the parasite. This might be as a result of good care attributed to the standards (their housing, feeding, medication and appropriate knowledge in keeping the birds) provided in keeping the birds. More so,

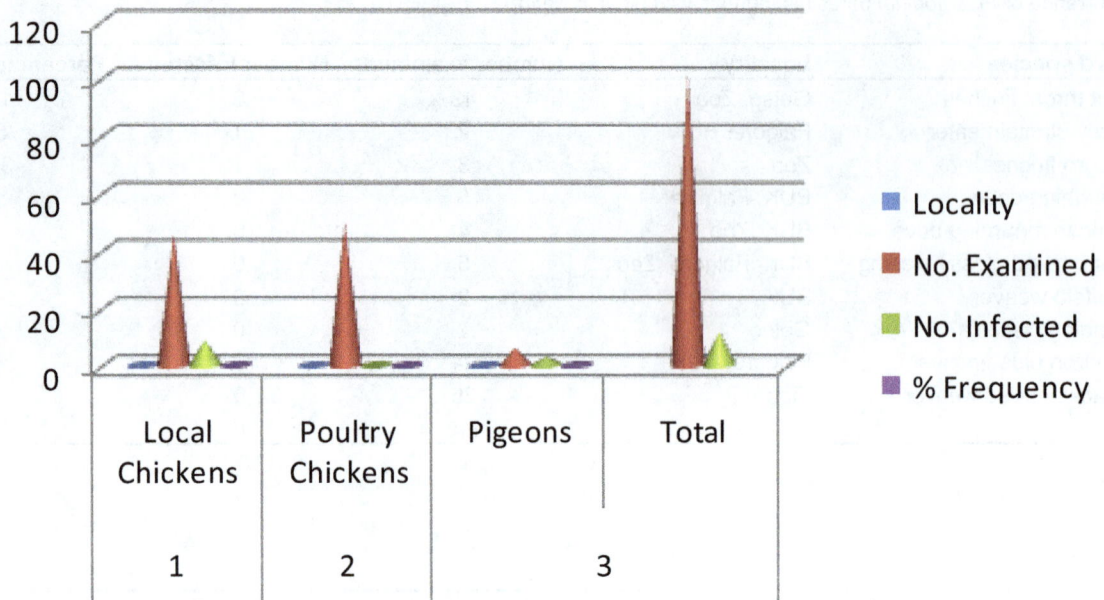

Figure 1. Domesticated birds examined against the infected birds.

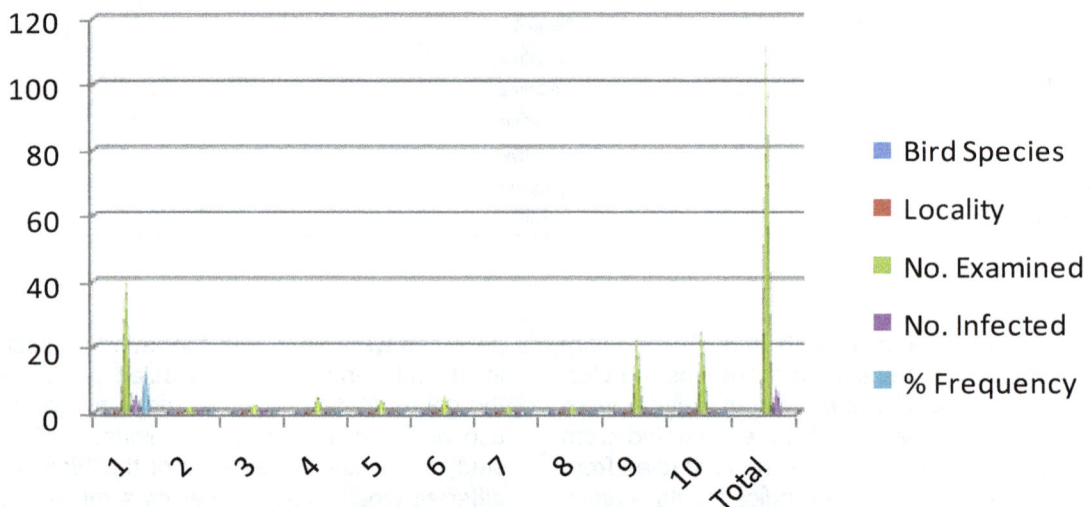

Figure 2. Wild birds examined against the infected birds.

the birds are known to have a very poor resistance for infection and are mostly kept for commercial purposes so proper care must be given to them to prevent break out of any epidemic or infection. As a result, no single bird was found to be infected with *Plasmodium*. The species of *Plasmodium* parasite were identified in both the domesticated and wild birds and with no statistically significant difference (P > 0.05). *Plasmodium gallinaecium* was found in the local chickens *G. gallus*, *H. columbae* was found in pigeons and *P. circumflexicum* was found in finches (Figure 2).

The parasitaemia value of the *Plasmodium* parasites in birds from the research conducted shows that Cut Throat Finches (*A. fasciata*) are more infected with the plasmodium parasite with mostly plus one degree of parasitaemia, that is, 1 to 10 parasites per 100 high power fields among the wild species of birds examined. While the domesticated birds pigeons and chickens were found to be infected more heavily than even the wild birds with degree of parasetaemia of plus two and three that is, 11 to 100 per 100 high-power field (Tables 2 and 3). The level of infection between the wild and domesticated species shows that there is significant difference (P < 0.05) in the parasitaemia value since most of the wild birds were in plus one degrees of infection and the domesticated are in the range of plus two and plus three parasitaemia value within the study area.

A possible new lineage identified in the present investigation

Results obtained from the molecular typing indicated a new variant of *Plasmodium* was found on certain blood samples of *Columba livia*. The results obtained on certain blood samples of *C. livia* (Pigeons) found new variant of *Plasmodium*, unknown to parasitologists before. The sample coded PE indicated perfect match to HAECOL1 (as in the MalAvi database) previously found in *C. livia* (Dranzoa et al., 1999) and identified as morphospecies *H. columbae* but sample PF was found to be a new lineage but only slightly different (4 changes) from HAECOL1. Most likely is a member of the morphospecies *H. columbae* and the lineage registered as COLIV03 in the MalAvi database.

Conclusion

The study confirms the presence of *Plasmodium* species among a variety of wild and domesticated avian species around Kano State. Infection rate was significantly higher in domesticated than the wild birds. The parasitaemia value was significantly (P < 0.05) higher in the local chickens than the wild birds and exotic chickens that had no parasitaemia. A new lineage of *Haemoproteus columbae* named as type COLIV03 in MalAvi data base was isolated from pigeons. This finding has thus stressed the need for active and emergent investigation for the new variant especially among the pigeons in Kano State.

RECOMMENDATION

Though bird's malaria is not well known in Nigeria, it is a well known disease of birds in many part of the world, and it has been a case study for decades in existing literatures. It is recommended that more research be carried out on this bird's disease in order to control it most especially in domesticated ones. This will prevent epidemic of the disease. In addition, obtaining a good data analysis on different States of Nigeria on bird's malaria could help in proper prevention of the disease.

ACKNOWLEDGEMENTS

Authors are grateful to Professor Staffan Bensch of Department of Animal Ecology, Lund University, Sweden for his Biotechnological Assistance especially with regards to identification of the parasites. The assistance rendered by Malam Nura Usman and Malam Isa A. Isa of School of Technology, Kano during the field and laboratory works is highly acknowledged.

REFERENCES

Baker JR (1976). Biology of the trypanosomes of birds. *In*: Biology of the Kinetoplastida. Volume 1. WHR Lumsden and DA Evans edition. Academic press, London, pp. 131-174.

Carlton MH (1938). Bird malaria and mosquito control. J. Protozool., 37: 25-31.

Cheesbrough M (2000). Protozoology. District laboratory practice in tropical countries. Low-price edition, UK. 1: 134-140.

Dranzoa C, Ocaido M, Katete P (1999). The ecto-gastrointestinal and haemo-parasites of life pigeons (Columba livia) in Kampala, Uganda. Avian Pathol., 28: 119-124.

Derraik JGB, Maguire T (2005). Mosquito-borne diseases in New Zealand: Has there ever been an indigenously-acquired infection? New Zealand Med. J., 118: 1670.

Jennings L, Julie W, Bruce EL (2006). Avian malaria. J. Vet. Clin. Pathol., 6: 1-4.

Laird M (1950a). Some blood parasites of New Zealand birds. J. Zool., 5: 1-20.

Laird M (1950b). Clinical episodes of *Plasmodium falciparum* malaria. J. Nat., 434(7030): 214-217.

Mukhtar FB (2005). An introduction to biostatistics. Spectrum Books Limited, Kano, Nigeria, pp. 92-104.

Rukhsana T (2005). Infectious Haematozoan parasites found in birds of Pakistan. Pak. J. Biol. Sci., 8(1): 1-5.

Shekarau MI (2010). Kano State Government of Nigeria. Environmental development, pp. 1-4.

Valkiunas G, Anthony C, Claire L, Tatjana I, Thomas BS, Ravinder NM (2005). Furtherobservations on the blood parasites of birds in Uganda. J. Wild Life Dis., 41(3): 580.

Valkiunas G, Anthony C, Claire L, Tatjana I, Thomas BS, Ravinder NM (2009a). Nested cytochrome B polymerase chain reaction diagnostics detect sporozoites of Haemosporidian parasites in peripheral blood of naturally-infected birds. J. Parasitol., 95(6): 1514.

Valkiunas G, Anthony C, Claire L, Staffan B, Asta K, Casmir VB (2009b). *Plasmodium relictum* (Lineage P-SGS1): Further observations of the effects on experimentally-infected Passeriform birds with remarks on treatment with malrone. J. Exp. Parasitol., 123(6): 134-135.

Wiersch SC, Maier WA, Kampen H (2005). *Haemamoeba cathemerium* gene sequences for phylogenetic analysis of malaria parasite. J. Parasitol., 96(2): 90-94.

Wikipedia E (2008). *Plasmodium* species infecting birds, pp. 2-8.

2

Effect of feeding malted foods on the nutritional status of pregnant women, lactating women and preschool children in Lepakshi Mandal of Ananthapur district, Andhra Pradesh, India

Vijaya Khader* and K. Uma Maheswari

Department of Foods and Nutrition, Post Graduate and Research Centre, Acharya N. G. Ranga Agricultural University, Rajendranagar, Hyderabad-500 030, India.

Information of preschool children (400), pregnant women (100) and lactating women (100) was collected. Anthropometric, hemoglobin, clinical and morbidity assessments were carried out before & after supplementation. Amylase Rich Malted Mixes (ARMMs) 2 types (Ragi/Wheat) were formulated and suitable products namely *laddu, roti, kheer* and *porridge* were prepared using formulated malted mixes. Malting decreased grain length, width, kernel weight (0.45 to 19.0g), volume (0.50 –31.2 ml) and hardness (1.12 to 5.9 kg/cm^2), thus reduced the bulk density of the malted mixes. Chemical composition revealed that, the significant increase (P<0.05) in fat (2.27 g), carbohydrate (98.0 g) and calorie (396 kcal) content of wheat malted mix. However significant increase was observed in calcium (440 mg), thiamine (0.7 mg) and riboflavin (0.9 mg) content of ragi malted mix. Germinated greengram had significantly higher protein (33.0 g), fibre (11.5 g), iron (8.0 g) and vitamin C (157.8 mg) content. The selected preschool children, pregnant women & lactating women were divided into 3 groups. Group II and III fed with ragi malted mix & wheat malted mix respectively served as the experimental groups and remaining group I served as the control group. Significant increase was observed in weight of preschool children and lactating women after supplementation. Hemoglobin level in pregnant and lactating women significantly increased (P<0.01) after supplementation. Considerable reduction (50%) in the incidence of PEM, vitamin A, B vitamins, vitamin C and iron deficiency symptoms in experimental groups. After supplementation, morbidity rate decreased to 50% both the Group II and III.

Key words: Malted foods, chemical composition, physical parameter.

INTRODUCTION

Malting is the controlled germination followed by controlled drying of the kernels. The main objective of malting is to promote the development of hydrolytic enzymes, which are not present in non-germinated grain (Dewar et al., 1997). Other benefits of the malting process include increased vitamin C content, phosphorus availability, and synthesis of lysine and tryptophan, calcium content (Dulby and Tsai, 1976 and (Sangita and Sarita, 2000). Furthermore, amylases are elaborated and as a result the viscosity of gelled starch decreases (Brandtzaeg et al., 1981). Malting also includes the inhibition of growth of pathogens through the fermentation process.

Amylase rich food (ARF) is germinated cereal flour, which is extremely rich in the enzyme alpha-amylase. The alpha-amylase cleaves the long carbohydrate chains in the cereal flour into shorter dextrins. It modifies the starch content of the cereals so that they do not thicken and would therefore not require dilutions resulting in enhanced digestibility (Inyang and Idoko, 2006). This remarkable property makes it possible to offer a low viscosity yet high energy dense preparation. Therefore, a study was conducted to assess the effect of feeding Amylase Rich Malted Mixes (ARMMs) on the nutritional status of vulnerable groups of population, that is, preschool children, pregnant and lactating women in backward and remote villages of a Lepakshi mandal, Ananthapur district, Andhra Pradesh, India.

MATERIALS AND METHODS

In Ananthapur district 3 mandals namely Hindupur, Parigi and Lepakshi were identified for the study after discussing with local officials. Baseline data was collected in the identified 14 villages from 3 mandals at Ananthapur district using a pretested, structured schedule. After collecting baseline information of the identified villages of 3 mandals, Lepakashi mandal was selected for implementing the study based on the availability of more subjects of low socio economic group with majority of families belonging to schedule caste and schedule tribes. ARMMs of two types (ragi/wheat) were formulated.

Raw ingredients, that is, whole ragi, wheat and greengram were subjected to physical parameters using standard procedures (Kumar et al., 1991) for colour, texture, length, width, kernel weight, kernel volume and hardness in native and malted grains. For the assessment of colour, texture, length and width, an average of 10 whole grains was taken. Proximate composition (moisture, protein, fat, fibre and ash) of whole grains, germinated grains and ARMMs were analyzed (AOAC, 1990). Vitamin C was assessed by colorimetric method (Association of vitamin chemist Inc, 1947). Thiamine was assessed by modified thiochrome method (AOAC, 1990), Riboflavin was assessed by Florescence technique (AOAC, 1990). Minerals such as iron and calcium were determined by atomic absorption spectrophotometer method (AOAC, 1990). Amylase activity was estimated by the standard procedure given by Bernfield (1955). Carbohydrate content was calculated by difference and energy values were computed.

Ragi malted mix was distributed in Siriravam, Gopindevarapally, Tirumaladevarapally and Sadlapally villages, whereas Wheat Malted Mix was distributed in Manempally, Venkatapuram, Gourganpally and Pulamathi Villages.

Individual information of preschool children (400), pregnant women (100) and lactating women (100) was collected on anthropometric, hemoglobin, clinical and morbidity assessments before and after supplementation for a period of 3 months. The purpose of the supplementation was to evaluate the effect of two types of malted mixes (ragi/wheat) on the growth and nutritional status. The selected preschool children, pregnant women and lactating women were divided into 3 groups. Group II fed with ragi malted mix and Group III fed with wheat malted mix served as the experimental groups and remaining group I served as the control group. Permission was obtained from the Director, ICDS project to feed the beneficiaries with ARMMs for a period of 3 months in place of regular RTE food supplied by ICDS.

Anthropometric measurements, which include height and weight of preschool children, pregnant women and lactating women and mid upper arm circumference of preschool children were measured. Clinical observation and morbidity pattern was collected by structured questionnaire developed by Seth et al. (1979). Hemoglobin level of preschool children, pregnant women and lactating women were assessed by Cyanometh hemoglobin method (Dacie and Lewis, 1991). The preschool children were selected based on the degree of malnutrition (<60% to below 90%) as per Gomez classification (Jelliffe, 1966) and pregnant and lactating women were selected with BMI <16 to <20 based on BMI classification (WHO Expert Consultation, 2004).

Clinical symptoms (nutritional deficiency symptoms) of the selected subjects were recorded before and after supplementation. Nutritional deficiency symptoms such as protein energy malnutrition (PEM), Vitamin 'A' deficiency, Vitamin 'D' deficiency, 'B' complex vitamins deficiencies, Vitamin 'C' deficiency, iodine deficiency and iron deficiency was observed.

RESULTS AND DISCUSSION

Physical parameters of native and malted grains

Native and malted grains (ragi, wheat and greengram) were assessed for the physical parameters. The results are given in Table 1. Malting resulted in decrease of 1000 kernel weight by 0.45 to 19.0 g and the volume by 0.50 to 31.2 ml. A noticeable level of decrease in grain hardness was also observed in all grains after malting because the malts of the grains had shrunken appearance and their hardness was considerably lower (1.1 to 5.9 kg/cm^2) than the native grains. Malting helped to reduce the bulk density of the malted mixes. Similar observations in weight, volume and hardness of native and malted grains were reported by Suhasini et al. (1995) and Sangita et al. (2000).

Chemical composition of native grains, germinated grains and malted mixes

The chemical compositions of native grains, germinated grains and malted mixes are given in Table 2. It was observed that wheat malted mix has significantly higher (p < 0.00) content of fat (2.27 g), carbohydrate (98.0 g) and calories (396 k.cal). Whereas, ragi malted mix has significantly higher (p < 0.05) content of calcium (440 mg), thiamine (0.7 mg), riboflavin (0.9 mg) and amylase activity (169 mg%) when compared to native and germinated grains. Aisien (1982) and Glennie (1984) reported significant decrease in fibre on malting due to cell wall degradation during sprouting process. The results in the present study can be comparable with the results reported by Malleshi et al. (1981). However, germinated greengram had significantly higher (p < 0.05) content of protein (33.0 g), fibre (11.5 g), iron (8.0 g) and vitamin C (157.8 mg). Thus, the incorporation of malted greengram mix and skimmed milk powder to the wheat malted mix and ragi malted mix helped to improve nutritional status of vulnerable segments of population with regard to protein, energy, iron, calcium and 'B' complex vitamins status.

Impact of supplementary feeding on the nutritional status of preschool children

Anthropometric assessment

Measurements of height, weight and midupper arm

Table 1. Physical parameters of native and malted grains (ragi, wheat and greengram).

S/N	Parameter	Ragi	Wheat	Greengram
1	Colour	Creamish White	Brown	Green
2	Texture	Hard	Hard	Hard
	Length (cm)			
3	Native	0.65	-	0.55
	Malted	0.75	-	0.5
	Width (mm)			
4	Native	0.35	-	0.45
	Malted	0.34	-	0.4
	1000 – Kernel wt (g)			
5	a) Native	52.00	2.87	44.0
	b) Malted	47.90	2.42	25.0
	1000 – Kernel volume (ml)			
6	Native	35.50	2.80	64.5
	Malted	33.70	2.30	33.3
	Hardness (kg/cm^2)			
7	Native	13.10	1.90	7.6
	Malted	7.20	0.80	3.5

Table 2. Nutrient composition of native, germinated grains and malted mixes (per 100 g).

S/N	Parameter	Native flour			Germinated grains			Malted mixes	
		Ragi	Wheat	Green gram	Ragi	Wheat	Green gram	Ragi malted mix	Wheat malted mix
1	Moisture (g)	11.9	11.5	9.4	17.0	43.5	90.10*	8.0	7.6
2	Protein (g)	7.8	12.25	23.75	10.3	14.0	33.0*	15.3	19.0
3	Fat (g)	1.5	1.2	1.6	1.3	2.0	2.1	1.8	2.27*
4	Fibre (g)	3.3	2.0	4.0	1.8	10.5	11.5*	2.0	8.0
5	Ash (g)	2.6	2.4	3.6	2.0	2.0	4.5*	2.9	1.8
6	Carbohydrates (g)[b]	72.6	71.0	57.0	75.0	82.1	60.8	94.51	98.0*
7	Energy (k.cal)[c]	336	348	228.0	390.0	403.0	380.0	376	396*
8	Calcium (mg)	350	30.0	320.0	242.0	54.0	109.0	440*	193.2
9	Iron (mg)	3.9	3.5	4.0	7.7	5.0	8.0*	4.4	5.9
10	Vitamin C (mg)	-	-	-	8.7	5.0	157.8*	25	14.1
11	Thiamine (mg)	0.19	0.17	0.28	-	0.9	0.37	0.7*	0.4
12	Riboflavine (mg)	0.42	0.45	0.43	-	1.23	0.31	0.9*	0.5
13	Amylase activity [@]	10.0	6.0	3.0	169*	145	132	117	102

@, mg maltose released by 1 g of malt flour when acted on 1 ml of 1% starch at 37°C for 30 min. *, Significant at 5% level; [b], calculated; [c] Computed.

circumference are commonly recognized as important indices of protein – energy malnutrition. The basic principle of anthropometry is that prolonged or severe nutrient depletion eventually leads to retardation of linear (skeletal) growth in children and to loss of, or failure to accumulate, muscle mass and fat in both children and

Table 3. Mean height of preschool children.

Group	Supplementation period (3 months)			
	Before	After	Increase in height (cm)	't' value
Control I	83.33	83.64	0.49	9.74**
Group II	77.97	84.18	4.24	78.19**
Group III	79.82	83.64	3.81	85.19**
NCHS standards	109.9	-	-	-
'F'value	-	0.153^{NS}	-	-
Level of significance between experimental groups	-	0.153^{NS}	0.43	

n = 400; NS, not significant; **, Significant at 1% level.

Table 4. Mean weight of preschool children.

Group	Supplementation period (3 months)			
	Before	After	Increase in weight (kg)	' t' value
Control I	11.54	12.14	0.6	16.63**
Group II	10.61	12.35	1.72	193.63**
Group III	10.74	12.49	1.74	85.19**
NCHS standards	18.7	-	-	
'F value	-	148.98**	-	
Level of significant between experimental groups	-	0.21^{NS}	0.02	

n = 400; NS, not significant; **, significant at 1% level.

adults. These problems can be detected by measuring body dimensions, such as standing height or upper-arm circumference or total body mass, that is, weight (Hoddinott, 2002). The mean increase in height of the preschool children was significant (p < 0.01) in both experimental and control groups after supplementation (Table 3). When compared to control, significant difference (P < 0.01) in height was not observed in both experimental groups after supplementation.

Sankhala et al. (2004) also observed an improvement in the height of malnourished children (6 to 10 years) after feeding paushtic laddu and mathari for a period of 90 days. The increment in height was of greater magnitude as compared to their control counterparts yet the difference was statistically non significant. Significant increase in height of preschool children in control group after 3 months period could be due to consumption of RTE food provided by the ICDS supplementary programme. Similar findings were observed by Bhagyalakshmi and Vijayalakshmi (2002), whereby effect of supplementation in ICDS beneficiaries (3 to 4 years old girls) was studied. A reduction in height deficit from 25 to 11% of National Centre for Health Statistics (NCHS) values after one year of feeding was found.

In the present study, increase in height was greater than the figures reported by Gopalan et al. (1996), that is, an increase of 2.3 cm in height was observed in the

preschool children for a period of four months supplementation. It was mainly due to providing additional calories and protein through malted mixes, which promoted better growth.

The increase in height observed in the present study in three months period is comparable to the values reported by Devadas et al. (1984), even though the quantity of protein and calorie provided by Devadas et al. (1984) was more as compared to the present study. The better impact was perhaps due to the more calorie gap in the present study.

Table 4 shows that the mean gain in weight by the preschool children increased significantly (P < 0.01) in both experimental and control groups after supplementation. A significant improvement in weight of the preschool children was observed in both the experimental groups after supplementation as compared to control due to consumption of amylase rich malted mixes which helped in decrease the bulk density and improving the digestibility and nutritional quality. Hossain et al. (2005) also observed higher weight gain and increments in length and weight-for-height in 65 malnourished children after receiving amylase rich flour added supplementary food for a period of 6 weeks.

The gain in weight by the preschool children in the present study was three times more than the gain in weight (0.56 kg) reported by Gopalan et al. (1996) for a

Table 5. Frequency distribution of Preschool Children as per Gomez classification*.

S/N	Gomez classification	Control I		Group II		Group III	
		Before	After	Before	After	Before	After
1	90% and above (Normal)	-	2 (10)	- (0)	38 (19)	- (0)	76 (38)
2	75-90% Grade I (Mild)	7 (35)	8 (40)	85 (43)	108 (54)	117 (59)	72 (36)
3	60-75% Grade II (Moderate)	7 (35)	5 (25)	65 (32)	49 (25)	33 (16)	52 (26)
4	<60% Grade III (Severe)	6 (30)	5 (25)	50 (25)	5 (2)	50 (25)	- (0)

n = 400, * Gomez classification (weight/age)

Table 6. Mean mid upper arm circumference of preschool children.

Group	Supplementation period (3 months)			
	Before	After	Increase in MUAC (cm)	't' value
Control I	12.92	13.23	0.31	18.44**
Group II	12.83	13.33	0.5	78.19**
Group III	12.70	13.3	0.68	142.25**
ICMR standard*	14.7	-	-	
'F' value	-	0.512^{NS}	-	
Level of significance between experimental groups	-	0.04^{NS}	0.18	

n = 400, NS- Not Significant; ** - Significant at 1% level; *-Gopalan et al. (1996)

Table 7. Mean hemoglobin level of preschool children

Group	Before	After	Increase in hemoglobin level (g/dl)	't' Value
Control I	9.32	9.46	0.14	11.77**
Group II	9.41	9.69	0.28^{NS}	35.41**
Group III	9.32	9.59	0.27^{NS}	79.91**
ICMR standard*	11 - 12	-	-	
'F' value	-	0.431^{NS}	-	
Level of significance between experimental groups	-	0.02^{NS}	0.1	

n = 400; NS, not significant ; **, Significant at 1% level; *, Gopalan et al. (1996).

period of four months supplementation. The additional calories and protein supplied through malted mixes must have increased the weight at a much rapid rate than normal weight.

The children in the present study showed greater gain in weight than the observations made by Upadhya (1970), that is, 15.4 kg who supplemented with wheat flour, bengal gram flour, peanut, skim milk powder, which provided 260 k cal and 10 g of protein for 6 weeks.

As per Gomez classification in experimental Group II, 2% of the preschool children were severely malnourished. Whereas none of the preschool children in the experimental Group III were severely malnourished after supplementation (Table 5).

On completion of three months duration feeding trials, the significant improvement (p < 0.01) in the mean mid upper arm circumference (MUAC) of both the experimental and control groups of preschool children was found (Table 6). However, when compared to control the increase was not found to be statistically significant in both the experimental groups. Supplementation of RTE food through ICDS supplementary programme could be a reason of significant increase in MUAC of preschool children in control group after 3 months period. Effect of *misola* (Porridge) consisted of millet, soybean, peanut, vitamins, minerals, and industrial amylase was studied on 300 malnourished children aged 6 to 48 months. After eight weeks of nutritional rehabilitation an improvement in

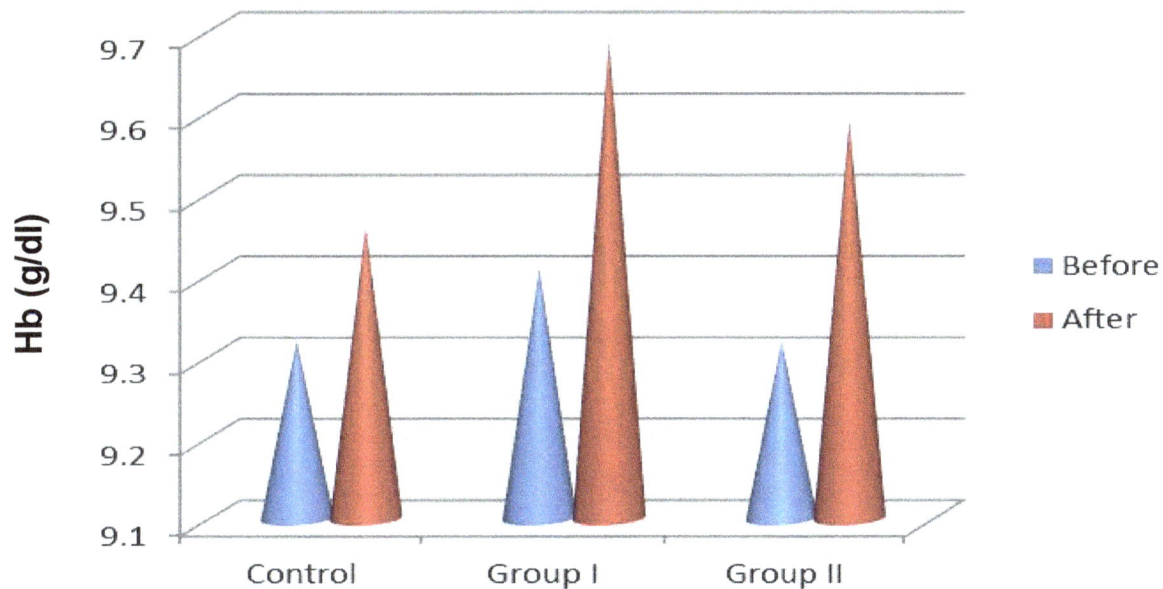

Figure 1. Mean hemoglobin level (g/dl) of preschool children (n = 400).

weight, height, head perimeter (HP) and MUAC was found (Zoenabo et al., 2012)

Biochemical estimation

Table 7 presents that, the mean increase in hemoglobinlevel of the preschool children was significant (p < 0.01) in both experimental and control groups after supplementation. Though, it was not statistically increased in both the experimental groups when compared to control. Significant increase in hemoglobin level of preschool children in control group after 3 months period could be due to consumption of RTE food supplied by the ICDS (Figure 1). The mean hemoglobin level of preschool children in both the experimental and control groups was lower than the WHO cut off values, that is, 11 to 12 g/dl for preschool children. This clearly indicated that the supplementation of malted mixes for at least 6 months with fortified vitamins and minerals would have better improvement.

Impact of supplementary feeding on the nutritional status of pregnant women

Anthropometric assessment

In both the experimental and control groups, significant difference in height of the pregnant women was not observed after supplementation. No significant gain inweight of the pregnant women was observed in both the experimental and control groups after supplementation (Table 8). The maximum increase in

weight was observed in Group III (5.32 kg), that is, pregnant women fed with wheat malted mix followed by Group II (5.29 kg), that is, pregnant women fed with ragi malted mix. An increase in weight (3.50 kg) was observed in the control group women. The results observed in the present study can be comparable to the values reported by Kumari and Singh (2004).

Biochemical estimation

The mean increase in hemoglobin level by the pregnant women increased significantly in both experimental and control groups after supplementation (Table 8, Figure 2). When compared to control, significant increase (P < 0.01) in hemoglobin level of pregnant women was observed in the experimental groups. Chaudhary (2004) also found an increase in Hb (1 to 2 g%) and weight gain (3 to 5 kg) of pregnant women of age 20 to 35 years, at 16 to 36 weeks of gestation, after supplementing a ladoo made with soya flour, wheat flour, roasted gram flour, groundnut, ghee and jaggery for a period of 12 weeks. An increase in Hb level (1.4 g/dl) of pregnant women was also reported by Kumari and Singh (2004) after supplementing quality protein from maize, green gram, ragi, gingerly seeds, amanranthus and jaggery.

Impact of supplementary feeding on the nutritional status of lactating women

Anthropometric assessment

The difference in height before and after the study of both

Table 8. Mean height, weight and hemoglobin level of pregnant women.

Group	Height (cm)				Weight (kg)				Hemoglobin level (g)			
	Before	After	Increase	't' Value	Before	After	Increase	't' Value	Before	After	Increase	't' Value
Control I	151.19	151.55	0.36	19.56**	43.14	46.64	3.50	15.98**	9.35	9.62	0.27	18.59**
Group II	151.19	151.69	0.50	8.28**	42.56	47.85	5.29	35.64**	9.35	11.00	1.67	40.01**
Group III	151.51	151.89	0.38	31.68**	43.16	48.48	5.32	45.28**	9.46	10.82	1.36	38.60**
Standard	-				-				12 - 14 g			
'F'Value		0.74 NS				1.24 NS				14.25**		
Level of significance between experimental groups		0.02 NS	0.28			0.16 NS	0.03			0.12 NS	0.31	

n = 100, NS, Not Significant; **, Significant at 1% level.

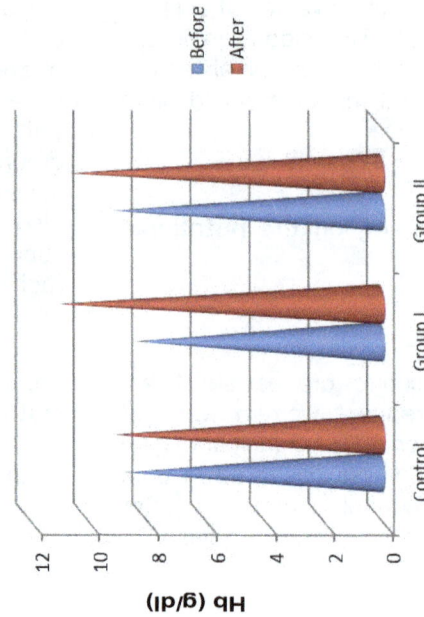

Figure 2. Mean hemoglobin level (g/dl) of pregnant women (n = 100).

the experimental and control groups was found to be statistically non significant (Table 9). The mean gain in weight of the lactating women increased significantly in both experimental and control groups after supplementation. However, when compared to control significant increase (P < 0.01)

in weight was observed in both experimental groups. The mean increase in hemoglobin level (Table 9, Figure 3) was significant in both experimental and control groups after supplementation. A significant difference (P < 0.01) in hemoglobin level was observed in both the experimental groups when compared to their control counterparts. Devadas et al. (1982) reported that the supplementation of indigenous low cost nutritive foods has a definite effect on the nutritional status of the nursing mothers.

BMI classification

The BMI classification for frequency distribution of pregnant (Table 10) and lactating women (Table 11) showed that 4% of both the pregnant and lactating women were severely malnourished in experimental Group II, where as none of them were severely malnourished in experimental Group III. Studies carried out at National Institute of Nutrition revealed that 75% of pregnant women in India are anaemic and anaemia remains to be a major factor responsible for maternal morbidity,

mortality and low birth weight. Continuous supplementation to those with energy gap helps to reduce under nutrition in pregnant and lactating women (http://wcd.nic.in/research/nti1947).

Prevalence of nutritional deficiency symptoms in preschool children

Assessment of the nutritional status was done based on the nutritional deficiency symptoms viz. protein energy malnutrition (PEM), vitamin A, vitamin C, vitamin D, B complex vitamin deficiency, iron deficiency and iodine deficiency symptoms in preschool children (Table 12). The results indicated that 60% of the preschool children in the control group were suffering from PEM before supplementation, whereas in both the experimental groups, that is, in Groups II and III, 16 and 17.5% of preschool children had PEM symptoms. There was a considerable reduction in the PEM deficiency symptoms of the experimental Group II (11%) and Group III (7%) than that of control group (2%). The reduction of protein energy malnutrition in the experimental groups

Table 9. Mean height, weight and hemoglobin level of lactating women.

Group	Height (cm)				Weight (kg)				Hemoglobin level (g)			
	Before	After	Increase	't' value	Before	After	Increase	't' value	Before	After	Increase	't' value
Control I	149.56	149.56	-	0.00 NS	40.50	41.78	1.25	6.08**	8.77	9.11	0.34	17.33**
Group II	150.76	150.95	0.19	20.97**	41.05	43.78	2.73	46.43**	8.37	10.98	2.61	46.38**
Group III	149.95	150.04	0.09	18.69**	40.65	43.35	2.70	46.24**	9.09	10.58	1.49	45.64**
Standard 'F'Value	1.365 NS				12.07**				77.86**			
Level of significance between experimental groups	0.03 NS	0.1			0.00 NS	0.03			0.1 NS	1.12		

n = 100, NS- Not Significant; ** - Significant at 1% level.

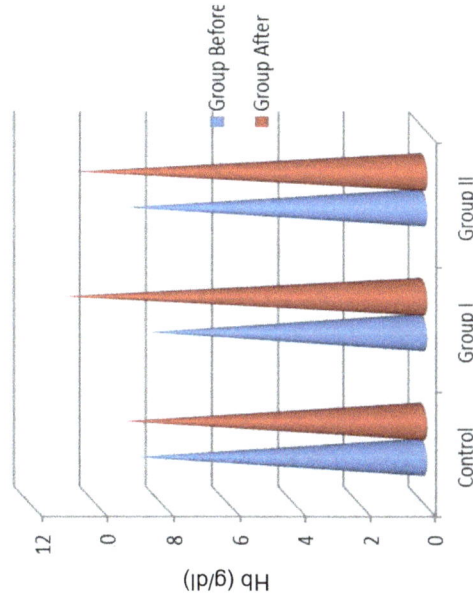

Figure 3. Mean hemoglobin level (g/dl) of lactating women (n = 100).

was mainly due to additional intake of calories and protein through the supplementary food.

Similarly, 50% reduction in vitamin A deficiency symptoms was observed in the preschool children of both the experimental and control groups after supplementation. This was mainly because of taking vitamin 'A' drops through primary health care centers.

None of the preschool children had vitamin 'D' and iodine deficiency symptoms during the experimental period. There was 50% improvement in the 'B' complex vitamin deficiency and vitamin C deficiency symptoms in preschool children of experimental groups. No change was observed in deficiency symptoms over 'B' complex vitamin and vitamin C in the control group by the end of the study period. This is mainly because malted grains are good sources of vitamin C and B complex vitamins.

Anemia is one of the common nutritional deficiencies seen in a large percentage of low-income population.

Lack of iron in the body leads to anemia. Considerable improvement (50%) in the iron nutritional status of preschool children of experimental groups was observed after supplementation. Iron deficiency symptoms increased in case of control group. Apart from iron deficiency in their diets, poor hygienic conditions and unsanitary surroundings may also be responsible for high incidence of anemia.

Prevalence of nutritional deficiency symptoms of pregnant and lactating women

Clinical examination of pregnant and lactating women was carried out before and after supplementation with the help of medical practitioner (Tables 13 and 14). Deficiency symptoms viz. PEM, vitamin A, vitamin C, vitamin D, B complex vitamins, iron and iodine in pregnant and lactating women of experimental as well as control groups were observed. 50% reduction in the deficiency symptoms of PEM by pregnant and lactating women in the experimental groups was observed; whereas, considerable reduction in deficiency symptoms of PEM was not observed in case of control group women. This was mainly because of supplementation of additional calories and protein through malted mixes.

Similarly, 'B' complex vitamin deficiency and vitamin C deficiency symptoms were lower (50%) in both the pregnant and lactating women of experimental groups. None of the pregnant and lactating women had vitamin A, vitamin D and iodine deficiency symptoms during the study period. Considerable reduction in the iron

Table 10. Frequency distribution of pregnant women as per BMI*classification.

BMI grade	Groups					
	Control I		Group II		Group III	
	Before	After	Before	After	Before	After
Grade III, Severe (<16.0)	2 (9)	1 (8)	10 (20)	2 (4)	10 (20)	(0)
Grade II, Moderate (16.0 - 17.0)	6 (30)	4 (18)	11 (22)	6 (12)	7 (14)	2 (4)
Grade I, Mild (17.0 - 18.5)	12 (61)	3 (13)	29 (58)	10 (20)	33 (66)	12(24)
Normal (18.5-25.0)	(0)	12 (61)	(0)	32 (64)	(0)	36 (72)

n=100; BMI*, body mass index = wt (kg/Ht^2 (mt).

Table 11. Frequency distribution of lactating women as per BMI* classification.

BMI grade	Groups					
	Control I		Group II		Group III	
	Before	After	Before	After	Before	After
Grade III, Severe (<16.0)	4 (22)	4 (22)	10 (20)	2 (4)	10 (20)	(0)
Grade II, Moderate (16.0 - 17.0)	7 (34)	8 (35)	25 (50)	17 (34)	24 (48)	16 (32)
Grade I, Mild (17.0 - 18.5)	9 (44)	6 (30)	15 (30)	16 (32)	16 (32)	21 (42)
Normal (18.5 - 25.0)	(0)	2 (13)	(0)	15 (30)	(0)	13 (26)

n = 100, BMI* - Body Mass Index = wt [kg/Ht^2 (mt)].

Table 12. Frequency distribution of preschool children in view of nutritional deficiency symptoms.

Symptoms of nutritional deficiency	Control I		Group II		Group III	
	Before	After	Before	After	Before	After
Protein energy malnutrition	12 (60)	10 (50)	32 (16)	10 (5)	35 (17.5)	21 (10.5)
Vitamin A deficiency	8 (10)	2 (10)	12 (6)	7 (3.5)	5 (2.5)	3 (1.5)
Vitamin D deficiency	-	-	-	-	-	-
B Complex vitamin deficiency	5 (25)	6 (30)	26 (13)	11 (5.5)	20 (10)	9 (4.5)
Vitamin C deficiency	3 (15)	3 (15)	29 (14.5)	19 (9.5)	32 (16)	12 (6)
Iron deficiency	9 (45)	9 (45)	48 (24)	23 (11.5)	52 (26)	25 (12.5)
Iodine deficiency	-	-	-	-	-	-

n = 400. Note: Figures in the parenthesis indicates percentages.

Table 13. Frequency distribution of pregnant women in view of nutritional deficiency symptoms.

Symptoms of nutritional deficiency	Control I		Group II		Group III	
	Before	After	Before	After	Before	After
Protein energy malnutrition	9.0 (45)	8.0 (40)	24.0 (48)	10.0 (20)	21.0 (42)	11.0 (22)
Vitamin A deficiency	-	-	-	-	-	-
Vitamin D deficiency	-	-	-	-	-	-
B Complex vitamin deficiency	7.0 (35)	8.0 (40)	19.0 (38)	9.0 (18)	21.0 (42)	11.0 (22)
Vitamin C deficiency	9.0 (45)	8.0 (40)	15.0 (30)	5.0 (10)	17.0 (34)	7.0 (14)
Iron deficiency	14.0 (70)	19.0 (95)	32.0 (64)	17.0 (34)	39.0 (7.8)	20.0 (40)
Iodine deficiency	-	-	-	-	-	-

n = 100; Figures in the parenthesis indicates percentages.

deficiency symptoms was also observed in the pregnant women (40%) and lactating women (50%) of experimental groups. Not much improvement in the iron nutritional status was observed in case of pregnant

Table 14. Frequency distribution of lactating women in view of nutritional deficiency symptoms.

Symptoms of nutritional deficiency	Control II		Group II		Group III	
	Before	After	Before	Before	After	Before
Protein energy malnutrition	13.0 (65)	12.0 (60)	28.0 (56)	13.0 (26)	25.0 (50)	12.0 (24)
Vitamin A deficiency	-	-	-	-	-	-
Vitamin D deficiency	-	-	-	-	-	-
B Complex vitamin deficiency	8.0 (55)	8.0 (40)	19.0 (38)	8.0 (16)	15.0 (30)	6.0 (12)
Vitamin C deficiency	11.0 (55)	8.0 (40)	12.0 (24)	6.0 (12)	16.0 (32)	6.0 (12)
Iron deficiency	12.0 (60)	15.0 (75)	39.0 (78)	21.0 (42)	35.0 (70)	18.0 (36)
Iodine deficiency	-	-	-	-	-	-

n = 100; Figures in the parenthesis indicates percentages.

Table 15. Frequency distribution of Preschool children as per Morbidity pattern.

S/N	Type of illness	Control I		Group II		Group III	
		Before	After	Before	After	Before	After
1	Diarrhea	4 (20)	5 (25)	23 (11.5)	15 (7.5)	22 (11)	14 (7)
2	URI	5 (25)	3 (15)	7 (8.5)	6 (3)	15 (7.5)	4 (2)
3	Bronchitis	-	-	-	-	-	-
4	Vomiting	8 (40)	7 (35)	26 (30)	18 (9)	24 (12)	16 (8)
5	Fever	4 (20)	3 (15)	13 (6.5)	6 (6.5)	11 (5.5)	4 (2)
6	Malaria	-	-	-	-	-	-
7	Measles	2 (10)	1 (5)	4 (2)	2 (1)	4 (2)	1 (0.5)
8	Skin infection	6 (30)	4 (20)	10 (5)	7 (3.5)	13 (6.5)	12 (6)
9	Dysentery	-	-	-	-	-	-
10	Ear infection	3 (15)	3 (15)	17 (8.5)	10 (5)	15 (7.5)	7 (3.5)
11	Eye infection	2 (10)	1 (5)	16 (8)	7 (3.5)	11 (5.5)	5 (2.5)
12	Scabies	-	-	-	-	-	-
13	Cough	7 (35)	7 (35)	18 (9)	8 (4)	13 (6.5)	6 (3)
14	Whooping cough	-	-	-	-	-	-
15	Chicken pox	-	-	-	-	-	-
16	Jaundice	2 (10)	3 (15)	11 (5.5)	2 (1)	9 (4.5)	3 (1.5)
17	Sour mouth	5 (25)	4 (20)	11 (5.5)	4 (2)	9 (4.5)	2 (1)
18	Edema	11 (55)	13 (65)	82 (41)	62 (31)	80 (40)	60(30)

n = 400; Figures in the parenthesis indicates percentages.

and lactating women in control group.

Morbidity pattern of preschool children, pregnant women and lactating women

A varied spectrum of illness were seen viz. diarrhea, upper respiratory infection (URI), vomiting, fever, meseals, skin infection, ear and eye infection, cough, jaundice, sour mouth and edema in the sample studied before and after supplementation in three groups of preschool children, pregnant women and lactating women of both the control and two experimental groups. Considerable decrease in the incidence of various morbidities was observed in preschool children, pregnant women and lactating women of both the experimental groups compared to control group after supplementation (Tables 15 to 17). Most of the important infections in the preschool children, pregnant women and lactating women render their consequences more seriously by the presence of malnutrition.

Conclusion

Supplementation of ARMMs incorporated food products to the vulnerable groups of population in Lepakshi mandal of Ananthapur district, Andhra Pradesh, India

Table 16. Frequency distribution of pregnant women as per morbidity pattern.

S/N	Type of illness	Control I		Group II		Group III	
		Before	After	Before	Before	After	Before
1	Diarrhea	4 (20)	4 (20)	11 (22)	5 (10)	19 (38)	12(24)
2	URI	3 (15)	3 (15)	5 (10)	2 (4)	6 (12)	2 (4)
3	Bronchitis	-	-	-	-	-	-
4	Vomiting	15 (75)	13 (65)	26 (52)	12 (24)	28 (56)	14 (28)
5	Fever	6 (30)	4 (20)	14 (28)	8 (16)	12 (24)	6 (12)
6	Malaria	-	-	-	-	-	-
7	Measles	-	-	-	-	-	-
8	Skin infection	6 (30)	5 (25)	5 (10)	2 (4)	6 (12)	3 (6)
9	Dysentery	-	-	-	-	-	-
10	Ear infection	3 (15)	2 (10)	4 (8)	2 (4)	3 (6)	1 (2)
11	Eye infection	2 (10)	2 (10)	1 (2)	1 (2)	1 (2)	-
12	Scabies	-	-	-	-	-	-
13	Cough	8 (40)	6 (30)	7 (14)	3 (6)	4 (8)	2 (4)
14	Whooping cough	-	-	-	-	-	-
15	Chicken pox	-	-	-	-	-	-
16	Jaundice	3 (15)	2 (10)	6 (12)	3 (6)	6 (12)	3 (6)
17	Sour mouth	5 (25)	7 (35)	15 (30)	3 (6)	13 (26)	5 (10)
18	Edema	9 (45)	9 (45)	33 (66)	18 (36)	39 (78)	11 (22)

n = 100; Figures in the parenthesis indicates percentages.

Table 17. Frequency distribution of lactating women as per morbidity pattern.

S/N	Type of illness	Control I		Group II		Group III	
		Before	After	Before	Before	After	Before
1	Diarrhea	9 (45)	8 (40)	10 (20)	5 (10)	9 (18)	5 (10)
2	URI	7 (35)	2 (10)	7 (14)	-	8 (16)	2 (4)
3	Bronchitis	-	-	-	-	-	-
4	Vomiting	9 (45)	6 (30)	10 (20)	4 (8)	9 (18)	5 (10)
5	Fever	6 (30)	4 (20)	13 (26)	8 (16)	11 (22)	6 (12)
6	Malaria	-	-	-	-	-	-
7	Measles	-	-	-	-	-	-
8	Skin infection	2 (10)	2 (10)	4 (8)	2 (4)	5 (10)	4 (8)
9	Dysentery	-	-	-	-	-	-
10	Ear infection	1 (5)	1 (5)	2 (4)	1 (2)	2 (4)	1 (2)
11	Eye infection	1 (5)	1 (5)	-	-	1 (2)	1 (2)
12	Scabies	-	-	-	-	-	-
13	Cough	5 (25)	1 (5)	2 (4)	1 (2)	2 (4)	1 (2)
14	Whooping cough	-	-	-	-	-	-
15	Chicken pox	-	-	-	-	-	-
16	Jaundice	-	-	-	-	-	-
17	Sour mouth	3 (15)	2 (10)	4 (8)	3 (6)	4 (8)	3 (6)
18	Edema	8(40)	8 (40)	26 (52)	14 (28)	39 (78)	15 (30)

n = 100, Note: Figures in the parenthesis indicates percentages.

showed that significant increase (P < 0.01) was observed in weight and height of preschool children, weight of lactating women and hemoglobin level of pregnant and lactating women after supplementation. The results of clinical assessment showed considerable reduction (50%) in nutritional deficiency symptoms and morbidity rate of preschool children, pregnant women and lactating women in experimental groups compared to control group at the end of the study.

Hence, these nutrient dense, easily digestible food products prepared with malted mixes will be beneficial to children, teenagers, anemic patients, pregnant women and lactating women. Further, these feeding trials could be carried out for geriatric population also.

ACKNOWLEDGEMENTS

The authors are thankful to the Department of Biotechnology, Ministry of Science and Technology, Government of India, New Delhi for the financial support. The authors are also thankful to Acharya N. G. Ranga Agricultural University, Hyderabad for successfully carrying out the project.

REFERENCES

AOAC (1990). Official methods of Analysis Association of Official Analytical chemists. Washington, DC.

Association of vitamin chemists (1947). Methods of vitamin Assay. Inc., (ed) Inter Science publishers. p. 159. (Association of vitamin chemists Inc, 1947).

Bernfield P (1955). Amylase alpha In calomick SP, Kaplan NO. Methods in enzymology, New York; Academic Press. 1:149-151.

Bhagyalakshmi G, Vijayalkshmi P (2002). Impact of ICDS on the health status of children. Ind. J. Nutr. Dietet. 39:519.

Brandtzaeg B, Malleshi NG, Svanberg U, Desikachar HSR, Mellander O (1981). Dietary bulk as a limiting factor for nutrient intake in pre-school children. III. Studies of malted flour from ragi, sorghum and green gram. J. Trop. Pediatrics 27:184-189.

Chaudhary R (2004). Biochemical Assessment of Nutritional Status of Pregnant Anemic Women after a Nutritional Supplement. Asian J. Exp. Sci. 18 (1&2):95-112

Dacie JV, Lewis SM (1991). Practical Hematology. 7th edn, Churchill livingstone, Edinburgh. p. 61

Devadas RP, Chandrasekar U, Bhooma N (1982). Acceptability of diets based on low cost locally available foods for various target groups. Ind. J. Nutr. Dietet. pp.19-23.

Devadas RP, Chandrasekhar U, Bhooma N (1984). Nutritional outcomes of a rural diet supplemented with low cost locally available foods – V, Impact on preschoolers followed over a period of four and a half years. Ind. J. Nutr. Dietet. 21:53

Dewar J, Taylor JRN, Berjak P (1997). Effects of germination conditions with optimized steeping on sorghum malt quality with particular reference to free amino nitrogen. J. Inst. Brew. 103:171-175

Dulby A, Tsai CY (1976). Lysine and Tryptophan increases during germination of cereal grains. Cereal Chem. 53:222-224

Gopalan C, Rama sastri BV, Balasubramanian SC (1996). Nutritive Value of Indian foods revised and updated by Narasingarao BS, Deosthale and Pant KC, National Institute of Nutrition, ICMR, Hyderabad, India.

Hoddinott J (2002). Food security in practice. Methods for rural development projects, International Food Policy Research Institute Washington, DC.

Hossain MI, Wahed MA, Ahmed S (2005) Increased food intake after the addition of amylase-rich flour to supplementary food for malnourished children in rural communities of Bangladesh. Food Nutr. Bull. 26(4):323-329

Inyang CU, Idoko CA (2006). Assessment of the quality of ogi made from malted millet. Afr. J. Biotechnol. 5(22):2334-2337

Jelliffe DB (1996). The assessment of Nutritional status of the community, WHO monograph series 53:1.

Kumar L S, Prakash H S, Shetty HS, Malleshi NG (1991). Influence of seed microflora and harvesting conditions on milling popping and malting qualities of sorghum. J. Sci. Food Agric. 55:617-625.

Kumari P, Singh U (2004). Impact of supplementation of quality protein maize based lactose on the nutritional status of pregnant women. Ind. J. Nutr. Dietet. 41:528.

Sangita K, Sarita S (2000). Nutritive value of malted flours of finger millet genotypes and their use in the preparation of burfi. J. Food Sci. Technol. 37(4):419-122.

Sankhala A, Sankhla AK, Bhatnagar B, Singh A (2004). Impact of Intervention Feeding Trial on Nutritional Status of 6-10 Years Old Malnourished Children. Anthropologist 6(3):185-189.

Seth V, Sundaram KR, Ghai OP, Gupta M (1979). Profile of morbidity and nutritional status and their effects on the growth promotion in Preschool Children in Delhi. Trop. Pediatrics Environ. Child Health 23:23-29

Upadhya G (1970). A Study with vegetable protein mixture in combating protein energy malnutrition at outdoor clinic, Thesis submitted to Osmania University for the degree of M. Sc (Appl. Nutr.).

WHO Expert Consultation (2004). Appropriate body-mass index for Asian populations and its implications for policy and intervention strategies. Lancet 363:157-63

Zoenabo D, Martinetto M, Pietra V, Pignatelli S, Schumacher F, Nikiema JB, Simpore J (2012). Effects of a Cereal and Soy Dietary Formula on Rehabilitation of Undernourished Children at Ouagadougou, in Burkina Faso Hindawi Publishing Corporation J. Nutr. Metab. Article ID 764504. http://wcd.nic.in/research/nti1947.

3

Assessment of antifungal effect of omeprazole on *Candida Albicans*

Fahriye Küçükaslan[1], Hasibe Cingilli Vural[2] , Didem Berber[1], Zeki Severoğlu[3], Sabri Sümer[3] and Meltem Doykun[2]

[1]Institute of Graduate Studies in Pure and Applied Sciences, Marmara University, 34722 Goztepe, Istanbul, Turkey.
[2]Department of Molecular Biology, Science Faculty, Selcuk University, Kampus Selcuklu, Konya, Turkey.
[3]Department of Biology, Faculty of Arts and Sciences, Marmara University, 34722 Goztepe, Istanbul, Turkey.

The goal of this study was to investigate the *in vitro* effect of omeprazole on *Candida albicans* and analyze the antifungal activity of omeprazole. A total of 150 samples were collected from the patients in Bakirköy Dr. Sadi Konuk Education and Research Hospital and samples were evaluated for *C. albicans*. After the microbiological analyses, fifty one patients (18 men and 33 women) between 0 and 78 of age were found to be *C. albicans* positive and they were included in the study. All consecutive isolates of *C. albicans* were recovered from blood, urine, sputum, oral cavity, vagina, catheter tip and ascitic fluid. Antifungal suspectibility test was carried out by microdilution assay according to the method outlined in the NCCLS document M27-A. It was determined that omeprazole is fairly effective in particular MIC range. Furthermore, it was observed that omeprazole in high concentrations support the growth of fungi.

Key words: *Candida albicans*, omeprazol, antifungal effect, antifungal agents.

INTRODUCTION

The incidence of fungal infections has been increasing dramatically over the last two decades in the world in immunocompetent and immunocompromised (premature newborns, elderly individuals, chemotherapy-treated patients, HIV patients and transplant recipients) patients as well as in patients with leukemia, lymphoma and organ transplantation (Banerjee et al., 2009; Beil and Sewing, 1984; Campbell et al., 1998; Clemons et al., 2006). Also, the incidence of fungal infections is associated with high morbidity and mortality in these patient groups (Beil and Sewing, 1984). The known predisposing factors for fungal infections are immature immune system, breakdown of cellular immunity and colonization seen during the broad-spectrum antibiotic therapy (Campbell et al., 1998; Costa et al., 2010). In particular, yeasts (such as *Candida* spp.) are the most frequently isolated fungi from human infections as well as dermatophytes (Banerjee et al., 2009; Fleischhacker et al., 2008; Garey et al., 2006). *Candida* spp. are commonly encountered polymorphic yeasts in the gut lumen and on cutaneous surfaces and they can exist as 2 to 5 μm round to oval cells in shape (blastospores) and can reproduce by budding. Also, they produce mostly pseudohyphae elongating from the cells but few species have ability to produce actual hyphae (Fleischhacker et al., 2008; Gatta et al., 2003). Of the 200 *Candida* spp., *Candida albicans* is one of the most frequently isolated fungal pathogen in humans (Campbell et al., 1998; Fleischhacker et al., 2008). This commensal yeast belongs to the normal flora of skin as well as gastro-intestinal and genital tracts of healthy individuals.

Candida albicans is a Gram-positive and opportunistic yeast that can cause life-threatening systemic infections by entering bloodstream and infecting the organs (Clemons et al., 2006). Their new forms in bloodstream are called chlamydospores. This unique property is a resting stage of yeast and specific to *C. albicans.* Chlamydospores have cylindirical extentions called germ tubes (Fleischhacker et al., 2008; Guery et al., 2009). *C. albicans* produce germ tubes at 33 to 42°C in pH 6 to 8 (Guery et al., 2009). The formation of germ tubes and chlamydospores are the distinctive features for *C. albicans* and these characterictics are rarely observed in other *Candida* species (Fleischhacker et al., 2008; Guery et al., 2009; Harrington et al., 2007). The germ tube test and controlling the ability to produce chlamydospores are the most helpful, rapid and reliable methods for the presumptive clinical identification of *C. albicans* isolates (Heelan et al., 1998).

It has been reported that the antifungal therapies recommended as the first-line treatment have usually been selected based on the clinical status of the patient (Gatta et al., 2003; Johnson et al., 2003; Katiyar et al., 2006). The narrow-spectrum antifungals may not supply sufficient treatment for the patients who are suffering from fungal infections (Gatta et al., 2003). In the recent studies, it has been suggested that early intervention by adequate antifungal agents may considerably reduce mortality in patients (Gatta et al., 2003; Keeling et al., 1985; Kumar et al., 2006; Larner and Lendrum, 1992). The first-line proposed antifungal agents are commonly azoles (especially fluconazole) for patients with fungemia (Johnson et al., 2003; Larsson et al., 1983). Other alternative antifungal agents such as lipid formulations of amphotericin and echinocandins with considerably broader activities have been preferred at a lower rate due to their high costs than azoles (Larsson et al., 1983). Given the increasing concerns on the resistant yeast species to the antifungal therapies administered in clinical practice (azoles and polyene antibiotics, etc.), newer antifungal agents like omeprazole could be useful in guiding the treatment of

fungemia.

Omeprazole, a substituted benzimidazole, is a potent proton pump inhibitor and has been reported in the treatment of acid-peptic diseases of the gastrointestinal tract, duodenal ulcer and Zollinger–Ellison syndrome over the past decade (Martínez et al., 1998; Merlino et al., 1998; Mogensen and Mühlschlegel, 2008; Molero et al., 1998; Monk and Perlin, 1994; Morrell et al., 2005).

The plasma membrane H+-ATPase inhibitor plays a major role in yeast cell physiology. This ion translocation enzyme is responsible for maintaining the electrochemical proton gradient for the adjustment of intracellular pH and nutrition uptake in the fungal cell. The inhibition of H+-ATPase activity by antagonists causes cell death. Therefore, utilization of the plasma membrane H+-ATPase as a molecular target appears to be more attractive approach for the antifungal drug therapies only if the connection is maintained between the inhibition of enzyme activity and suppression of cell growth (Mogensen and Mühlschlegel, 2008).

The aim of this study was to investigate the antifungal activity of omeprazole on *C. albicans* isolated from the clinical samples of blood, urine, sputum, oral cavity, vagina, catheter tip and ascitic fluid of inpatients and outpatients by evaluating its resistance, sensitivity and applicability in candidiasis.

MATERIALS AND METHODS

Sample collection

A total of 150 samples were collected from the inpatients and outpatients in Bakirköy Dr. Sadi Konuk Education and Research Hospital. These samples were collected under the aseptic conditions and placed in sterile sample bags and then, they were immediately transported to the laboratory. The samples were evaluated for *C. albicans*. After the microbiological analyses, fifty one patients (18 men and 33 women) between 0 and 78 of age were found to be *C. albicans* positive and they were included in the study. All consecutive isolates of *C. albicans* were recovered from blood, urine, sputum, oral cavity, vagina, catheter tip and ascitic fluid. The reference strains were supplied from The American Type Culture Collection for quality control. These ATCC strains were *C. albicans* ATCC 90028, *Candida krusei* ATCC 6258 and *Candida parapsilosis* ATCC 22019). These strains were supplied and grown as described for *Candida* isolates. The reference strains and *C. albicans* isolates were identified in microbiologically.

Omeprazol activation

Omeprazol was gotten from Eczacibaşi Holding Co. Since antifungal agents like omeprazole become active in acidic environment, the pH of omeprazol was adjusted to pH 2 with HCl after dissolving the agent in dimethyl sulfoxide (DMSO) in accordance with The

National Commmittee for Clinical Laboratory Standards (NCCLS). After an hour, the activation of omeprazol was confirmed by the color change (orange) on the slides. Then, the pH was again increased to pH 7.

Microbiological identification of *Candida* species

All the collected strains were analysed microscopically for the identification of *Candida* by 10% KOH or sterile physiological saline solution. The clinical isolates were streaked onto the Saboraud-Dextrose agar (SDA) plates which is a selective media for the growth and identification of fungi (Shindo et al., 1998). They were incubated at 35°C for 24 h. The growing yeast colonies on SDA were examined for *C. albicans* by germ tube test in serum, colony colour on *CANDIDA* ID2 and identification by API Candida (BioMerieux, France). Firstly, yeast cultures were incubated with human serum placed into blood culture bottles for 2-2.5 h at 37°C and then the presence or absence of germ tubes recorded. All germ tube positive yeast isolates were accepted as identified as *Candida albicans*. The same samples were cultured on the chromogenic agar plates of *CANDIDA* ID2 which is a commercially a ready-to-use medium allowing the specific identification of *Candida* spp. The plates were read and results interpreted after the incubation for 24 h at 37°C according to the manufacturer's instructions. *C. albicans*, *Candida dubliniensis* and *C. krusei* were characterized with blue, turquoise and pink colonies in the *CANDIDA* ID2 agar plates, respectively. The blue colored colonies, identified as *C. albicans* on the basis of their typical appearances, were enrolled in the study. Then, definitive identification of *C. albicans* was made with the API Candida test kit (BioMerieux, France) on the basis of biochemical reactions from the microbiological aspect. Each pure isolate of *C. albicans*, identified microbiologically before, were stored in the medium at -20°C until use (Stevens et al., 2006). The inoculums of yeast strains were prepared based on the NCCLS document. The stored isolates of *C. albicans* and reference strains were re-streaked onto SDA and incubated for 24 h at 35°C. Then, three or five colonies, which were in similar morphology and ≥1 mm in diameter, were picked up and suspended in 5 ml of 0.85% sterile physiological saline solution by vortexing. The density and turbidity of the suspension were adjusted to 1.5×10^6 CFU/ml and a Mc Farland standard of 0.5. Then, the suspensions were diluted (1:100 and 1:20) in RPMI 1640 medium. In this way, the final concentrations of the inoculums were adjusted to 0.5 to 2.5×10^3 CFU/ml.

Antifungal suspectibility testing

Antifungal suspectibility testing was carried out by microdilution assay according to the method outlined in the NCCLS document M27-A (1997) (Stevens et al., 2006). In the antifungal suspectibility testing, RPMI 1640 medium, which was supplemented with both glutamine and pH indicator, without sodium bicarbonate, was used. The pH of this medium was adjusted to pH 7 at 25°C with morpholinepropanesulfonic acid (MOPS) until the final concentration was 0.165 mol/L and sterilized by filtering. *C. albicans* ATCC 90028, *Candida parapsilosis* ATCC 22019 and *C. krusei* ATCC 6258 were used as reference strains. During the experiments, these reference strains were periodically streaked onto SDA plates and checked for their purity to prevent contamination.

The dilution intervals of omeprazole

The powdered omeprazole was diluted in DMSO in quantities specified as in Tables 1 to 3 and three omperazole suspensions in different concentrations were prepared.

The sterile microdilution plates (96-well and U-bottom) were used for the microdilution technique M27-A recommended by the NCCLS. During all the experiments, freshly prepared samples, which were incubated for 48 h at 35°C, were used. The omperazole suspensions ranging from 32 to 6 µg/L in concentration were distributed in 96-well microdilution plates in amounts of 100 µl per well. Then, the *Candida* suspensions, which were adjusted to a Mc Farland standard of 0.5, were distributed to the wells in amounts of 100 µl loaded with omeprazole. One well was used as control for the yeast growth and only yeast cells were placed and omeprazole was not added to the wells. The last wells were loaded only with medium to control the contamination of the test medium. The plates were incubated for 48 h at 35°C. Same procedure was used for the second and third concentrations of omperazole given in Tables 2 and 3.

RESULTS AND DISCUSSION

Since *Candida* spp. are commonly encountered opportunistic yeasts in normal flora of healthy individiuals, they have critically important role in terms of high morbidity and mortality. There are several studies aiming to reduce the rates of fungal morbidity and mortality but antifungal drug discovery are still needed to be developed (Sümer et al., 2005). The $(H^+ + K^+)$-ATPase inhibitor omeprazole is an effective treatment with a favourable safety profile for acid-peptic disease of the gastrointestinal tract, duodenal ulser and Zollinger–Ellison syndrome over the past decade (Martínez et al., 1998; Merlino et al., 1998; Mogensen and Mühlschlegel, 2008; Molero et al., 1998; Monk and Perlin, 1994; Morrell et al., 2005).

Johnson et al. (2003) showed that omeprazole increased the serum concentration of itrakonazole and positively affected its antifungal effect. In the other preliminary studies, it was determined that omeprazole interacts with fungal ATPase and inhibits this enzyme as well as gastric $(H^+ + K^+)$ATPase in a similar pattern (Sümer et al., 2005; Thomas et al., 2001).

Keeling and coworkers (1985) reported that the inhibition by omeprazole, degraded by acid, was more pronounced in Na^+, K^+-ATPase rather than H^+, K^+-ATPase. Furthermore, they emphasized that H^+, K^+-ATPase could be inhibited by considerably high omeprazole concentrations (Mogensen and Mühlschlegel, 2008).

Beil and Sewing have reported that omeprazole inhibited the $(H^+ + K^+)$-ATPase activity of preparations isolated from parietal cells of guinea pig (National Committee for Clinical Laboratory Standards, 1997; Reboli et al., 2007). In the other study, it was showed that omperazole was unstable in the acidic solutions and the inhibition of

Table 1. The first concentration range of omeprazole (32 to 0.06 µg/ml).

Potency	100%
Weight (mg)	100
Desired stock concentration (µg/ml)	12800
DMSO quantity	7.8125
Used microplaque number	10
Initial concentration (mcg/ml)	32

Step	Required concentration	Total volume	Sample concentration	Sample quantity	DMSO quantity	Proportion	Final volume	Diluted (1:50) RPMI
1	3200	0.4	12800	0.1	0.3	0.25	0.2	9.8
2	1600	0.2	3200	0.1	0.1	0.5	0.2	9.8
3	800	0.2	3200	0.05	0.15	0.25	0.2	9.8
4	400	0.4	3200	0.05	0.35	0.125	0.2	9.8
5	200	0.2	400	0.1	0.1	0.5	0.2	9.8
6	100	0.2	400	0.05	0.15	0.25	0.2	9.8
7	50	0.4	400	0.05	0.35	0.125	0.2	9.8
8	25	0.2	50	0.1	0.1	0.5	0.2	9.8
9	12.5	0.2	50	0.05	0.15	0.25	0.2	9.8
10	6.25	0.4	50	0.05	0.35	0.125	0.2	9.8
11	ATIK		6.25	0.2				

($H^+ + K^+$)-ATPase activity by omeprazole was highly dependent upon pH (Reboli et al., 2007; Rex and Sobel, 2001; Richardson and Elewski, 2000).

Resistance to antifungal drugs has become a major problem worldwide and has become a significant problem increasingly in pathogenic mycology. In general, the antifungal-resistant strains arised from haphazardly use of antifungal drugs in repeated dosages. It is known that some strains of *Candida* are resistant to the antifungal agents such as azoles and flukanazole. In recent studies, it was reported that this resistance problem can occur due to the developing mutations depending on their origin, living conditions and its host. Nowadays, our knowledge on the mechanism of the resistant strains of *C. albicans* to antifungal drugs has increased. On the other hand, some antifungal drugs are mostly effective on both pathogenic fungi and host due to the high similarities of eucaryotic cells. Since many antifungal drugs have side-effects on humans, it is important to discoverfairly new drugs targeting the non-shared features between host and fungi or application of antifungal drugs with less side-effects. Thus, it has been considered that omperazole, a proton pomp inhibitor in humans, can be effective on fungal plasma membrane ATPases and might be used in fungal diseases.

In our study, the patient age range of study group (19 males and 32 females) was from 0 to 78 years of age. There was no significant difference in terms of age distribution within the patient groups ($p > 0.05$).

The minimum inhibitory concentration (MIC) of omeprazole was evaluated by the antifungal suspectibility test in a total of 51 isolates of *C. albicans* which were isolated from blood, urine, dental plaque, oral cavity, sputum, vagina, catheter tip and ascitic fluid and reference strains of *C. albicans* ATCC 90028, *C. parapsilosis* ATCC 22019 and *C. krusei* ATCC 6258 in accordance with the NCCLS M27-A microdilution method. MICs were defined as the highest concentration that showed a sharp decline in the density of growth. 30 of 51 *C. albicans* strains (59%) were susceptible to 0.06 µg/ml of omeprazole in the first

Table 2. The second concentration range of omeprazole (512-1 µg/ml).

Potency	100%
Weight (mg)	100
Desired stock concentration (µg/ml)	51200
DMSO quantity	1.95
Used microplaque number	10
Initial concentration (mcg/ml)	512

Step	Required concentration	Total volume	Sample concentration	Sample quantity	DMSO quantity	Proportion	Final volume	Diluted (1:50) RPMI
1	51200	0.4	51200	0.4	0	1	0.2	9.8
2	25600	0.2	51200	0.1	0.1	0.5	0.2	9.8
3	12800	0.2	51200	0.05	0.15	0.25	0.2	9.8
4	6400	0.4	51200	0.05	0.35	0.125	0.2	9.8
5	3200	0.2	6400	0.1	0.1	0.5	0.2	9.8
6	1600	0.2	6400	0.05	0.15	0.25	0.2	9.8
7	800	0.4	6400	0.05	0.35	0.125	0.2	9.8
8	400	0.2	800	0.1	0.1	0.5	0.2	9.8
9	200	0.2	800	0.05	0.15	0.25	0.2	9.8
10	100	0.4	800	0.05	0.35	0.125	0.2	9.8
11	ATIK		100	0.2				

concentration range of omeprazole (32 to 0.06 µg/ml), while 16 (31.7%) were susceptible to 1 µg/ml of omeprazole in the second concentration range of omeprazole (512 to 1 µg/ml). On the other hand, almost all these strains (except 3 strains) were fairly resistant to omeprazole in the third concentration range of omeprazole (1 to 0.001 µg/ml). As a result, the effective MIC range of omperazole was found to be 1 to 0.06 µg/ml and this efficacy disappeared in higher or lower concentrations. This phenomenon is called "the eagle effect" or "paradoxical effect" in the literature. There are several reports analyzing the effects of high concentrations of three antifungal substances, including caspofungin, on the growth of *Candida* spp. and it was demonstrated that *in vitro* efficiency on *Candida* spp. was reduced by increasing antifungal dosesm (Wallmark et al., 1983; Whitley-Williams, 2006; Zomorodi and Houston, 1996).

Sümer and her colleagues (2005) examined the efficiency of omeprazole on *C. albicans* and they observed that most of the strains of *Candida* spp. showed increasing suspectibility to the high concentration of omeprazole (320 µg/ml). In the other parallel studies examining the different antifungal drugs, it was observed that this concentration was considerably high. Therefore, it was concluded that utilization of different antifungal drug combinations or experiencing different preparations was more efficacious to reduce the concentration of antifungal drugs, and also local application instead of systemic utilization was necessary.

In this study, we determined that omeprazole is fairly effective on *C. albicans* in 1 to 0.06 µg/ml and surely appliciable in candidiasis. On the other hand, we observed that omeprazole application in high concentrations supported the growth of fungi. Especially, to chose and apply the most effective antifungal therapy is essential for controlling the nosocomial infections and epidemiological studies and also for maintaining public health. In conclusion, we believe that omeprazole therapy in *Candida*

Table 3. The third concentration range of omeprazole (1 to 0.001 µg/ml).

Potency	100%
Weight (mg)	100
Desired stock concentration (µg/ml)	12800
DMSO quantity	7.8125
Used microplaque number	10
Initial concentration (mcg/ml)	32

Step	Required concentration	Total volume	Sample concentration	Sample quantity	DMSO quantity	Proportion	Final volume	Diluted (1:50) RPMI
1	51200	0.4	51200	0.4	0	1	0.2	9.8
2	25600	0.2	51200	0.1	0.1	0.5	0.2	9.8
3	12800	0.2	51200	0.05	0.15	0.25	0.2	9.8
4	6400	0.4	51200	0.05	0.35	0.125	0.2	9.8
5	3200	0.2	6400	0.1	0.1	0.5	0.2	9.8
6	1600	0.2	6400	0.05	0.15	0.25	0.2	9.8
7	800	0.4	6400	0.05	0.35	0.125	0.2	9.8
8	400	0.2	800	0.1	0.1	0.5	0.2	9.8
9	200	0.2	800	0.05	0.15	0.25	0.2	9.8
10	100	0.4	800	0.05	0.35	0.125	0.2	9.8
11	ATIK		100	0.2				

infections is clearly effective and promising application, and also can solve the major antifungal drug-resistance problem.

REFERENCES

Banerjee U, Satyanarayana T, Kunze G (2009). Opportunistic Pathogenic Yeasts . Yeast Biotechnology: Divers. and Appl. pp. 215-236.

Beil W, Sewing KF (1984). Inhibition of partially purified K+/H+ -ATPase from guinea-pig isolated and enriched parietal cells by substituted benzimidazoles, Br. J. Pharmac. 82:651-657.

Campbell CK, Holmes AD, Davey KG, Szekely A, Warnock DW (1998). Comparison of a New Chromogenic Agar with the Germ Tube Method for Presumptive Identification of Candida albicans. Eur. J. Clin. Microbiol. Infect. Dis. 17(5):367-368.

Clemons KV, Espiritu M, Parmar R, Stevens DA (2006). Assessment of the Paradoxical Effect of Caspofungin in Therapy of Candidiasis. Antimicrob. Agents Chemother. 50(4):1293–1297.

Costa AR, Silva F, Henriques M, Azeredo J, Oliveira R, Faustino A (2010). Candida Clinical Species İdentification: Mol. Biochem. Methods Ann. Microbiol. 60:105–112.

Fleischhacker M, Radecke C, Schulz B, Ruhnke M (2008). Paradoxical Growth *Effects of The Echinocandins Caspofungin and Micafungin, But Not of Anidulafungin, On Clinical Isolates of Candida albicans and C. Dubliniensis. Eur. J. Clin. Microbiol. Infect. Dis. 27(2):127-131.

Garey KW, Rege M, Pai MP, Mingo DE, Suda KJ, Turpin RS, Bearden DT (2006). Time to Initiation of Fluconazole Therapy Impacts Mortality in Patients with Candidemia: a Multi-Institutional Study. Clin. Infect. Dis. 43:25–31.

Gatta L, Perna F, Figura N, Ricci C, Holton J, D'Anna L, Miglioli M, Vaira D (2003). Antimicrobial activity of esomeprazole versus omeprazole against Helicobacter pylori. J. Antimicrob. Chemother. 51:439-442.

Guery BP, Arendrup MC, Auzinger G, Azoulay E, Borges Sa´ M, Johnson EM, Müller E, Putensen C, Rotstein C, Sganga G, Venditti M, Crespo RZ, Kullberg BJ (2009). Management of Invasive Candidiasis and Candidemia in Adult Non-neutropenic Intensive Care Unit Patients: Part II. Treatment Intensive Care Med. 35:206–214.

Harrington A, McCourtney K, Nowowiejski D, Limaye A (2007).

Differentiation of Candida albicans from Non-Albicans Yeast Directly From Blood Cultures By Gram Stain Morphology. Eur. J. Clin. Microbiol. Infect. Dis. 26:325–329.

Heelan JS, Sotomayor ER, Coon K, D'Arezzo JB (1998). Comparison of The Rapid Yeast Plus Panel with the API20C Yeast System for Identification of Clinically Significant Isolates of Candida Species. J. Clin. Microbiol. 36(5):1443–1445.

Johnson MD, Hamilton CD, Drew RH, Sanders LL, Pennick GJ, Perfect JR (2003). A randomized comparative study to determine the effect of omeprazole on the peak serum concentration of itraconazole oral solution. J. Antimicrob. Chemother. 51:453-457.

Katiyar S, Pfaller M, Edlind T (2006). Candida albicans and Candida glabrata Clinical Isolates Exhibiting Reduced Echinocandin Susceptibility. Antimicrob. Agents Chemother. 50(8):2892–2894.

Keeling DJ, Fallowfield C, Milliner KJ, Tingley SK., Ife RJ, Underwood AH (1985). Studies On The Mechanism of Action of Omeprazole. Biochem. Pharmacol. 34(16):2967-2973.

Kumar A, Roberts D, Wood KE, Light B, Parrillo JE, Sharma S, Suppes R, Feinstein D, Zanotti S, Taiberg L, Gurka D, Kumar A, Cheang M (2006). Duration of Hypotension Before Initiation of Effective Antimicrobial Therapy Is the Critical Determinant of Survival in Human Septic Shock. Crit. Care Med. 34:1589–1596.

Larner AJ, Lendrum R (1992). Oesophageal Candidiasis after Omeprazole Therapy. Gut, 33:860-861.

Larsson H, Carlsson E, Junggren U, Olbe L, Siostrand S, Skanberg I, Sundell G. (1983). Gastroenrerol. 85: 900.

Martínez JP, Gil M L, López-Ribot JL, Chaffin WL (1998). Serologic Response to Cell Wall Mannoproteins and Proteins of Candida albicans. Clin. Microbiol. Rev. 11(1):121–141.

Merlino J, Tambosis E, Veal D (1998). Chromogenic Tube Test for Presumptive Identification or Confirmation of Isolates as Candida albicans. J. Clin. Microbiol. 1157-1159.

Mogensen E, Mühlschlegel FA (2008). CO2 Sensing and Virulence of Candida albicans. Human and Animal Relationships, 2nd Edition. The Mycota VI A.A. Brakhage and P.F. Zipfel (Eds.), Springer-Verlag Berlin Heidelberg.

Molero G, Díez-Orejas, R, Navarro-García F, Monteoliva L, Pla J, Gil C, Sánchez-Pérez M, Nombela C (1998). Candida albicans: genetics, dimorphism and pathogenicity, Int. Microbiol. 1:95–106.

Monk BC, Perlin DS (1994). Fungal Plasma Membrane Proton Pumps as Promising New Antifungal Targets. Crit. Rev. Microbiol. 20(3):209-23.

Morrell M, Fraser VJ, Kollef MH (2005). Delaying the Empiric Treatment of Candida Bloodstream Infection Until Positive Blood Culture Results Are Obtained: A Potential Risk Factor For Hospital Mortality. Antimicrob. Agents Chemother. 49:3640–3645.

National Committee for Clinical Laboratory Standards (1997). Reference Method for broth Dilution Antimicrobial Susceptibility Testing of yeasts. Approved Standard M27-A. NCCLS, Wayne, PA, USA.

Reboli AC, Rotstein C, Pappas PG, Chapman SW, Kett DH, Kumar D, Betts R, Wible M, Goldstein BP, Schranz J, Krause DS, Walsh TJ (2007). Anidulafungin versus Fluconazole for Invasive Candidiasis. N. Engl. J. Med. 356:2472–2482.

Rex JH, Sobel JD (2001). Prophylactic Antifungal Therapy in the Intensive Care Unit. Clin. Infect. Dis. 32:1191–1200.

Richardson M, Elewski B (1998). Superficial Fungal Infections. Med. 33:89-90.

Shindo K, Machida M, Fukumura M, Koide K, Yamazaki R. Omeprazole Induces Altered Bile Acid Metabolism. Gut. 42:266–271.

Stevens DA, Ichinomiya M, Koshi Y, Horiuchi H (2006). Escape of Candida from Caspofungin Inhibition at Concentrations above the MIC (Paradoxical Effect) Accomplished by Increased Cell Wall Chitin; Evidence for β-1,6-Glucan Synthesis Inhibition by Caspofungin. Antimicrob Agents Chemother. 50(9):3160–3161.

Sümer Z, Kaya S, Çetin A, Hakgüdener Y (2005). In Vitro Antifungal Effect of Omeprazole in Candida Albicans and its Comparison with Fluconazole. C.Ü. Tıp Fakültesi Dergisi, 27(2):74 – 78.

Thomas G.A, Williams DL, Soper SA (2001). Capillary Electrophoresis - Based Heteroduplex Analysis With A Universal Heteroduplex Generator For Detection of Point Mutations Associated With Rifampin Resistance in Tuberculosis. Clin. Chem. 47(7):1195-203.

Wallmark B, Jaresten B, Lasod H, Ryberg B, Brandstrom A, Fellenius E (1983). Am. J. Physiol. pp.45-264.

Whitley-Williams P (2006). Candida Chapter. Infectious Disease. Congenital and Perinatal Infections A Concise Guide to Diagnosis. Humana Pres Inc., Totowa, NJ.

Zomorodi K, Houston JB (1996). Diazepam-Omeprazole Inhibition Interaction: an in vitro Investigation Using Human Liver Microsomes. Br. J. Clin. Pharmacol. 42(2):157-62.

4

Extraction, purification and characterization of L-asparaginase from *Penicillium* sp. by submerged fermentation

Krishna Raju Patro and Nibha Gupta*

Regional Plant Resource Centre, Bhubaneswar, 751 015 Odisha, India.

A fungal strain identified as *Penicillium* sp. was evaluated for its L–asparaginase enzyme production. The L-asparaginase enzyme was purified to homogeneity from *Penicillium* sp. that was grown on submerged fermentation. Different purification steps including salt precipitation, followed by separation on sephadex G-100-120 gel filtration and DEAE were applied to obtain pure enzyme preparation. The purified enzyme showed 13.97 IU/mg specific activity and 36.204% yield. The polyacrylmide gel electrophoresis of the pure enzyme exhibited one protein of 66 kDa. The enzyme showed maximum activity at 7.0 pH and 37°C and K_m value 4.00×10^{-3} M.

Key words: L – Asparaginase, fungi, *Penicillium*.

INTRODUCTION

Amino acid degrading enzymes are important chemotherapeutic agents for the cure of some types of cancers. Among them, L-asparaginase also emerged as potent health care agent for the treatment of acute lymphocytic leukemia (Bessoumi et al., 2004; Sahu et al., 2007; Gupta et al., 2009a, b; Shah et al., 2010). It is fact that tumor cells take L-asparagine from blood circulation or body fluid as it cannot synthesize L –asparagines. The presence of L-asparaginase enzyme as chemothera-peutic agents may degrade the L-asparagine present in blood circulation and indirectly starve tumor cells and lead to cell death.

Microbial enzymes are preferred over plant or animal sources due to their economic production, consistency, ease of process modification, optimization and purification. They are relatively more stable than corresponding enzymes derived from plants or animals (Savitri et al., 2003). L-asparaginase production using microbial systems has attracted considerable attention owing to the cost-effective and eco-friendly nature. A wide range of microorganisms such as filamentous fungi, yeasts, and bacteria have proved to be beneficial sources of this enzyme (Verma et al., 2007). L-asparaginase has been use as a chemotherapeutic agent for over 30 years, mainly from the bacterial strains of *Escherichia coli* and *Erwinia chrysanthemi* (Aghaiypour et al., 1999, 2001; Krasotkina et al., 2004). L-asparaginases from bacterial sources sometimes cause allergic reactions and anaphylaxis (Keating et al., 1993; Bessoumy et al., 2004; Sarquis et al., 2004). The search for other asparaginase sources, like eukaryotic microorganisms, can lead to an enzyme with less adverse effects. It has been observed that some eukaryotic microbes like yeast and filamentous fungi have a potential for L- asparaginase production (Theantana et al., 2007; Sukumaran et al., 1979; Nakahama et al., 1973). Some fungi such as *Aspergillus tamari* and *Aspergillus terreus* have proved to be beneficial sources of this enzyme (Soni, 1989). For the commercial production of enzyme, selection of superior strain and harvesting protocol is a crucial step. Besides source organisms, sufficient quantity of enzyme is also important for clinical trails. To overcome such type of

---

*Corresponding author. E-mail: nguc2003@yahoo.co.in.

difficulty, other sources of enzyme should be explored.

MATERIALS AND METHODS

Fungal strains used

Previously screened fungal isolates from the Microbiology Laboratory of Regional Plant Resources Centre having potentiality for L-asparaginase production were selected for experimentations.

Characteristics and identification

The isolates were identified on the basis of their morphology. Identification of the fungal cultures was done from Agharkar Research Institution, Pune.

Pure culture and inoculum preparation

Pure cultures were maintained on Czapeckdox agar slants and preserved at 4°C with sub-culturing in regular intervals. 5 to 6 days old cultures grown on Czapeck dox agar plates were used as inoculum.

Media used

Glucose-aspargine broth with pH 4.5 was prepared; 25 ml of the same was distributed in 150 ml flasks. The flasks were sterilized at 121°C for 15 to 20 min.

Inoculation

A single 5 mm disc of inoculum derived from the culture plates was inoculated into the flask containing broth .The culture flasks were incubated at 30°C for 10 days in static condition.

L-Asparaginase enzyme assay and protein estimation

The enzyme L-asparaginase was assayed by estimating the amount of ammonia released in the reaction (Imada et al., 1973). The amount of ammonia released by the test sample was calculated with reference to the standard graph. The enzyme activity was expressed in terms of enzyme units (IU/ml). The protein estimation was done by Bradford's method (Bradford, 1976).

Large scale production and extraction, purification, and characterization

The organisms were inoculated in the aforementioned broth media and incubated for a period of 10 days in static condition at 30°C. They were harvested after due incubation period. The biomass was weighed and was macerated with 0.05 M Tris-HCL buffer, pH 8.5 in the ratio 1:5. It was centrifuged at 600 rpm for 20 min at 4°C and the supernatant was collected. This was treated as the crude preparation of the enzyme. It was estimated for enzyme activity and protein, and was subjected for first purification step with ammonium sulphate precipitation.

Ammonium sulphate precipitation

The crude extract was precipitated with finely powdered ammonium

sulphate with 80% saturation. It was left at 4°C over night followed with centrifugation at 8000 rpm for 20 min. The pellet was collected and was dissolved in 0.05 M Tris-HCl buffer, pH 8.5.

Sephadex G-100-120 gel filtration

Ammonium sulphate precipitated samples were tested for enzyme activity and protein and were further subjected to gel filtration-using sephadex G-100-120, with bead size 40 to 120 µl.

Preparation of gel-column and application of sample

A chromatography column made up of glass tubing having a diameter of 2.2 cm and a height of 60 cm was used. 0.05 M Tris-HCl buffer, pH 8.5 was used as eluent. The eluent was stored in tightly stoppered brown bottles and brought at the same temperature as that of the gel bed, to prevent bubble formation within the gel bed. For the preparation of gel slurry, 10 g of sephadex was suspended in 400 ml of 0.05 M Tris-HCl buffer and left for 24 h to swell at room temperature. The column was packed with sephadex and stabilized. Ammonium sulphate precipitated samples were poured continuously into the column and fractions were collected in vials (5 ml each). The collected fractions were tested for enzyme activity and protein randomly, and the fractions showing better enzyme activity were pooled together.

Ion exchange chromatography

Required amount of DEAE-cellulose was dissolved in Tris-HCl buffer and was left over night and was used to make the column. The column was packed; initially it was washed with 5 M NaOH to remove ionic charges from the ion exchanger then with 5 M KCl to generate desired form of weak ion exchange material. Finally, it was washed with distilled water followed by 0.05 M Tris-HCl buffer (pH 8.5). Finally, pooled peak fractions collected from sephadex filtration were applied to the ion exchange column. 5 ml fractions were collected. Samples were analyzed for enzyme activity and protein; fractions having better enzyme activity were pooled together and stored in deep freezer for later use.

Electrophoretic separation of L-asparaginase

Gel electrophoresis was performed for determining the homogeneity of the pure enzyme and to estimate the molecular weight. The resolving gel (dist. water 3.75 ml, resolving buffer 2.5 ml, acrylamide solution 3.6, 10% SDS 100 µl, 10% APS 57.1 µl, TEMED 42.85 µl) was prepared and was pipetted into the Sandwich. 100 ml of water-saturated butanol was added to the gel and was allowed to polymerize for 30 TO 40 min. After the due time, the butanol solution was decanted out and washed with ddH$_2$O and soaked with tissue paper. Stacking gel was prepared (dist. water 3.0 ml, stacking buffer 0.65 ml, acrylamide solution 1.2 ml, 10% SDS 50 µl, 10% APS 50 µl, TEMED 6 µl) added above the resolving gel. The comb was inserted and allowed to polymerize for 30 to 45 min). After the polymerization, the gel was clamped in the electrophoresis unit and the comb was removed carefully after adding the tank buffer. A standard marker and denatured sample to each well were loaded; electrophoresis was run at 40 V for 1 h and then 55 V for 3 h. When the dye reaches to the bottom of the gel the supplied power was turned up and the gel unit was removed from the electrophoresis unit. The plates were gently separated with the help of a spatula and the gel was carefully transferred into the staining solution for staining overnight. Next, it was distained for 7 h or overnight. Then, it was paced in the fixer solution and the bands

Table 1. Analysis of L-asparaginase in different purification steps.

Sample	Enzyme (IU/ml)	Protein (mg/ml)
Crude	2.058	0.280
Am.Sulphate precipitate	10.000	1.220
Sephadex gel filtration f3	0.000	0.400
6	0.000	0.870
9	10.294	0.800
12	12.941	0.790
15	10.117	0.820
18	9.705	0.840
21	10.235	0.850
24	0.000	0.270
27	0.000	0.000
30	0.000	0.000
Biologic LP-Ion exchange fraction No- 4	0.588	0.006
5	17.058	0.160
6	34.117	0.320
7	10.00	0.156
8	2.058	0.010
Precipitatedsample-fraction-6 of biologic LP	33.529	2.400

were observed under Gel Doc. The molecular weight of the protein was measured with reference to the standard marker used.

Characterization of the partially purified enzyme

Substrate specificity

Substituting L-asparagine with different substrates like L-arginine, L-Phenylalanine, L-Histidine, L-glutamine and L-aspartic acid were used in the assay mixture to determine the substrate specificity of the enzyme. The substrates had a concentration of 0.04 M, keeping L-asparagine as control.

pH optima

pH of the Tris-HCl buffer added in the reaction mixture for enzyme activity was adjusted to 3, 4, 5, 6, 7, 8, 9 and 10 respectively, and the pH optima was determined by detecting the enzyme activity at each level.

Temperature tolerance

The enzyme was kept at a varying temperature of 30, 37, 45 and 50°C before adding into the reaction mixer for assay.

Determination of K_m

The K_m of the enzyme was determined by making the reaction mixture containing fixed volume (0.25 ml) of the partially purified enzyme and varying concentration of the substrate L-asparagine. The total volume of the mixture was made up to 2 ml with 0.5 M Tris-HCl buffer of pH 7.2 and the enzyme activity was estimated. A graph of the substrate concentration was plotted against the reaction velocity.

RESULTS AND DISCUSSION

The *Penicillium* sp. was grown in basal glucose asparagines medium (pH 4.5 at 30°C for 10 days in static conditions). The fungus exhibited the L-asparaginase production under submerged culture developed in basal glucose asparagine medium. The partial purification of L-asparaginases crude preparation was affected by ammonium sulphate precipitation (80%) and showed that most of the enzyme activity was preserved in the precipitate (Table 1). The total protein decreased from 126 to 14 mg in ammonium sulphate precipitation. The specific activity increased from 7.350 to 13.970 IU/mg after sephadex gel filtration and ion-exchange chromatography respectively (Table 2).

The pH optimization of the enzyme was studied using a 0.05 M Tris-HCl buffer of different pH values ranging from 3 to 10 showed maximum enzyme activity at pH 7 (Table 3). The purified enzyme showed maximum activity 37°C. The analysis of substrate specificity showed that the enzyme catalyzes L-asparagine as a substrate. A Lineweaver-Burk analysis of the enzyme showed the K_m of 4.00×10^{-3} M. Molecular weight of L-asparaginase-SDS-PAGE of the enzyme showed the presence of the single peptide chain of ~66.00 kDa (Figure 1). This is a preliminary study on L-asparaginase from *Penicillium* sp. Further exploratory research is required to optimize the nutritional and cultural amendments in order to develop

Table 2. Purification profile of L-asparaginase from *Penicillium* sp.

Steps	Collected volume (ml)	Total activity (IU)	Total protein (mg)	Specific activity (IU/mg)	Purification (fold)	Yield (%)
Crude extract	450	926.10	126.00	7.350	0.000	100.00
Ammonium sulphate precipitation	90	900.00	109.8.	8.196	1.115	97.181
Sephadex gel filtration	65	692.77	53.30	12.997	1.768	74.850
Ion-exchange chromatography	10	335.29	24.00	13.970	1.900	36.204

Table 3. Effect pH, temperature and substrate on partially purified L-asparaginase.

Parameter		Enzyme activity (IU/ml)
pH	4	20.882
	5	26.470
	6	24.705
	7	35.882
	8	16.176
	9	13.529
	10	7.941
Temperature	30	20.294
	37	33.529
	45	19.411
	50	12.352
Substrates	L-asparagine	33.529
	L-aspartic acid	0
	L-Phenylalanine	0
	L-Glutamine	0
	L-histidine	0

bioprocess technology.

Several fungi are reported as producer of L–asparaginase enzyme (Mohapatra et al. 1997; Drainas and Draines 1985; Raha et al., 1990; Gupta et al., 2009a; Siddalingeswara and Lingappa, 2010b). The *Penicillium* sp. showed good enzyme activity in cell biomass.

L-asparaginase of different microbes has different substrate affinity and species play different ecophysiological roles in the enzyme activity (Warangkar and khobragade, 2010). Our fungal strain produces enzyme of K_m 4.0×10^{-3} M. It is comparatively higher than other organisms like *vibrio succinogens* (0.0745 mM) and *Pseudomonas aeruginosa* (0.147 mM) (Willis and Woolfolk, 1974; Bessouny et al., 2004). Higher K_m value exhibited its high affinity with the L-asparagine as substrate. The enzyme production is the complex chain reactions and is supported and induced by suitable substrates. *Penicillium sp.* preferred L–asparagines as substrate. This characteristic phenomenon of *Penicillium*

sp. was corroborated with the Dunlope and Roon (1975) who noted the increment in enzyme production due to the addition of L glutamine or glutamate in the fermentation medium. The purification steps followed for the L-asparaginase from this fungal species achieved a protein of single peptide chain of 66 KDa. The temperature tolerance of the enzyme showed that it had maximum activity at 37°C and may be quite stable at high temperature too. Enzymatic activity was optimum at pH 7. It clearly indicates that L- asparaginase from this fungal species is pH dependant. This observation is corroborated with the earlier studies done by Sahu et al. (2007). Since, the work has been done under batch culture; further experimentation for the development of fermentation technology in mega scale production of this enzyme is needed. Any drug discovery and development program necessitate the standardization of small scale protocols, their exploration and standardization for mega scale, *in vitro* and *in vivo* trials and clinical trials etc, the

Figure 1. Gel electrophoresis analysis of L-asparaginase. Lane 1: BSA (66 kDa), Lane 2: pure enzyme (66 kDa).

present work and its outcome may be a mile stone in drug development program.

ACKNOWLEDGEMENT

The financial support received from the Ministry of Earth Sciences, Govt. of India (sanction No. MoES/11/MRDF/1/30/P/08) is gratefully acknowledged.

REFERENCES

Aghaiypour K, Wlodawer A, Lubkowski J (2001). Structuralbasis for the activity and the substrate specificity of *Erwinia chrysantemi* L-Asparaginase. Biochemistry 40:5655-5664.

Aghaiypour K , Farzami B, Mohammadi AA,Vand-Yoosefi J, Pasalar P, Safavieh S (1999). Comparative study of L-asparaginase activity in different bacterial species. Arch. Razi Inst. 50:51-58.

Bradford MM (1976). A Rapid and Sensitive Method for the Quantification of Microgram Quantities of Protein Utilizing the Principle of Protein-Dye Binding, Anal. Biochem. 72:248-254.

Bessoumy AA, Sarhan M, Mansour J (2004). Production, Isolation, and Purification of L-Asparaginase from *Pseudomonas aeruginosa* 50071 Using Solid-state Fermentation, J. Biochem. Mol. Biol. 37(4):387-393.

Dunlop PC, Roon RJ (1975). L-asparaginase of *Sacharomyces cerevisiae*: an Extracellular Enzyme. J. Bacteriol. 122(3): 1017-1024.

Gupta N, Dash SJ, Basak UC (2009a). L- asparaginases from fungi of Bhitarkanika mangrove ecosystem, AsPac. J. Mol. Biol. Biotech., 17(1):27-30.

Gupta N, Sahoo D, Basak UC (2009b). Screening of Streptomyces for L-asparaginase Production. Acad. J. Can. Res. 2(2):92-93.

Imada A, Igarasi S, Nakahama K, Isono M (1973). Asparaginase and Glutaminase activities of microorganisms, J. Gen. Microbiol. 76:85-99.

Keating MJ, Holmes R, Lerner SH (1993). L-asparaginase and PEG asparaginase past, present and future. Leuk. Lymp. 10:153-157.

Krasotkina J, Borisova AA, Gevaziev YV, Sokolov N (2004). One-step purification and kinetic properties of the recombinant L-Asparaginase from *Erwinia Carotovora*. Biotech. Appl. Biochem. 39:215-221.

Nakahama K, Imada A, Igarasi S, Tubaki K (1973). Formation of L Asparaginase by Fusarium Species. J. Gen. Microbiol. 75:269-273.

Sahu MK, Sivakumar K, Poorani E, Thangaradjou T, Kannan L (2007). Studies on L-asparaginase enzyme of *actinomycetes* isolated from estuarine fishes. J. Environ. Biol. 28(2):465-474.

Sarquis M, Oliveira EMM, Santos AS, Costa GLD (2004). Production of L-asparaginase by Filamentous Fungi, *Mem Inst Oswaldo Cruz*, Rio de Janeiro, 99(5):489-492.

Savitri NA, Azmi W (2003). Microbial L-asparaginase: A Potent Antitumour Enzyme. Ind. J. Biotechnol. 2:184-194.

Shah AJ, Karadi RV, Parekh PP (2010). Isolation, optimization and production of L-asparaginase from *Coliform* Bacteria. Asian J. Biotechnol. 2(3):169-177.

Soni K (1989). Isolation and Characterization of L-Asparaginases from microorganisms. Ph.D. Thesis. Department of Biosciences. Pt. Ravishankar University, India.

Sukumaran CP, Singh DV, Mahadevan PR (1979). Synthesis of L-asparaginase by *Serratia marcescens* (Nima). J. Biosci. 1(3):263-269.

Theantana T, Hyde KD, Lumyong S (2007). Asparaginase production by Endophytic fungi isolated from some Thai medicinal plants. KMITL Sci. Tech. J. 7(S1):13-18.

Verma N, Kumar K, Kaur G, Anand S (2007). L-asparaginase: a promising chemotherapeutic agent. Crit. Rev. Biotechnol. 27:45-62.

Willis RC, Woolfolk R (1974). Aspraginase utilization in *E. coli*. J. Bacteriol. 118:231-235.

5

Theileria parva apical membrane antigen-1 (AMA-1) shares conserved sequences with apicomplexan homologs

Nyerhovwo J. Tonukari

Department of Biochemistry, Delta State University, P. M. B. 1, Abraka, Nigeria. E-mail: tonukari@gmail.com.

The apical membrane antigen-1 (AMA-1) is a leading malaria vaccine candidate that is expressed in mature stage parasites and is thought to be essential for invasion. In *Plasmodium falciparum*, AMA-1 is localized initially to the micronemes, apical organelles of the parasite. A similar apicomplexan parasite, *Theileria parva* infects and transforms lymphocytes of cattle and African buffalo causing the disease called East Coast fever (ECF). The AMA-1 homolog in *T. parva* was isolated, cloned and sequenced. The predicted amino acid sequence was further analyzed. The partial gene sequence, which is 1422 bp long and encoding 473 amino acid residues, is located on chromosome 1 of the *T. parva* genome. The AMA-1 homolog in *T. parva* shares conserved sequences with apicomplexan homologs. It is 64% similar to the AMA-1 in *Babesia bigemina*, and 56% similar to the homolog in *P. falciparum*. Given the importance of AMA-1 in invasion and the central role invasion plays in pathogenesis, the *T. parva* AMA-1 may likely have implications for vaccine design in the East Coast fever disease.

Key words: *Theileria parva*, East Coast fever, apical membrane antigen-1, apicomplexan parasite.

INTRODUCTION

Apicomplexa is a protozoan class of obligate intracellular parasites that includes many important human and animal pathogens. Proteins with constitutive or transient localization on the surface of *Apicomplexa* parasites are of particular interest for their potential role in the invasion of host cells (Healer et al., 2004). Apical membrane antigen-1 (AMA-1) is a microneme protein that is highly conserved among apicomplexan parasites (Waters et al., 1990; Donahue et al., 2000; Hehl et al., 2000; Gaffar et al., 2004). AMA-1 is a leading malaria vaccine candidate that is expressed in mature stage parasites and is thought to be essential for invasion (Peterson et al., 1989; Waters et al., 1990; Remarque et al., 2008). *Plasmodium falciparum* AMA-1 (*Pf*AMA-1) is a type I integral membrane protein that is produced as an 83-kDa precursor and is localized initially to the micronemes, apical organelles of the parasite (Bannister et al., 2003; Healer et al., 2002). Six to eight conserved intramolecular

Abbreviations: AMA-1, Apical membrane antigen-1; *Pf*AMA-1, *Plasmodium falciparum* apical membrane antigen-1; **BLAST,** basic local alignment search tool; **PCR,** polymerase chain reaction.

disulfide bonds constrain this protein into three distinct domains (Waters et al., 1990; Hodder et al., 1996; Donahue et al., 2000; Hehl et al., 2000; Gaffar et al., 2004; Pizarro et al., 2005). AMA-1 has been shown to elicit a protective immune response against merozoites dependent on the correct pairing of its numerous disulfide bonds. Shortly after synthesis, the AMA-1 precursor is proteolytically cleaved to a 66-kDa product which is translocated onto the merozoite surface where much of the ectodomain is shed during invasion (Narum and Thomas, 1994; Donahue et al., 2000; Hehl et al., 2000; Howell et al., 2003; Crewther et al., 1996; Remarque et al., 2008).

These biological features of AMA-1, together with its relative conservation within the genus, suggest a role for this molecule in erythrocyte invasion and thus support its candidacy as a vaccine antigen. AMA-1 has also been identified in *Plasmodium knowlensi* and *Plasmodium fragile*. This antigen can induce protection in mice and monkeys with monoclonal antibodies blocking the invasion of rhesus erythrocytes by *P. knowlensi* merozoites *in vitro* (Xu et al., 2000). Using a different simian system (Collins et al., 1994), it was possible to obtain partial protection by immunization with recombinant *P.*

fragile AMA-1 produced in a baculovirus expression system. Several observations indicate a role for apical membrane antigen-specific antibody-independent T cell-mediated immunity against *Plasmodium chaubadi adami* infection in mice (Xu et al., 2000; Harris et al., 2005). Also, *Plasmodium vivax* AMA-1 was immunogenic during natural malaria infections in Sri Lanka, where low transmission and unstable malaria conditions prevailed (Wickramarachchi et al., 2006).

Apical membrane antigens and several other known apicomplexan genes act as target for membrane anchorage due to the presence of a signal peptide or characteristic peptide structure of glycosylphosphatidylinositol (GPI) anchored proteins. Gardner et al. (2005) reported the genome sequence of *Theileria parva*, an apicomplexan pathogen causing economic losses to smallholder farmers. *T. parva* infects and transforms lymphocytes of cattle and African buffalo causing the disease called East Coast fever (ECF). Transmitted by *Rhipicephalus appendiculatus* ticks, the parasite causes a severe lymphoproliferative disease of cattle in eastern, central, and southern Africa (Katzer et al., 2006). It is an intracellular parasite that infects and transforms bovine lymphocytes. Gene homologs encoding antigens from other apicomplexan parasites constitute a source of possible vaccine candidate antigens (Tonukari and Kangethe, 2009a). Here, the identification and characterization of the *T. parva* TpAMA-1 homolog is presented. Its relatedness with other known apicomplexan genes from related organisms was also evaluated.

MATERIALS AND METHODS

Identification of antigen's homolog

The homologs of apicomplexan antigens were selected from the *T. parva* genome sequence database (http://www.tigr.org/tdb/e2k1/tpa1/), in the following manner:

1. Apicomplexan gene products that have been described as antigenic in literature were selected for analysis.
2. The selected gene sequences were used to perform a basic local alignment search tool (BLAST) search (Altschul et al., 1990; Altschul et al., 1997) of the *T. parva* genome database to identify homologous genes.
3. Those identified in the *T. parva* database were translated to their amino acid sequences and analyzed using SignalP-2.0 and TMHMM software for the presence of signal peptides and transmembrane domains. This would identify secreted proteins or proteins located on the surface of the parasite.
4. The selected *T. parva* sequences were then confirmed in the GenBank database:

a) *T. parva* DNA sequences were subjected to BLASTn to seek identical DNA sequences and BLASTx on non redundant databases that compare amino acid sequences to confirm the homologous apicomplexan antigens originally selected in 1.
b) *T. parva* translated amino acid sequences were subjected to BLASTp to also confirm the homologous apicomplexan antigens originally identified to 1. Further analysis was carried out on

identified homologs e.g. predicted isoelectric point, protein motifs, etc.

Polymerase chain reaction (PCR) and cloning

PCR was performed with a thermocycler (MJ Research, Watertown, MA) using *Taq* DNA polymerase (Promega) and two primers based on the sequences identified in *T. parva* using cDNA library as template (Graham et al., 2006). The PCR product generated above was cloned into pGEM T-easy vector (Promega, Madison, WI). Vector specific primers, both forward and reverse were synthesized as indicated; 5' GCCGCCACCATGAGTTTTAGCCCTAACACTGCTGA 3' (forward primer) and 5' TTATATGAATGGTCCTGAAGAAACGGGT 3' (reverse primer). These primers were then used to amplify the TCP-1 ortholog gene from cDNA (Graham et al., 2006). PCR was performed in the following conditions: initial denaturation at 94°C for 3 min, 35 cycles of denaturation, 94°C for 1 min; annealing at 55°C for 1 min and polymerization at 72°C for 2 min. A final round of polymerization at 72°C for 10 min was performed at the end of the 35 cycles. Aliquots (10 μl) of PCR products with 5 μl loading buffer were loaded onto a 0.8% TAE agarose gel, stained with ethidium bromide and visualized on a UV transilluminator.

PCR products were extracted and purified using QIA quick Gel DNA Extraction Kit protocol (QIAGEN Co.) and the purified PCR products ligated into pGEM-T Vector (Promega Co., USA) according to the manufacturer's instructions. 1 μl of ligation reaction was transformed into *Escherichia coli* strain DH5-α competent cells. DNA nucleotide sequences were determined by gel based sequencing at the International Livestock Research Institute sequencing unit in Nairobi, Kenya.

Sequence analysis

The translated *T. parva* amino acid sequence was compared with sequences present in the GenBank and analyzed using BLAST (Altschul et al., 1997; http://ncbi.nlm.nih.gov/blast/) and ClustalW multiple sequence alignment program (http://www.ebi.ac.uk/Tools/clustalw2/).). Clustal W is a general tool used for multiple alignments of DNA sequences. The *T. parva* protein hydrophobicity was also plotted; this characterizes its hydrophobic and hydrophilic character, which may be useful in predicting membrane-spanning domains, potential antigenic sites and regions that are likely exposed on the protein surface (Kyte and Doolittle, 1982; Hopp and Woods, 1981; Gomase and Changbhale, 2007).

RESULTS AND DISCUSSION

Characterization of the AMA-1 homolog in *T. parva* was carried out because of the similarity in peptide structure with other apicomplexan homologs and for its potential as a T cell target. The coding sequence of the *T. parva* partial AMA-1 is 1422 bp (Figure 1) and the translated amino acid sequence is 473 residues. The hydrophobicity plot reveals several transmembrane domains in *T. parva* AMA-1 (Figure 2). A BLAST search (NCBI) indicates that *T. parva* AMA-1 is 100% identical to the amino acid sequence obtained from the translated genome sequence (Gardner et al., 2005) with a conspicuous AMA-1 superfamily motif (Figure 3). It is also 64% similar to the AMA-1 in *Babesia bigemina*, and 56% similar to

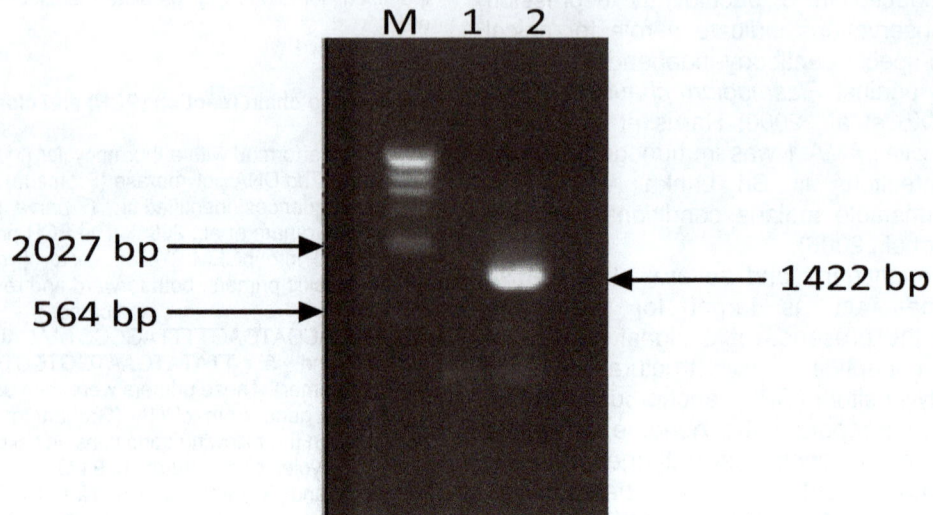

Figure 1. PCR product of *T. Parva* AMA-1 using the *T. parva* cDNA library as a template. Lane M, Molecular size markers; lane 1, negative control; lane 2, *T. Parva* AMA-1 PCR product.

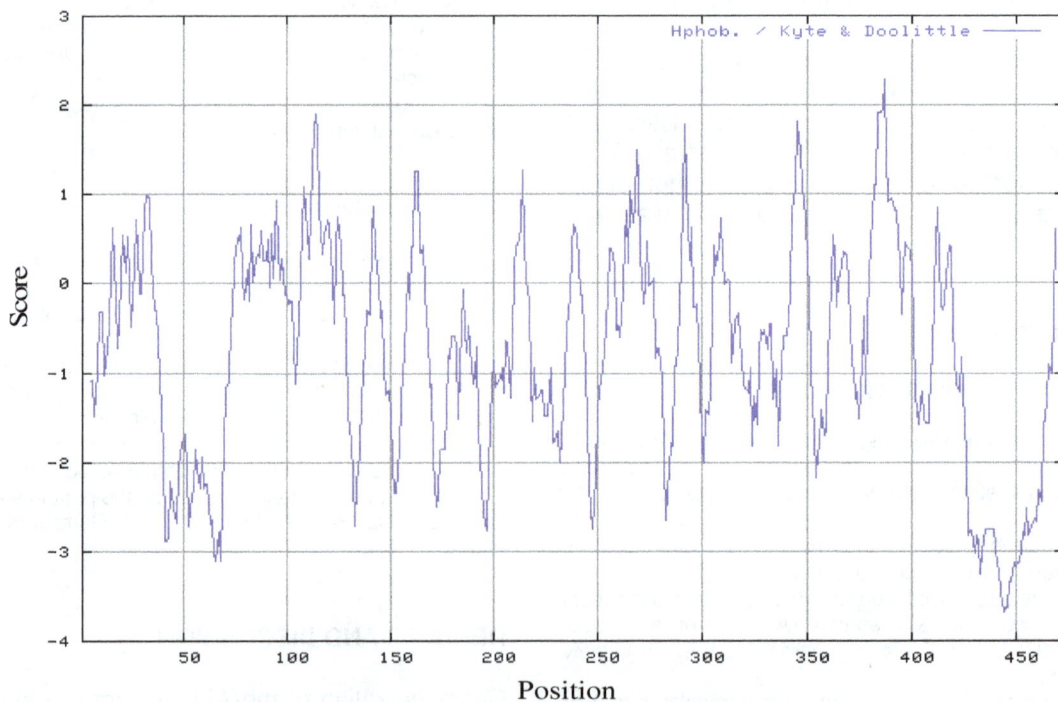

Figure 2. Kyte-Doolittle hydrophobicity plot of AMA-1 protein of *T. parva*. The hydrophobic character may be useful in predicting membrane-spanning domains, potential antigenic sites and regions that are likely exposed on the protein's surface.

the homolog in *P. falciparum*. Comparison of the *T. parva* AMA-1 with protein homologs from *P. falciparum* and *B. bigemina* is shown in Figure 4. The conserved domains are indicated. Specific BLAST search of the *T. parva* genome (TIGR) using the nucleotide sequence as query shows that AMA-1 is located on chromosome 1.

Other putative *T. parva* antigenic genes have been cloned and expressed in *E. coli* (Tonukari and Kangethe, 2009a, b). Identification of the AMA-1 ortholog in *T. parva* and other apicomplexan parasites suggests a conserved function across species (Hehl et al., 2000). An antigen such as AMA-1 indicates a role for apical membrane

```
                  1        75        150        225       300          375      450  473
Query seq.       |------------------------------------------------------------------------|
Superfamilies                    |                AMA-1 superfamily                    |
```

Figure 3. AMA-1 superfamily motif in the *T. parva* AMA-1 amino acid sequence obtained using protein BLAST.

```
T. parva       -SFSPNTADNTSLVET-------RSSVLNTTLGRFGSFLQSGLITSKSEKKKRTGANRRA 52
B. bigemina    ---MQCILNRITLLATPVIFFLWLSTEISPAGCAFVTFQNEPTSTRGTRRSSRSSRHQQA 57
P. falciparum  NSDVYHPINEHSEHPT------EYEYPLHQEHTYQQEDSGEDENTLQHAYPIDHEGAEPA 54
                      :. :    *       .  :      .      *          . *

T. parva       P--KGKKGKGGETEEK-----------RNKWTEFMAKFDIAKVHGSGVYVDLGESATVGI 99
B. bigemina    TSSTSQAGAGEATERTGGRTAGSKLIPQTPWTRYMIKYDIARCHGSGIFVDIGGYEAVGN 117
P. falciparum  PQEQNLFSSNEIVERSN--------YMGNPWTEYMAKYDIEEVHGSGIRVDLGEDAEVAG 106
                .       .*..        .   **.:* *:** . ****: **:*     *.

T. parva       YDYRMPIGKCPVVGKAIILENG-ADFLSSITHHDP--KERGLGFPATKVASN-SSKQDME 155
B. bigemina    KYYRMPTGKCPVMGKVISLASG-ADFLEPISADNP--RYRGLAFPETVIKHTGASAGALT 174
P. falciparum  TQYRLPSGKCPVFGKGIIENSNTTFLTPVATGNQDLKDGGFAFPPTNPLMSPMTLDDMR 166
                 **:* *****.** *   : .. :  ** .:: :    :   *:.** *   . :   :

T. parva       NQ-----LLSPISAQVLRSWNYKHESDLSNCAEYSRNIVPGSNRNSKYRYPFVYDESEKL 210
B. bigemina    NAGNIHGNLSPVSAADLRKWGYKGN-AVTNCAEYASNIVPGSDQRTKYRYPFVYDGKEEM 233
P. falciparum  RFYK-----------DNEYVKNLDELTLCSRHAGNMIPDNDKNSNYKYPAVYDDKDKK 213
                .                *    :: *:.:: *::*..::.::*:** *** .::

T. parva       CYILYSPMQYNQGVKYCDKD-SADEGTSSLACMYPDKSKDDSHLFYGTSGLHMDWPVVCP 269
B. bigemina    CYILYSPMQYNQGTRYCDEDGSAKEGPSSLLCMKPYKSEADAHLYYGSARIDPKWDQNCP 293
P. falciparum  CHILYIAAQENNGPRYCNKDQSKRN---SMFCFRPAKDKSFQNYTYLSKNVVDNWEKVCP 270
                *:***  . *  *:* :**::** :    *: * *.: :  * : .*    **

T. parva       VYPIRDSIFGSYDDEKDECVPIDPIFVEDADDYEECAKIIFEYSPSDVDISTNNQKLSDV 329
B. bigemina    MKPIKDAIFGTWVSG--ACVALESAFEEYVDSAEECAAILFEHSAADVDIDIESERYNEI 351
P. falciparum  RKNLQNAKFGLWVDG--NCEDIPHVDEFSANDLFECNKLVFELSASDQPK-QYEQHLTDY 327
                ::::  **  :.   *    :    .:.  ** ::** *.:*     .:: .:

T. parva       DLYKEAMNNGKLSTALSIMFAPR-YSEDRPIYTKGVGINWATYSVEEKKCNILDVVPTCL 388
B. bigemina    SELYNGLKNLKLQQIAFSLFAPMAKSAASATLSKGVGMNWANYESETGVCRILNATPTCL 411
P. falciparum  EKIKEGFKN-KTASMIKSAFLPTGAFKADRYKSHGKGYNWGNYTAETQKCEIFNVKPTCL 386
                .    :.::*  *      * *            ::*  * **..*   *    *.*::. ****

T. parva       IISNGYYALTSLSSPNEDDAINYPCNI--------------------------------- 415
B. bigemina    IINAGSLAMTALGSPLESDAINYPCHIDTLGYVEPRKRDSREDGDRNSGITTALNMKTLK 471
P. falciparum  INNSSYIATTALSHPIEVEHN-FPCSLYKDEIMKEIERESKRFKLNDNDDEGNKKIIAPR 445
                *   . .  *  *:*. * *  :   :** :

T. parva       ----VHGK---------------GFLK--------------------------------- 423
B. bigemina    CTKYVHSKYSESCGTYYYCSEEKSGYLSRLYQFMCSHNVKKAAVISTALVLLCLAIYWIY 531
P. falciparum  IFISDDKDSLKCPCDPEIVSNSTCNFFVCKCVERRAEVTSNN------------------ 487
                     . .                .  ::

T. parva       ----NPNTGKKKDQKPPEPPKDEKQNKKEEESKPKEKGKTEQ----------TNKTPVSS 469
B. bigemina    QRLWSTKKGRQHDDYDRLMSKYEYDDVSHDNIEPEHQLRTDAYIWGEAAARPSDITPVHL 591
P. falciparum  EVVVKEEYKDEYADIPEHKPTYDNMKIIIASSAAVAVLATILMVYLYKRKGNAEKYDKMD 547
                   . :  :       .. :     .     .    .     *          ::

T. parva       GPFI--------------- 473
B. bigemina    TKLN--------------- 595
P. falciparum  QPQDYGKSKSRNDEMLDPE 566
```

Figure 4. Amino acid sequence alignment showing the similarities between AMA-1 in *T. Parva* and its homologs in *B. Bigemina* and *P. falciparum*. Conserved amino acid residues are indicated by "*" while similar residues are indicated by "." or ":".

antigen-specific Ab-independent T cell-mediated immunity (with CD4[+] T cells acting independently of antibodies to contribute to protective immunity) against *P.chaubadi adami* infection in mice (Xu et al., 2000), whereas the same antigen in *P. knowlensi* can induce protection in monkeys with monoclonal antibodies blocking the invasion of rhesus erythrocytes by *P. knowlensi* merozoites *in-vitro* (Xu et al., 2000). It has also been reported that AMA-1 is a cross-reactive antigen between *Neospora caninum* and *Toxoplasma gondii* and a potential common vaccine candidate to control two parasites (Zhang et al., 2007).

Immunological studies have shown that anti-AMA-1 polyclonal and monoclonal antibodies block parasite invasion *in vitro* (Kennedy et al., 2002; Kocken et al., 2000; Taylor et al., 2002; Collins et al., 2007; Sabo et al., 2006). AMA-1 is expressed in sporozoites, and anti-AMA-1 antibodies block parasite invasion of liver cells, suggesting a role for AMA-1 in both erythrocyte and heaptocyte invasion (Silvie et al., 2004). AMA-1 is also a target of the naturally acquired immune response, and anti-AMA-1 antibodies isolated from individuals exposed to malaria block parasite invasion *in vitro* (Hodder et al., 2001; Thomas et al., 1994; Remarque et al., 2008). Moreover, stage-specific expression and localization of AMA-1 indicate a role in invasion, and AMA-1 knockouts are not viable, implying that AMA-1 is critical in the parasite life cycle (Triglia et al., 2000). However, allelic polymorphisms in apical membrane antigen-1 are responsible for evasion of antibody-mediated inhibition in *P. falciparum* (Healer et al., 2004), and are maintained by balancing selection arising from host immune recognition (Verra and Hughes, 1999). This may pose a problem for a vaccine based on this antigen (Bai et al., 2005). Nevertheless, Narum et al. (2006) have shown that passive immunization with a multicomponent vaccine against conserved domains of apical membrane antigen 1 and 235-kilodalton rhoptry proteins protects mice against *Plasmodium yoelii* blood-stage challenge infection.

Two of the three distinct AMA-1 domains (Waters et al., 1990; Hodder et al., 1996; Donahue et al., 2000; Hehl et al., 2000; Gaffar et al., 2004; Pizarro et al., 2005) (I and II) belong to the PAN module superfamily, suggesting that they may function in adhesion to protein or carbohydrate receptors (Pizarro et al., 2005). While Kato et al. (2005) determined that domain III of AMA-1 binds to the erythrocyte membrane protein, Kx. Trans-species complementation experiments (Triglia et al., 2000) and heterologous expression experiments (Fraser et al., 2001; Kato et al., 2005) suggest that AMA-1 may function as an adhesin during host cell invasion.

Host cell invasion is a critical step in the pathogenesis of the diseases caused by apicomplexan parasites. Given the importance of AMA-1 in invasion and the central role invasion plays in pathogenesis, the *T. parva* AMA-1 may likely have implications for vaccine design in East Coast fever disease. *T. gondii* tachyzoites carrying a conditional knockout of apical membrane antigen-1 are severely compromised in their ability to invade host cells (Mital et al., 2005), providing direct genetic evidence that AMA-1 functions during invasion.

ACKNOWLEDGEMENTS

Part of this study was carried out at the International Livestock Research Institute, Nairobi, Kenya. The author is grateful to Richard T. Kangethe for his assistance.

REFERENCES

Altschul SF, Gish W, Miller W, Myers EW, Lipman DJ (1990). Basic local alignment search tool. J. Mol. Biol. 215(3): 403-410.

Altschul SF, Madden TL, Schäffer AA, Zhang J, Zhang Z, Miller W, Lipman DJ (1997). Gapped BLAST and PSI-BLAST: a new generation of protein database search programs. Nucleic Acids Res. (17): 3389-3402.

Bai T, Michael Becker, Aditi Gupta, Phillip Strike, Vince J. Murphy, Robin F. Anders, Adrian HB (2005). Structure of AMA-1 from *Plasmodium falciparum* reveals a clustering of polymorphisms that surround a conserved hydrophobic pocket. PNAS. 102(36): 12736-12741.

Bannister LH, Hopkins JM, Dluzewski AR, Margos G, Williams IT, Blackman MJ, Kocken CH, Thomas AW, Mitchell GH (2003). *Plasmodium falciparum* apical membrane antigen 1 (PfAMA-1) is translocated within micronemes along subpellicular microtubules during merozoite development. J. Cell Sci. 116: 3825-3834.

Collins CR, Withers-Martinez C, Bentley GA, Batchelor AH, Thomas AW, Blackman MJ (2007). Fine mapping of an epitope recognized by an invasion-inhibitory monoclonal antibody on the malaria vaccine candidate apical membrane antigen 1. J. Biol. Chem. (10): 7431-7441.

Collins WE, David P, Crewther PE, Kirsten LV, Galland GG, Alexander JS, David JK, Stirling JE, Ross LC, Joann SS, Carla LM, Robin FA (1994). Protective immunity induced in squirrel monkeys with recombinant apical membrane antigen-1 of *Plasmodium fragile*. Am. J. Trop. Med. Hyg. 51: 711-719.

Crewther PE, Mary LSMM, Robert HF, Robin FA (1996). Protective Immune Responses to Apical Membrane Antigen 1 of *Plasmodium chabaudi* Involve Recognition of Strain-Specific Epitopes. Infect. Immun. 64(8): 3310–3317.

Donahue CG, Carruthers VB, Gilk SD, Ward GE (2000). The *Toxoplasma* homolog of *Plasmodium* apical membrane antigen-1 (AMA-1) is a microneme protein secreted in response to elevated intracellular calcium levels. Mol. Biochem. Parasitol. 111: 15-30.

Fraser TS, Kappe SH, Narum DL, VanBuskirk KM, Adams JH (2001). Erythrocyte-binding activity of *Plasmodium yoelii* apical membrane antigen-1 expressed on the surface of transfected COS-7 cells. Mol. Biochem. Parasitol. 117: 49-59.

Gaffar FR, Yatsuda AP, Franssen FF, de Vries E (2004). Erythrocyte invasion by *Babesia bovis* merozoites is inhibited by polyclonal antisera directed against peptides derived from a homologue of *Plasmodium falciparum* apical membrane antigen 1. Infect. Immun. 72, 2947-2955.

Gardner MJ, Richard B, Trushar S, de Villiers EP, Jane MCarlton, Neil H, Qinghu R, Ian TP, Arnab P, Matthew B, Robert JMW, Shigeharu S, Stuart AR, David JM, Zikai X, Shamira JS, Janice W, Lingxia J, Jeffery L, Bruce W, Azadeh S, Alexander RD, Delia W, Jonathan C, Jennifer RW, Brian H, Samuel VA, Todd HC, Charles L, Bernard S, Joana CS, Teresa RU, Tamara VF, Mihaela P, Jonathan A, William CN, Evans LNT, Steven LS, Owen RW, Henry AF, Subhash M, Venter JC, Claire MF, Vishvanath N (2005). Genome Sequence of *Theileria parva*, a Bovine Pathogen that Transforms Lymphocytes. Science. 309(5731): 134-137.

Gomase VS, Changbhale SS (2007) Antigenicity Prediction in Melittin: Possibilities of in Drug Development from Apisdorsata. Current Proteomics. 4:107-114.

Graham SP, Pelle R, Honda Y, Mwangi DM, Tonukari NJ, Yamage M, Glew EJ, De Villiers EP, Shah T, Bishop R, Abuya E, Awino E, Gachanja J, Luyai AE, Mbwika F, Muthiani AM, Ndegwa DM, Njahira M, Nyanjui JK, Onono FO, Osaso J, Saya RM, Wildmann C, Fraser CM, Maudlin I, Gardner MJ, Morzaria SP, Loosmore S, Gilbert SC, Audonnet JC, van der Bruggen P, Nene V, Taracha EL (2006). *Theileria parva* candidate vaccine antigens recognized by immune bovine cytotoxic T lymphocytes. Proc. Natl. Acad. Sci. USA 103: 3288–3291.

Harris KS, Joanne LC, Andrew MC, Rosella M, Sabo JK, David WK, Erinna FL, Andrew M, Raymond SN, Robin FA, Michael F (2005). Binding Hot Spot for Invasion Inhibitory Molecules on *Plasmodium falciparum* Apical Membrane Antigen. Infect. Immun. 73(10): 6981-6989.

Healer J, Crawford S, Ralph S, McFadden G, Cowman AF (2002). Independent translocation of two micronemal proteins in developing *Plasmodium falciparum* merozoites. Infect. Immun. 70: 5751-5758.

Healer J, Vince M, Anthony NH, Rosella M, Alan WG, Robin F, Alan FC, Drian B (2004). Allelic polymorphisms in apical membrane antigen-1 are responsible for evasion of antibody-mediated inhibition in *Plasmodium falciparum*. Mol. Microbiol. 52(1): 159-168.

Hehl AB, Lekutis C, Grigg ME, Bradley PJ, Dubremetz JF, Ortega-Barria E, Boothroyd JC (2000). *Toxoplasma gondii* homologue of *Plasmodium* apical membrane antigen 1 is involved in invasion of host cells. Infect. Immun. 68: 7078-7086.

Hodder AN, Crewther PE, Anders RF (2001). Specificity of the protective antibody response to apical membrane antigen 1. Infect. Immun. 69: 3286-3294.

Hodder AN, Crewther PE, Matthew ML, Reid GE, Moritz RL, Simpson RJ, Anders RF (1996). The disulfide bond structure of *Plasmodium* apical membrane antigen-1. J. Biol. Chem. 271: 29446-29452.

Hopp TP, Woods KR (1981). Prediction of protein antigenic determinants from amino acid sequences. Proc. Natl. Acad. Sci. 78(6): 3824-3828.

Howell SA, Well I, Fleck SL, Kettleborough C, Collins CR, Blackman MJ (2003). A single malaria merozoite serine protease mediates shedding of multiple surface proteins by juxtamembrane cleavage. J. Biol. Chem. 278: 23890-23898.

Kato K, Mayer DCG, Singh S, Reid M, Miller LH (2005). Domain III of *Plasmodium falciparum* apical membrane antigen 1 binds to the erythrocyte membrane protein Kx. Proc. Natl. Acad. Sci. USA 102: 5552-5557.

Katzer F, Daniel N, Chris O, Richard PB, Evans LNT, Alan RW, Declan JM (2006). Extensive Genotypic Diversity in a Recombining Population of the Apicomplexan Parasite Theileria parva. Infect. Immun. 74(10): 5456-5464.

Kennedy MC, Wang J, Zhang Y, Miles AP, Chitsaz F, Saul A, Long CA, Miller LH, Stowers AW (2002). *In vitro* studies with recombinant *Plasmodium falciparum* apical membrane antigen 1 (AMA-1): production and activity of an AMA-1 vaccine and generation of a multiallelic response. Infect. Immun. 70: 6948-6960.

Kocken CH, Narum DL, Massougbodji A, Ayivi B, Dubbeld MA, van der WA, Conway DJ, Sanni A, Thomas AW (2000). Molecular characterisation of *Plasmodium reichenowi* apical membrane antigen-1 (AMA-1), comparison with *P. falciparum* AMA-1, and antibody-mediated inhibition of red cell invasion. Mol. Biochem. Parasitol. 109: 147-156.

Kyte J, Doolittle RF (1982). A simple method for displaying the hydropathic character of a protein. J. Mol. Biol. 157(1): 105-132.

Mital J, Markus M, Dominique S, Gary EW (2005). Conditional Expression of *Toxoplasma gondii* Apical Membrane Antigen-1 (TgAMA-1) Demonstrates That TgAMA-1 Plays a Critical Role in Host Cell Invasion. Mol. Biol. 16(9): 4341-4349.

Narum DL, Ogun SA, Batchelor AH, Holder AA (2006). Passive immunization with a multicomponent vaccine against conserved domains of apical membrane antigen 1 and 235-kilodalton rhoptry proteins protects mice against *Plasmodium yoelii* blood-stage challenge infection. Infect Immun. 74(10): 5529-5536.

Narum DL, Thomas AW (1994). Differential localization of full-length and processed forms of *Pf*83/AMA-1 an apical membrane antigen of *Plasmodium falciparum* merozoites. Mol. Biochem. Parasitol. 67: 59-68.

Peterson MG, Marshall VM, Smythe JA, Crewther PE, Lew A, Silva A, Anders RF, Kemp DJ (1989). Integral membrane protein located in the apical complex of *Plasmodium falciparum*. Mol. Cell. Biol. 9: 3151-3154.

Pizarro JC, Brigitte VN, Marie-Laure C, Christine RC, Chrislaine W, Fiona H, Michael JB, Bart WF, Edmond JR, Clemens HMK, Thomas WA, Graham AB (2005). Crystal structure of the malaria vaccine candidate apical membrane antigen 1. Science. 308: 408-411.

Remarque EJ, Faber BW, Kocken CH, Thomas AW (2008). Apical membrane antigen 1: a malaria vaccine candidate in review, Trends in Parasitology. 24(2): 74-84.

Sabo JK, David WK, Zhi-Ping F, Joanne LC, Kathy P, Andrew MC, Michael F, Raymond SN (2006). Mimotopes of apical membrane antigen 1: structures of phage-derived peptides recognized by the inhibitory monoclonal antibody 4G2dc1 and design of a more active analogue. Infect. Immun. (75): 61-73.

Silvie O, Franetich JF, Charrin S, Mueller MS, Siau A, Bodescot M, Rubinstein E, Hannoun L, Charoenvit Y, Kocken CH, Thomas AW, van Gemert GJ, Sauerwein RW, Blackman MJ, Anders RF, Pluschke G, Mazier D (2004). A role for apical membrane antigen 1 during invasion of hepatocytes by *Plasmodium falciparum* sporozoites. J. Biol. Chem. 279: 9490-9496.

Taylor HM, Grainger M, Holder AA (2002). Variation in the expression of a *Plasmodium falciparum* protein family implicated in erythrocyte invasion. Infect. Immun. 70: 5779-5789.

Thomas AW, Trape JF, Rogier C, Goncalves A, Rosario VE, Narum DL (1994). High prevalence of natural antibodies against *Plasmodium falciparum* 83-kilodalton apical membrane antigen (*Pf*83/AMA-1) as detected by capture-enzyme-linked immunosorbent assay using full-length baculovirus recombinant *Pf*83/AMA-1. Am. J. Trop. Med. Hyg. 51:730-740.

Tonukari NJ, Richard TK (2009a). Cloning and expression of *Theileria parva* T-complex 1 protein zeta subunit ortholog. J. Cell Ani. Biol. 10: 183-187.

Tonukari NJ, Richard TK (2009b). Isolation of *Theileria parva* ring-infected erythrocyte surface antigen (RESA) homolog. J. Cell Ani. Biol. 3(10): 179-182.

Triglia T, Healer J, Caruana SR, Hodder AN, Anders RF, Crabb BS, Cowman AF (2000). Apical membrane antigen 1 plays a central role in erythrocyte invasion by *Plasmodium* species. Mol. Microbiol. 38: 706-718.

Verra F, Hughes AL (1999). Natural selection on apical membrane antigen-1 of Plasmodium falciparum. Parassitologia 1999 41(1-3):93-95.

Waters AP, Thomas AW, Deans JA, Mitchell GH, Hudson DE, Miller LH, McCutchan TF, Cohen S (1990). A merozoite receptor protein from *Plasmodium knowlesi* is highly conserved and distributed throughout *Plasmodium*. J. Biol. Chem. 265: 17974-17979.

Wickramarachchi T, Prasad HP, Lakshman Perera KLR, Sumith B, Clemens HMK, Thomas AW, Shiroma MH, Udagama-Randeniya PV (2006). Natural Human Antibody Responses to *Plasmodium vivax* Apical Membrane Antigen 1 under Low Transmission and Unstable Malaria Conditions in Sri Lanka Infect. Immun. 74(1): 798–801.

Xu H, Hodder AN, Yan H, Crewther PE, Anders RF, Good MF (2000). CD4+ T cells acting independently of antibody contribute to protective immunity to *P.chaubadi* infection after apical membrane 1 immunisation. J. Immunol. 165: 389-396.

Zhang H, Compaore MK, Lee EG, Liao M, Zhang G, Sugimoto C, Fujisaki K, Nishikawa Y, Xuan X (2007). Apical membrane antigen 1 is a cross-reactive antigen between *Neospora caninum* and *Toxoplasma gondii*, and the anti-NcAMA-1 antibody inhibits host cell invasion by both parasites. Mol. Biochem. Parasitol. 151(2): 205-212.

6

A quick bud breaking response of a surface model for rapid clonal propagation in *Centella asiatica* (L.)

AK Bhandari[1] , M Baunthiyal[2], VK Bisht[1], Narayan Singh[1] and JS Negi[1]

[1]Herbal Research and Development Institute (HRDI) - Mandal, Gopeshwar, Chamoli, Uttarakhand, India.
[2]Department of Biotechnology, G. B. Pant Engineering College- Ghurdauri, Pauri-Garhwal, Uttarakhand, India.

Present investigation was planned to evaluate time period of bud breaking in *Centella asiatica* with different concentration of plant growth regulators, a medicinal herb distributed throughout the worldwide. For the study, concentrations were designed for response surface model describing bud breaking growth in optimum conditions. A combination of BAP (2 mg/L) + gibberellic acid (GA$_3$, 0.5 mg/L) was achieved at a best initial bud breaking at 8th hour. Longest time period taken for bud breaking was shown in combination of BAP (0.5 mg/L) + naphthalene acetic acid (NAA, 0.5 g / L) and BAP (0.1 mg/L) + adenine sulphate (0.5 mg/L) which was recorded at 84th hour. Half strength MS media was supplemented with IBA alone (2 mg/L) and in combination with IAA (0.5 mg/L) to attain an early *in vitro* rooting. Their interactions observed were statistically significant ($P < 0.05$).

Key words: *Centella asiatica*, bud, plant growth regulator, medicinal plant.

INTRODUCTION

Centella asiatica (L.) Urban, synonym *Hydrocotyle asiatica* (Family: Apiaceae) is a small perennial herb, commonly known as Mandukparni. In India, this species is mostly found in the swampy areas up to an altitude of 600 to 1800 m asl (Patra et al., 1998). Medicinally, *C. asiatica* used as memory enhancer and in the treatment of chronic diseases, mental disorders and neuropharma-cological disorders like insomnia, insanity, depression, psychosis, epilepsy and stress (Chopra et al., 1980). The major bio-active ingredients in the plant are the triterpenes, asiatic acid, madecassic acid and their glycosides such as asiaticoside and madecassoside (Zheng and Qin, 2007). Due to the presence of these active ingredients, it possesses antileprotic, antifilarial, antibacterial, antifeedant, adaptogenic and antiviral properties (Warrier et al., 1994). The roots contain many polyacetylenic compounds, the major compound being 8-acetoxyfalcarinol (Loc and Tam, 2010).

Over-exploitation of *C. asiatica* from natural habitats for medicinal purposes causes depletion of plant population. There has been an increase interest in *in-vitro* culture techniques for mass multiplication of important species to overcome the pressure of over-exploitation and to restore species diversity (Patra et al., 1998; Tiwari et al., 2000; Bhandari et al., 2010). However, till date *in-vitro* technique has been applied only for < 20% of medicinally important species (Shukla et al., 1999). *In-vitro* propagation of *C. asiatica* was also carried out through leaf explants (Banerjee et al., 1999), axillary buds (George et al., 2004), stolons (Sampath et al., 2001), shoot tips

(Sangeetha et al., 2003), callus cultures (Patra et al., 1998; Rao et al., 1999) and somatic embryogenesis (Martin, 2004). Besides all, there is information available on the methods to initiate early bud breaking, shoot formation and root initiation. Therefore, present study was designed to understand the effect of different PGRs in alone and in combination for bud breaking, shoot formation and root initiation. Present study will be useful in producing quality planting material with in short duration.

MATERIALS AND METHODS

Ex- plant selection

For the Rapid clonal propagation of *C. asiatica,* explants were collected from the Herbal garden (1545 m asl) of Herbal Research and Development Institute, Mandal, Gopeshwar (Chamoli) Uttarakhand, India; It is bounded by North Latitude 30° 27' 13.40" and East Longitude 79° 16' 21.61".

The media for clonal propagation was prepared by following Murashige and Skoog (1962), All Chemicals used for the research purpose were purchased from HiMedia Laboratories (Mumbai, India), and growth regulators were purchased from Sigma Chemical Co. (St Louis, MO) and HiMedia Laboratories, India. Cultures were established for the bud explants on MS medium (Murashige and Skoog, 1962) containing 58 mM sucrose and gelled with 0.7% (w/v) agar. The pH of medium was adjusted between 5.6 to 5.8 using 0.1 N HCL or 0.1 N NaOH solution prior to the autoclaving at 121°C and a pressure of 15 psi for 20 min then allowed to cool at room temperature. The explants thoroughly washed with running tap water for 15 to 20 min, then treated with 1% (v/v), Tween 20 solution and subsequently for 15 min with a sodium hypochlorite solution (0.5% active chlorine) in laminar air flow cabinet and finally the explants were washed thoroughly with autoclaved distilled water for several times to remove the traces of sodium hypochlorite. In support of surface disinfection, bud segments were trimmed from the cut ends in appropriate size, and cultured.

Culture conditions and *in-vitro* establishment of plantlets

For establishment of cultures, the surface disinfected explants were inoculated on full strength MS (Murashige and Skoog, 1962) basal medium having 3% of sucrose, semi-solidified with 0.7% (w/v) agar and supplemented with different concentrations of plant growth regulator viz. BAP (0.1 to 2 mg/L), adenine sulphate (0.1 to 0.5 mg/L) and gibberellic acid (0.5 mg/L). Half strength of MS medium supplemented with growth regulator IBA (0.5 to 2.0 mg/L) and NAA 0.1 to 0.5 mg/L was attempt for rooting. Each hormonal combination was tried in three replicates. 250 ml (Borosil, India) Conical flasks containing 20 ml of medium were used. Cultures vessels were used for incubated at 25 ± 1°C under a 16/8 h light/dark photoperiod with light provided by cool-white fluorescent lamps (Philips India, Mumbai, India) at a light intensity of 1000 lux. The multiplied cultures were taken out; every single shoot was excised and kept in small plastic cup filled with a mixture of soil: sand (1:1) for *ex-vitro* rooting.

Multiple shoots from bud induction

The explants were inoculated in semi-solid MS medium with

concentrations (0.1, 0.2, 1.5 and 2 mg/L) of BAP in alone, with combination of BAP (0.5 and 2 mg/L) along with gibberellic acid in 0.5 mg/L, BAP (0.1, 0.5,1 and 2 mg/L) with adenine sulphate 0.5 mg/L and BAP (0.5 to 2 mg/L) with the combination of NAA (0.5 mg/L). Sub culturing was carried out at periodic intervals of three weeks.

Rooting of microshoots

Developed shoots having one or two nodes were excised and transferred to root induction medium comprising of ½ strength MS medium with 3% sucrose and supplemented with different concentrations of IBA (0.5, 1, 1.5 and 2.0 mg/L) in alone and (0.1, 0.2, 0.5, 1.5 and 2.0 mg/L) with IAA (0.5 mg/L) in combination. Number of roots per shoot and root length was score in alternate day.

Hardening and acclimatization

For *in vitro* hardening, rooted shoots were transferred to ¼ MS strength medium having 3% sucrose devoid of PGR for seven days in flasks. Thereafter, they were transferred to polybags containing a mixture of soil: sand: FYM manure (1:1:1) and kept for two weeks in mist-chamber under controlled condition (tamp-25°C ± 2°C), humidity (65% ± 5%). Acclimatized plants were later shifted to soil in pots in agronet-shade house for one week and after that in field.

Statistical analysis

The data collected was subjected to the analysis of variance (ANOVA); using MS Excel 2007 for calculating the significance among different treatments and time of bud breaking and time of root initiation values at $P < 0.05$ were computed to compare means from various treatments.

RESULTS AND DISCUSSION

Initial study on *C. asiatica* was undertaken by Patra et al. (1998), Banerjee et al. (1999) and Tiwari et al. (2000). Tiwari et al. (2000) reported that initiation of nodal culture is better using different combination of plant growth regulators. The results of the present study on bud initiation, bud establishment and root initiation in MS medium supplemented with various combinations of growth regulators are presented in Tables 1 to 3. The earliest bud breaking in this study was achieved in BAP (2 mg/L) + GA_3 (0.5 mg/L). Initiation of bud breaking within 8 h of *in-vitro* culture in *C. asiatica* was reported first time in present study. Achieving early bud breaking is of importance as it produces quality planting material vis a vis reduces time and efforts. Different combinations of PGRs have also been reported to initiate bud breaking (Sen and Sharma, 1991). The longest time period (84 h) taken for bud breaking in present study was noticed in combination of BAP (0.5 mg/L) + NAA (0.5 mg/ L) and BAP (0.1mg/L) + adenine (0.5 mg/L) which are presented in Tables 2 and 3.

BAP alone at higher concentration (2 mg/L) seems to initiate early bud breaking. Similar observations in

Table 1. Morphogenetic response of *C. asiatica* buds cultured on MS medium supplemented with different concentrations of BAP and Gibberellic acid (Bud breaking is found positively significant LSD = 24.25 ($P < 0.05$).

BAP (mg/L)	Gibberellic acid (mg/L)	Time of bud breaking (h)
Control	-	0
0.1		72
0.1		60
0.1		66
0.2		60
0.2		72
0.2		72
0.5	0.5	36
0.5	0.5	38
0.5	0.5	40
1.5		40
1.5		46
1.5		42
2	0.5	8
2	0.5	12
2	0.5	14
2		24
2		18
2		16

t-value= 2.02, LSD = 24.25 ($P < 0.05$)

Table 2. Morphogenetic response of *C. asiatica* buds cultured on MS medium supplemented with different concentrations of BAP and Adinine (Bud breaking is found positively significant LSD=19.36; $P < 0.05$).

BAP (mg/L)	Adinine sulphate (mg/L)	Time of bud breaking (h)
Control	-	0
0.1	0.5	84
0.1	0.5	80
0.1	0.5	78
0.5	0.5	72
0.5	0.5	68
0.5	0.5	66
1	0.5	48
1	0.5	44
1	0.5	40
2	0.5	24
2	0.5	20
2	0.5	18

t-value = 2.13, LSD =19.36; ($P < 0.05$).

Table 3. Morphogenetic response of *C. asiatica* buds cultured on MS medium supplemented with different concentrations of BAP and NAA (Bud breaking is found positively significant LSD=19.36; $P < 0.05$).

BAP (mg/L)	NAA (mg/L)	Time of bud breaking (h)
Control	-	0
0.5	0.5	84
0.5	0.5	80
0.5	0.5	76
2	0.5	48
2	0.5	46
2	0.5	42

t-value= 2.03, LSD=19.36; $P < 0.05$

Ocimum basilicum was found by Pattnaik and Chand (1996). In addition, BAP was found more efficient over kinetine (Kn) in *in-vitro* shoot proliferation in different species (Purohit, 1994; Martin, 2003). In *Swertia chirata*, BAP with higher concentration have optimal response for shoot proliferation (Chaudhuri, 2007; Pant et al., 2010).

In present study, BAP in combination with GA_3 (with different concentrations) significantly ($P < 0.05$) enhance the rate of bud breaking, shoot proliferation and root initiation (Table 1, Figure 1A-E). Sharma and Sharma (2010) attributed this to the stimulating effects of various hydrolytic enzymes activities thus increasing availability of nutrients for growth. The result of the effects of BAP and GA_3 on shoot proliferation in present study was found comparable to the earlier reports. However, Tiwari et al. (2013) reported improved bud breaking using high concentration of BAP (5 mg/L) and improved root formation in combinations of BAP (4.0 mg/L) and IBA (0.5 mg/L). Karthikeyan et al. (2009) described the rapid clonal propagation through auxiliary shoot proliferation in *C. asiatica*. The shoot elongation with the treatment of BAP and GA_3 might be due to cell enlargement and increase in normal cell division (Karivartharaju and Ramakrishnan, 1985). Earliest root initiation was achieved alone in IBA (2 mg/L) and in combination of IBA (2 mg/L) and IAA (0.5 mg/L) (Tables 4 and 5).

Thus, it is concluded that *in-vitro* micro-propagation offer rapid clonal multiplication of elite clones and further helps in dissemination fulfilling the need of vis a vis to quality planting material. BAP (2 mg/L) in combination with GA_3 (0.5 mg/L) is recommended for effective and earliest bud breaking. Likewise, IBA (2 mg/L) is recommended for earliest rooting in *C. asiatica*.

ACKNOWLEDGEMENTS

The authors are grateful to technical staff Megha Sati and Shweta Semwal for assistance in carrying out the research

Figure 1. In vitro regeneration of *Centella asiatica,* via bud explants; **(A)** axillary bud induction on nodal segment MS medium+BAP (2 mg/L)+gibberellic acid (0.5 mg/L), **(B)** bud induction on nodal segment, **(C)** culture establishment, **(D)** multiplication of shoots in BAP (0.5 mg/L)+NAA 0.1 mg/L, **(E)** rooted plantlet containing a mixture of soil : sand : manure (1:1:1), **(F)** plantlet in soil for hardening containing a mixture of soil : sand : manure (1:1:1).

Table 4. Morphogenetic response of root initiation explants of *C. asiatica* cultured on half strength MS medium supplemented with different concentrations of IBA (Root initiation is found positively significant LSD=3.16; ($P < 0.05$).

IBA (mg/L)	Time of root initiation (Days)
Control	0
0.5	16
0.5	14
0.5	15
1	15
1	14
1	13
1.5	12
1.5	11
1.5	11
2	10
2	9
2	8

t-value= 2.07, LSD=3.16; ($P < 0.05$)

Table 5. Morphogenetic response of root initiation explants of C. asiatica cultured on half strength MS medium supplemented with different concentrations of IBA and IAA (Bud breaking is found positively significant LSD=3.12; ($P < 0.05$).

IBA (mg/L)	IAA (mg/L)	Time of root initiation (Days)
Control	0	0
0.1	0.5	18
0.1	0.5	16
0.1	0.5	16
0.2	0.5	18
0.2	0.5	16
0.2	0.5	16
0.5	0.5	15
0.5	0.5	14
0.5	0.5	13
1.5	0.5	12
1.5	0.5	11
1.5	0.5	10
2	0.5	10
2	0.5	9
2	0.5	9

t-value= 2.02, LSD=3.12; ($P < 0.05$)

work and Agriculture and Processed Food Products Export Development Authority, Ministry of Commerce and Industry, Government of India to provide financial support for instrument facility to the Institute under Herbal Analytical Laboratory Project (Grant no. FLR/059/2006-07/13692).

REFERENCES

Banerjee S, Zehra M, Kumar S (1999). *In vitro* multiplication of *Centella asiatica*, a medicinal herb from leaf explants. Curr. Sci. 76:147-148.

Bhandari AK, Negi JS, Bisht VK, Bharti MK (2010). *In vitro* culture of *Aloe vera* - A plant with medicinal property. Nat. and Sci. 8(8):174-176.

Chaudhuri RK (2007). Production of genetically uniform plants from nodal explants of *Swertia chirata* Buch.-Ham.ex Wall.: an endangered medicinal herb. *In-vitro* Cell. Div. Biol. Plant 43:467-472.

Chopra RN, Nayar SL, Chopra IC (1980). Glossary of Indian Medicinal Plants (Including the Supplement). CSIR, New Delhi.

George S, Remashree AB, Sebastian D, Hariharan M (2004). Micropropagation of *Centella asiatica* L. through axillary bud multiplication. Phytomorphology 54:31-34.

Karivartharaju TV, Ramakrishnan V (1985) Seed hardening studiesin two varieties of ragi. Indian J. Pl. Physiol. 28:243-248.

Karthikeyan K, Chandran C, Kulothungan S (2009). Rapid clonal multiplication through *in- vitro* axillary shoot proliferation of *Centella asiatica* L. Ind. J Biotechnol. 8:232-235.

Loc NH, Tam NT (2010). An Asiaticoside production from Centella (*Centella asiatica* L. Urban) cell culture. Biotech. Bioproc. Eng. 15:1065-1070.

Martin KP (2003). Rapid *in-vitro* multiplication and ex-vitro rooting of *Rotula aquatic* Lour., a rare rhoeophytic woody medicinal plant. Pl. Cell Rep. 21:415-420.

Martin KP (2004). Plant regeneration through somatic embryogenesis in medicinally important *Centella asiatica* L. *In vitro* Cell. Dev. Biol. Plant. 40:586-591.

Murashige T, Skoog F (1962). A revised medium for rapid growth and bioassays for tobacco tissue cultures. Physiol. Plant. 15:473-97.

Pant M, Bisht P, Gusain PM (2010). *In vitro* propagation through axillary bud culture of *Swertia chirata* Buch.-Ham ex Wall: an endangered medicinal herb. Int. J. Int. Bio.10:48-53.

Patra A, Rai B, Rout GR, Das P (1998). Successful plant regeneration from callus cultures of *Centella asiatica*. Pl. Growth Regul. 24:13-16.

Pattnaik SK, Chand PK (1996). *In vitro* propagation of the medicinal herbs *Ocimum americanus* Pl. Cell Reports 15:846-850.

Purohit SD (1994). Micropropagation of safed musli (*Chlorophytum borivilianum*), a rare medicinal herb. Plant Cell Tiss. Org. Cult. 39:93-96.

Rao KP, Rao SS, Sadanandam M (1999). Tissue culture studies of *Centella asiatica*. Indian J. Pharm. Sci. 61:392-394.

Sampath P, Muthuraman G, Jayaraman P (2001). Tissue culture studies in *Bacopa monnieri* and *Centella asiatica*.National Research Seminar on Herbal Conservation, Cultivation, Marketing and Utilization with Special Emphasis on Chattisgarh, The Herbal State, Raipur, Chattisgarh, India. 12:18-21.

Sangeetha N, Buragohain AK (2003). *In vitro* method for propagation of *Centella asiatica* (L.) urban by shoot tip culture. J. Plant Biochem. Biotechnol. 12:167-169.

Sen J, Sharma AK (1991). Micropropagation of *Withania somnifera* from germinating seeds and shoot tips. Plant Cell Tiss. Org. Cult. 26:71-73.

Sharma S, Sharma RK (2010). Seed physiological aspects of pushkarmool (*Inula racemosa*), a threatened medicinal herb: response to storage, cold stratification, light and gibberellic acid. Curr. Sci. 99(12):1801-1806.

Shukla A, Rasik AM, Jain GK, Shankar R, Kulshrestha DK, Dhawan BN (1999). *In vitro* and *in vivo* wound healing activity of asiaticoside isolated from *Centella asiatica*. J. Ethnopharmacol. 65:1-11.

Tiwari C, Bakshi M, Vichitra A (2013). A rapid two step protocol of *in vitro* propagation of an important medicinal herb *Centella asiatica* Linn. Afr. J. Biot.12:1084-1090.

Tiwari KN, Sharma NC, Tiwari V, Singh BD (2000). Micropropagation of *Centella asiatica* (L.) a valuable medicinal herb. Plant Cell Tiss. Organ Cult. 63:179-185.

Warrier PK, Nambiar VP, Ramankutty C (1994). Indian medicinal plants. A compendium of 500 species.Vol-I, Orient Lonhman Pvt Ltd, Chennai, India: pp.52-55.

Zheng CJ, Qin LP (2007). Chemical components of *Centella asiatica* and their bioactivities. J. Chin. Integ. Med. 5:348-351.

7

Micropropagation of *Litsea glutinosa* (Lour) C.B

Syed Naseer Shah[1], Amjad M. Husaini[2] and S. A. Ansari[1]

[1]Genetics and Plant Propagation Division, Tropical Forest Research Institute, Mandla Road, Jabalpur 482 021, India.
[2]Centre for Plant Biotechnology, Division of Biotechnology, SKUAST-K, Shalimar, Srinagar-191121, J&K, India.

Litsea glutinosa (Lour) C.B (Hindi: Maida lakri) is a medicinal plant of immense pharmaceutical value. The species is critically endangered due to its indiscriminate collection as raw material for pharmaceutical industry, where it is used for manufacturing drugs for pain, arousing sexual power and in treatment of diarrhea and dysentery etc. An attempt has been made for development of *in vitro* propagation procedure for the species, involving four steps, namely: culture establishment, shoot multiplication, rooting and hardening. Aseptic cultures were established on Murashige and Skoog (MS) medium supplemented 10.0 µM N^6-benzyladenine (BA) using nodal segments (1 cm). Four sets of simple randomized experiment were carried out on MS medium to study the effect of four doses of each BA, GA_3, IAA, (0, 2.5, 5.0 and 10 µM) and ascorbic acid (0, 284, 852 and 1136 µM) for *in vitro* shoot multiplication. MS medium supplemented with 5.0 µM BA with 852 µM ascorbic acid significantly proved optimum for *in vitro* shoot multiplication and resulted in 1.05 shoot number explant^{-1}, 1.72 node number shoot^{-1} and 1.79 node number explant^{-1} at one month after inoculation. The *in vitro* multiplied shoots were tested for *in vitro* root induction on MS culture media containing auxin IBA (Indole-3-butaric acid) treatments (0, 2.5, 5.0 and 10.0 µM) in simple randomized designs experiment. MS media supplemented with 10.0 µM IBA, screened out to be significantly excellent for induction and growth of adventitious roots, resulting in 72.2% rooting and 0.72 root number explant^{-1} at 30 days after inoculation. The *in vitro* propagated plants exhibited excellent growth. Therefore, the present study recommends a four step micropropagation procedure for *in vitro* production of *L. glutinosa* plants on a commercial scale to meet the requirement of pharmaceutical industries and save the species from extinction.

Key words: *Litsea glutinosa* (Lour) C.B, ascorbic acid, nodal segments.

INTRODUCTION

Litsea glutinosa (Family Lauraceae) is an evergreen tree of medium size, which grows to a height of about 25 m. Found in mixed primary and secondary forest and thickets throughout india and in the outer Himalayas' (Kirtikar and Basu, 1981). *L. glutinosa* contain photoconstituents like alkaloids, glycosides, flavonoids, saponins, tannins, phenolic compounds etc. The bark of *L. glutinosa*, "is one of the most popular of native drugs", is considered to be capable of relieving pain, arousing sexual power and good for stomach in treatment of diarrohea and dysentery. *L. glutinosa* is widely used as a demulcent and as an emollient. The phytochemical constituents of bark of *L. glutinosa* have been shown to possess effective antibacterial and antifungal activity (Hosamath, 2011). This species is critically endangered (Reddy and Reddy 2008). The conventional propagation is hampered due to low seed viability and no rooting of vegetative cuttings (Rabena, 2010).

Thus there is need for alternative *in vitro* propagation method for large scale multiplication, improvement and conservation of the species. The objective of the study was to develop a procedure for its micropropagation.

Figure 1. Explant collection, culture establishment and shoot multiplication in *Litsea glutinosa,* (a) mother plant, (b) a twig, (c) nodal explants, (d) *in vitro* shoot establishment on MS medium + 10 µM BA.

MATERIALS AND METHODS

The selected (mother) plants (Figure 1a) were used to collect twig (s) (Figure 1b), which were brought in laboratory and washed thoroughly for 15 min under running water for removing the debris from the surface. The washed twigs were defoliated and cut into nodal explants (approximately 1 to 1.5 cm long and 0.5 to 0.6 cm diameter) (Figure 1c). These explants were washed with 2% Cetrimide® and kept for 10 min with constant vigorous shaking (150 rpm) on an orbital shaker incubator. The explants were rewashed 4 to 5 times with distilled water to remove traces of Cetrimide®. The washed explants were sterilized for 5 min with a composite sterilization treatment comprising $HgCl_2$ (0.1%), Bavistin® (1.0%) and Streptomycin® (0.2%) in the laminar flow cabinet. Finally, the surface sterilized nodal explants were rinsed 4 to 5 times with sterile distilled water for removal of sterilizing agent under laminar flow cabinet. The nodal segments were inoculated on MS medium (Murashige and Skoog, 1962) supplemented with 10.0 µM BA for culture establishment (Figure 1d).

Shoot multiplication

Four sets of simple randomized experiment were carried out on MS medium to study the effect of four doses of each BA, GA_3, IAA, (0, 2.5, 5.0 and 10 µM) and ascorbic acid (0, 284, 852 and 1136 µM) for *in vitro* shoot multiplication. Shoot number explant^{-1}, node number shoot^{-1} and node number explant^{-1} were recorded (Figure 2a-c).

Root induction

A simple randomized experiment was carried out to study the effect of four doses of IBA (0, 2.5, 5.0 and 10 µM) on root induction at 30 days after inoculation.

Culture conditions and statistical analysis

The inorganic salts used for preparation of culture medium were obtained from Qualigens Pvt. Ltd., India and phytohormones and B vitamins from Sigma Chemicals Pvt. Ltd., MS medium was used in all the experiments with 5.0 µM BA. India. The medium contained 3% (w/v) sucrose, 0.65% (w/v) agar (Hi-media chemical Ltd., India). The pH of the medium was adjusted to 6.0 before autoclaving for 15 min at 1.06 kg cm^{-2} (121°C). Explants were cultured in 150 ml culture bottles containing 40 ml semi-solid medium. For *in vitro*

Figure 2. *In vitro* shoot multiplication in *Litsea glutinosa,* (a) MS medium supplemented 5 µM BA + 5 µM IAA, (b) MS medium 5 µM BA + 5 µM GA_3, (c) MS medium + 5 µM BA + 825 µM ascorbic acid at 30 days after inoculation.

shoot multiplication experiments, the cultures were incubated at 25 ± 2°C under 16 h illuminations with fluorescent light (50 $\mu Em^{-2} s^{-1}$). The experiments had thee replicates for *in vitro* shoot multiplication and three replicates for *in vitro* rooting (Figure 3). Each replicate had 10 propagules. In all five experiments the data were recorded at 30 days after inoculation. The data were subjected to one way (factor) analysis of variance for all the experiments with "F" test for ascertaining level of significance. If the data were found significant

at $p \leq 0.05$, $LSD_{0.05}$ was computed for comparison of treatment means.

Hardening and transplantation

The *in vitro* raised plantlets were removed from rooting medium, washed with distilled water and the plantlets were transferred to

Figure 3. The *in vitro* adventitious root induction *Litsea glutinosa,* root formation on semi-solid MS medium with 10 µM IBA at 30 days after inoculation.

sand beds in the mistchamber and the plantlets were covered with culture bottles to maintain humidity (Figure 4a).

Subsequently, they were transferred to perforated polythene bags and kept initially in mistchamber for 10 days and finally transferred to natural environmental conditions (Figure 4b).

RESULTS

In vitro shoot multiplication

Experiment 1

Shoot number explant^{-1}: Effect of different doses of BA at 0 to 10 µM produced statistically similar value for shoot number explant^{-1} at 30 days after inoculation. The effect of BA was found to be non significant on shoot number explant^{-1} at the stage of sampling (Table 1).

Node number shoot^{-1}: Use of BA significantly influenced node number shoot^{-1} at 30 days after inoculation. 5 µM BA enhanced node number shoot^{-1} by 25% in comparison to the control at the stage of sampling.

Node number explant^{-1}: Various doses of BA significantly influenced node number explant^{-1} at 30 days after inoculation. 5.0 BA µM significantly enhanced node number explant^{-1} by 27% in comparison to the control at the stage of sampling.

Experiment 2

Shoot number explant^{-1}: Effect of different doses of IAA significantly influenced shoot number explant^{-1} at 30 days after inoculation. 5 µM IAA significantly enhanced shoot number explant^{-1} by 22% in comparison to the control at the stage of sampling (Table 2).

Node number shoot^{-1}: Use of IAA significantly influenced node number shoot^{-1} at 30 days after inoculation. 5.0 µM IAA significantly enhanced node number shoot^{-1} by 33% in comparison to the control at the stage of sampling.

Node number explant^{-1}: Various doses of IAA significantly influenced node number explant^{-1} at 30 days after inoculation. 5.0 µM IAA significantly enhanced node number explant^{-1} by 55% in comparison to the control at the stage of sampling.

Experiment 3

Shoot number explant^{-1}: Effect of different doses of GA$_3$ significantly influenced shoot number explant^{-1} at 30 days after inoculation. Maximum shoot number explant^{-1} were obtained on 5.0 µM GA$_3$ which was significantly higher than shoot number explant^{-1} into other doses of GA$_3$. 5.0

Figure 4. Hardening and acclimatization of the *in vitro* raised plantlets of *Litsea glutinosa*, transferred into (a) sand bed covered with culture bottles in mistchamber, (b) growth of plantlets in the open environment.

Table 1. Effect of different doses of BA on shoot number explant^{-1}, node number shoot^{-1} and node number explant^{-1} at 30 days after inoculation. MS medium supplemented with uniform dose of 5.0 μM BA.

Doses BA (μM)	Shoot number explant^{-1}	Node number shoot^{-1}	Node number explant^{-1}
0	1.0	1.0	1.0
2.5	1.0	1.0	1.0
5.0	1.0	1.25	1.27
10	1.0	1.0	1.0
LSD$_{(0.05)}$	NS	0.09	0.01

Table 2. Effect of different doses of IAA on shoot number explant^{-1}, node number shoot^{-1} and node number explant^{-1} at 30 days after inoculation. MS medium supplemented with uniform dose of 5.0 μM BA.

Doses IAA (μM)	Shoot number explant^{-1}	Node number shoot^{-1}	Node number explant^{-1}
0	1.0	1.0	1.0
2.5	1.0	1.0	1.05
5.0	1.22	1.31	1.55
10	1.0	1.0	1.0
LSD$_{(0.05)}$	0.10	0.03	0.25

μM GA$_3$ enhanced shoot number explant^{-1} by 44% in comparison to the control at the stage of sampling (Table 3).

Node number shoot^{-1}: Effect of different doses of GA$_3$ was found to be non significant at 30 days after inoculation. 5.0 to 10.0 μM GA$_3$ produced statistically

Table 3. Effect of different doses of GA_3 on shoot number explant^{-1}, node number shoot^{-1} and node number explant^{-1} at 30 days after inoculation. MS medium supplemented with uniform dose of 5.0 μM BA.

Doses GA_3 (μM)	Shoot number explant^{-1}	Node number shoot^{-1}	Node number explant^{-1}
0	1.0	1.0	1.0
2.5	1.0	1.0	1.11
5.0	1.44	1.11	1.41
10	1.0	1.16	1.11
LSD$_{(0.05)}$	0.19	NS	0.25

Table 4. Effect of different doses of ascorbic acid on shoot number explant^{-1}, node number shoot^{-1} and node number explant^{-1} at 30 days after inoculation. MS medium supplemented with uniform dose of 5.0 μM BA.

Doses Ascorbic acid (μM)	Shoot number explant^{-1}	Node number shoot^{-1}	Node number explant^{-1}
0	0.38	1.0	0.38
282	0.50	1.16	0.58
852	1.05	1.72	1.79
1136	0.55	0.94	0.53
LSD$_{(0.05)}$	0.44	0.73	0.83

equal value for node number shoot^{-1} at the stage of sampling.

Node number explant^{-1}: Various doses of GA_3 significantly influenced node number explant^{-1} at 30 days after inoculation. 5.0 μM GA_3 significantly enhanced node number explant^{-1} by 41% in comparison to the control at the stage of sampling.

Experiment 4

Shoot number explant^{-1}: Effect of different doses of ascorbic acid significantly influenced shoot number explant^{-1} at 30 days after inoculation. 852 μM ascorbic acid significantly enhanced shoot number explant^{-1} by 176% in comparison to the control at the stage of sampling (Table 4).

Node number shoot^{-1}: Use of ascorbic acid significantly influenced node number shoot^{-1} at 30 days after inoculation. 852 μM ascorbic acid significantly enhanced node number shoot^{-1} by 72% in comparison to the control at the stage of sampling.

Node number explant^{-1}: Various doses of ascorbic acid significantly influenced node number explant^{-1} at 30 days after inoculation. Ascorbic acid at 100 μM significantly enhanced node number explant^{-1} by 371% in comparison

to the control at the stage of sampling.

Experiment 5

***In vitro* adventitious rooting:** Various doses of IBA (0, 2.5, 5.0 and 10.0 μM) induced significant rooting and root number explant^{-1} at 30 days after inoculation (Table 5).

Adventitious rooting (%): MS medium supplemented with various doses of IBA significantly enhanced rooting (%) at 30 days after inoculation. 10.0 μM IBA enhanced rooting (%) by 767% in comparison to the control and maximum rooting of 72.22% was obtained on 10.0 μM IBA supplemented medium.

Root number explant^{-1}: Various doses of IBA significantly influenced root number explant^{-1} at 30 days after inoculation. 10.0 μM IBA significantly enhanced root number explant^{-1} by 134% in comparison to the control and maximum root number explant^{-1} (0.72) was obtained on 10.0 μM IBA supplemented medium.

DISCUSSION

The micropropagation of *L. glutinosa* comprises four steps namely: establishment of culture from nodal explants, shoot multiplication, root induction and harden-

Table 5. Effect of different doses of IBA on MS medium for *in vitro* rooting % and root number explant[-1] at 30 days after inoculation. Values in the parentheses are arc sine transformation.

Doses (μM) IBA	Rooting (%)	Root number explant[-1]
0	8.33(10.97)	0.05
2.5	12.49(17.41)	0.11
5.0	33.33(35.24)	0.33
10	72.22(66.55)	0.72
LSD$_{(0.05)}$	14.59	0.18

ing and acclimatization. The present investigations pertain to standardization of BA, GA$_3$, IAA and ascorbic acid doses for shoot multiplications and different concentrations of IBA for root induction. The best combination for *in vitro* shoot multiplication in four experiments emerged to be MS medium supplemented with 5.0 μM BA + 852 μM ascorbic acid. There is no published report on the *in vitro* shoot multiplication in the species. BA exhibiting superiority over other sources of cytokinins for differentiation and growth of new shoots is well documented in other species. The possible reason could be that BA is much closely related to natural cytokinins as far as the structures of the latter is concerned.

Amendment of culture medium with ascorbic acid significantly influenced shoot number explant[-1], node number shoot[-1] and node number explant[-1] at the stage of sampling. The increase in shoot number in the presence of ascorbic acid has also been reported in tobacco callus culture (Richard et al., 1988). According to Sharma and Chandel (1992) addition of ascorbic acid to the hormone supplemented medium was essential for bud break and further shoots multiplication. Mechanism of action of ascorbic acid, a common antioxidant/ antibrowning agent, is presently not known. Ascorbic acid or some product of its oxidation may possibly be increasing shoot number through ascorbate protection of endogenous phytohormones as implicated for tobacco, *Pinus* and *Picea* (Berlyn and Beck, 1980; Rumary and Thorpe, 1984; Richard et al., 1988). The auxins stimulate root development by inducing root initials that differentiate cells of the young secondary phloem, cambium and pith tissue (Gianfagna, 1995). Roots formed *de novo* from differentiated cells other than radical are defined as adventitious roots (Casson and Lindsey, 2003). A key stage in adventitious rooting is the *de novo* formation of root meristem. The *in vitro* rooting is complex process and is controlled by several factors, major being hormonal and nutritional status of media (Jarvis and Booth, 1981). A successful rooting procedure with high rooting percentage is essential for a competent micropropagation protocol.

In the present study 10 μM IBA stimulated adventitious

root formation and was found to have significant effect on rooting percentage and root number. IBA is preferred over NAA as it produces strong fibrous root system and is less toxic than NAA (Ahuja, 1991). There are earlier reports by many workers supporting our finding of rooting response with IBA. In medicinal plant species like *Gentiana lutea*, (Petrova et al., 2011) 1 mg L[-1] IBA was found effective for rooting. The best rooting was obtained on MS medium containing 0.5 mg/l IBA in *Solanum nigrum* (Kolar et al., 2008). On 3 mg L[-1] IBA, rooting was obtained in *Chlorophytum borivilianum* (Bathoju and Giri, 2012). Maximum number of roots was obtained on 1.5 IBA mg L[-1] IBA in *Centella asiatica* (Karthtikeyan et al., 2008). 100% *in vitro* rooting was obtained when shoot clusters were cultured on MS medium supplemented with 0.15 mg/L IBA in *Bacopa monnieri* (Sharma et al., 2010). IBA has been found effective for rooting of *in vitro* raised shoots of many tree species also. In *Dalbergia sissoo*, 1.0 mg/L of indole butyric acid (IBA) was reported to be the best treatment for rooting (Ali et al., 2012).

Conclusion

To the best of our knowledge this is the first report on micropropagation of *L. glutinosa*, a critically endangered medicinal tree species. As cultures have been established using nodal segments of mature trees from field, this method will be very useful for cloning of mature trees of this species. The high rooting frequency (72.2 %) obtained in the present study will help in its propagation in large numbers. The study will also be helpful for *ex situ* conservation of this endangered species in the form of *in vitro* cultures.

REFERENCES

Ahuja MR (1991). Biotechnology in forest trees. Plant Res. Dev. 33:106-120.

Ali A, Rizwan M, Majid A, Saleem A, Naveed NH (2012). Effect of media type and explants source on micropropagation of *Dalbergia sissoo*: A tree of medicinal importance. J. Med. Plants Res. 9:1742-1751.

Bathoju G, Giri A (2012). Production of medicinally important secondary metabolites (stigmasterol andhecogenin) from root cultures of *Chlorophytum borivilianum* (Safed musli), Recent Res. Sci. Technol. 5:45-48.

Berlyn GP, Beck RC (1980). Tissue culture as a technique for studying meristematic activity. In Control of Shoot Growth in Trees. Proceedings of the Joint Workshop of IUFRO Working Parties in Xylem and Shoot Growth Physiology. pp.305-324.

Casson SA, Lindsey K (2003). Genes and signaling in root development. New Phytol. 158:11-38.

Gianfagna T (1995). Natural and synthetic growth regulators and their use in horticultural and agronomic crops. In Plant Hormones: Physiology, Biochemistry and Molecular Biology, Davies, P.J. (ed.) Kluwer Academic Publishers, Dordrecht. pp.751-773.

Hosamath PV (2011). Evaluation of Antimicrobial activity of *Litsea glutinosa*, Int. J. Pharm. Appl. 1:105-114.

Jarvis BC, Booth A (1981). Influence of IBA, boron, myoinositol, vitamin D and seedling age on adventitious root development in cuttings of *Phaseolus aureus*. Physiology Plant. 53:213- 218.

Karthtikeyan K, Chandran C, Kulothungan S (2008) Rapid clonal multiplication through *in vitro* axillary shoot proliferation of *Centella asiatica* L. Indian Journal of Biotechnology. 8:232-235.

Kolar AB, Vivekanandan L, Ghouse BM (2008). *In vitro* regeneration and flower induction on *Solanum nigrum* L. from pachamalai hills of eastern ghats, Plant Tissue Cult. Biotech. 1:43-48.

Kritikar K, Basu BD (1981). Indian Medicinal Plant, Periodic Book Agency, Delhi, India, III. pp. 2158-2160.

Petrova M, Zayova E, Vitkova A (2011). Effect of silver nitrate on *in vitro* root formation of *Gentiana lutea* AC Romanian Biotechnol. Lett. 16:53-58.

Rabena AR (2010). Propagation Techniques of Endangered Sablot (Litsea gultinosa) Lour. C.B. Rob. Nat. Peer Rev. J. 5:56-77.

Reddy KN, Reddy CS (2008). First red list of medicinal plants of Andhra Pradesh, India- Conservation Assessment and Management Planing. Ethnobot. Leafl.12:103-107.

Richard WJ, Patel KR, Thorpe TA (1988). Ascorbic acid enhancement of organogenesis in tobacco callus, Plant Cell, Tiss. Organ Cult. 13:219-228.

Rumary C, Thorpe TA (1984). Plantlet formation in black and white spruce. Canadian J. For. Res. 14:10-16.

Sharma N, Chandel KPS (1992). Effects of ascorbic acid on axillary shoot induction in *Tylophora indica* (Burm. F.) Merrill. Plant Cell, Tiss. Organ Cult. 29:109-113.

Sharma S, Kamal B, Rathi N, Chauhan S, Jadon V, Vats N, Gehlot A, Arya S (2010). *In vitro* rapid and mass multiplication of highly valuable medicinal plant *Bacopa monnieri* (L.) Wettst. Afr. J. Biotechnol. 9:8318-8322.

8

Pathogenic variability within biochemical groups of *Pectobacterium carotovorum* isolated in Algeria from seed potato tubers

R. Yahiaoui-Zaidi*, R. Ladjouzi and S. Benallaoua

Université de Béjaia, Faculté des Sciences de la Nature et de la Vie, Département de Biologie physico-Chimique, 06000 Béjaia, Algérie. Laboratoire de Microbiologie Appliquée, Algeria.

One hundred *Pectobacterium carotovorum*, isolates, recovered from soft rot symptoms on potato tubers in Algeria and previously characterised taxonomically, were assessed in a half-tuber test for differences in ability to cause tuber rotting on two cultivars (Désirée and Bintje) respectively considered to be moderately and very susceptible to soft rot. A first trial at 20°C, involving the complete collection of isolates at two inoculum concentrations (2.10^5 and 2.10^7 cel. ml^{-1}), showed significant effects of inoculum dose, host cultivar and biochemical and molecular groups on pathogenicity. A significant interaction between pathogen groups and cultivars was also apparent. *P. carotovorum.* subsp. *atrosepticum* (*Pca*) was more pathogenic on cv. Bintje than on cv. Désirée, while the susceptibility of these two cultivars to *P. carotovorum.* subsp *carotovorum* (*Pcc*) was the opposite. Some *Pcc* isolates were non-pathogenic to both cultivars, and others were pathogenic on cv. Bintje but not on cv. Désirée. A second trial, conducted at 20 and 25°C with a high inoculum concentration (2.10^7 cel. ml^{-1}) of forty isolates representative from the collection, confirmed the previous findings, and showed a significant effect of temperature on pathogenicity. *Pca* isolates were more aggressive than *Pcc* isolates at both temperatures, but the difference was greater at 20°C. Our data therefore suggest that cultivar resistance rankings depend on the *Pectobacterium* subspecies considered, and should therefore be assessed separately for the various *Pectobacterium* subspecies.

Key words: Sensitivity, soft rot, aggressiveness, cultivars, resistance.

INTRODUCTION

Pectobacterium carotovorum subsp *carotovorum* (Vanhall) Dye and (*Pcc*), *Pectobacterium carotovorum* subsp. *atrosepticum* (*Pca*) (Jones) Dye (*Eca*) and *Pectobacterium chrysanthemi* (Burkholder) MacFadder and Dimmock (*Pch*) have a wide geographical distribution and are economically important pathogens. These bacteria can cause serious damage and are involved in soft rot diseases of several major crops, including potato, cucumber, tomato and carrot. (Pérombelon and Kelman, 1980; Stead, 1999; Farrar et al., 2000 Yishay et al., 2008). Soft rot on potato tubers can be a severe problem and affects potato industry, because of the long storage period between harvest and processing. Due to the vegetative mode of propagation of the plant, the inoculum concentration in the tubers seed is well correlated with the amount and severity of disease in the ware crop (Stead, 1999). Therefore, it is important to ensure that seed tubers must be pathogen-free to minimise the risk of disease occurrence in ware crops.

Infection by *Erwinia* usually results in extensive maceration and rotting of parenchymatous tissue in the organs affected due to the production of large amounts of proteases and pectic enzymes (Kotoujansky, 1987). In some cases, maceration is directly correlated with cell death (Garibaldi and Battman, 1971). The ability to cause soft rot varies among *Pcc* and *Pca* strains (Gregg, 1952; Johnson et al., 1989). It also depends on temperature, a factor essential for bacterial growth (Pérombelon and Kelman, 1980; Pérombelon, 2002). Disease potential is greatest when temperatures are in the range of (25 - 30°C). Temperature regimes may

*Corresponding author. E-mail: rachida_zaidi@yahoo.fr.

therefore determine the subspecies that will prevail when more than one of them is present in a region (Pérombelon and Salmon, 1995). Under cool and moist conditions, *Pca* is the main causal agent of blackleg, which originates from the mother tuber (Pérombelon and Kelman, 1987) and causes extensive decay of tubers when storage conditions are favourable, since the pathogen is capable of spreading rapidly from one tuber to another (Waterer and Pritchard, 1984). Under warmer and drier climates (Mediterranean or continental areas for instance), *Pcc* and *Pch* are commonly found and responsible for rotting of tubers under storage (Lumb et al., 1986; Cazelles et al., 1995).

Recently, Yishay et al. (2008) observed relationship between pathogenicity and genetic diversity among *P. carotovorum* subsp. *carotovorum* isolates which reveals a co-evolutionary specialization trend in the interaction between this pathogen and its hosts.

Several procedures have been described to determine the rate of maceration of tissue caused by *Pectobacterium* species and variation in the pathogenicity of these isolates (Smith and Bartz, 1990). Soft rot is routinely assessed by quantifying the decay of inoculated tubers through the measurement of lesion diameter (De Boer and Kelman, 1978), the volume of the rotted cavity in a half-tuber test (Ibrahim et al., 1978; Priou, 1992; Rabot et al., 1994), or the weight of rotted tissue (Bourne et al., 1981). Another possibility is to monitor the concentration of volatile metabolites (Waterer and Pritchard, 1984). A number of potato cultivars and Potato hybrids, assessed for susceptibility to rotting by *P. carotovorum*, proved more resistant to *Pcc* than to *Pca* (Lapwood et al., 1984; Carputo et al., 2000).

Over the last years, the presence of *Pectobacterium* causing soft rot was detected in Algeria on seed potato imported from the Netherlands, the British Isles, Canada and France, as well as on seed stocks multiplied locally (Yahiaoui-Zaidi et al., 2003). A large biochemical and molecular diversity was detected among these isolates, which could be assigned taxonomically to *Pca*, a number of groups related to *Pcc*, and *Pectobacterium carotovorum* subsp. *odoriferum* (*Pco*) (Yahiaoui-Zaidi et al., 2003). The purpose of the present study is to investigate the pathogenic variability within the biochemical groups on the two potato cultivars Bintje and Désirée, focusing on 1) inoculum concentration, and 2) temperature dependence.

MATERIALS AND METHODS

Pathogenicity tests

Plant material

Tubers of potato (*Solanum tuberosum*) cultivars Bintje and Désirée selected for uniform size (ca. 40 mm) and not effected by soft rot causing bacteria used for pathogenicity tests. These Algeria (Désirée); both have been regarded as susceptible to soft rot (Pérombelon, 1979). Tubers were stored at 4°C from harvest

until used and they were taken out of the cold storage and kept for 24 h at room temperature before inoculation.

Bacterial strains and growth conditions

Pectobacterium isolates used in this study were collected from rotted seed potato tubers in Algeria, during 1994 - 2000 and have been previously characterised taxonomically) (Yahiaoui-Zaidi et al., 2003). All isolates were maintained for extended periods as deep-frozen cultures (- 80°C) in Luria-Bertani (LB) medium (10 g.l^{-1} tryptone, 5 g.l^{-1} yeast extract and 10 g.l^{-1} NaCl; pH 7.3 (Sambrook et al., 1989) supplemented with 30% glycerol. The type strains for *Pca* (CFBP 1526 -NCPPB 549, Graham), *Pcc* (CFBP 2046 -NCPPB 312, Jones), and *Pch* (CFBP 2048 -ICPB EC17, Burkholder), obtained from the French Collection of Phytopathogenic Bacteria (CFBP, INRA Angers, France) were used as standards.

Identification of PCR/RFLP groups

To assess possible relationships between pathogenicity and genetic diversity among *P. carotovorum* sp., isolates were assigned to RFLP groups after the analysis of the restriction pattern of the PCR products obtained with Y1 and Y2 Primers. The DNA from 40 µl samples was digested with *Alu*1 *Hae* II et *Hpa*II and *Sau*3a. Restriction profiles with all four enzymes were then combined to identify the unique RFLP groups According to Yahiaoui-Zaidi et al. (2003).

Pathogenicity tests on half tubers

Before inoculation, a bacterial suspension of each strain was prepared in sterile distilled water (SDW) from cultures on King's B medium and incubated at 26°C for 24 h. Initial suspension corresponding to 4.10^8 (cel. ml^{-1}) each isolate was prepared and further dilutions in SDW were made to obtain the concentrations of 4.10^6 (cel. ml^{-1}), then, 50 µl of each suspension were used as inoculum. The half-tuber method given by Ibrahim et al. (1978) was followed for the study of the pathogenic variability of the hundred *Pectobacterium* strains on the two cultivars. Tubers of each test cultivar were washed, surface-sterilised by dipping in 10% ethanol for 5 min, rinsed in SDW and air-dried overnight before cutting. Each tuber was cut from the rose end to the heel end in two roughly equal parts with a sterile knife. A hole (5 mm diameter x 5 mm depth) was drilled with a cork borer in the center of each half-tuber, which was then placed on tissue paper soaked with 100 ml of water in plastic containers and allowed to dry for one hour before inoculation. Each half-tuber was inoculated by depositing 50 µl of the previously prepared bacterial suspensions in the hole. For the control, SDW was used. The plastic boxes containing the inoculated tubers were covered with a plastic sheet sealed with a rubber band to maintain a saturating relative humidity. Disease development was scored after five days of inoculation in a dark moist chamber at 20°C. A second trial was conducted at 20 and 25°C with a high inoculum concentration (2.10^7 cel. ml^{-1}) of forty isolates representative from the collection. Disease development was scored after five days of inoculation at either 20 or 25°C. Symptoms were assessed visually, before all rotted tissue was removed with a spatula and the volume of water necessary to fill up the hole was recorded (Ibrahim et al., 1978). In each test, 10 half-tubers per cultivar were inoculated with each inoculum concentration of each strain.

Data analysis

The pathogenic ability of *Pectobacterium carotovorum* on each cultivar was assessed as the means of the volume of water necessary to fill up the hole caused by rotting. Effects of cultivars, isolates (or groups thereof), temperature, and interactions between those parameters on pathogenicity were analysed by ANOVA using the GLM module of the SAS statistical package.

RESULTS

Biochemical characteristics

Tests for pectolytic activity, growth at 37°C, acid production from methyl α-D-glucoside, production of reducing substances from sucrose, utilisation of melibiose and citrate, showed that all isolates had the characteristics of soft rot *Pectobacterium*. The collection included 40 typical *Pcc* isolates, which grew at 37°C and did not utilise α-methylglucoside or produce reducing substances from sucrose, and 14 typical *Pca* isolates, which did not grow at 37°C, utilised α-methylglucoside and produced reducing substances from sucrose. The remaining isolates were clustered into seven groups and named as (*Pcc1, Pcc2, Pcc3, Pcc4, Pcc5, and Pcc6*) which were genetically similar or identical to the *Pcc* type strain but did not produce all the typical biochemical or physiological responses of this subspecies and one atypical *P. c.* subsp *odorifera* (*Pco1*)

Molecular typing

The RFLP analysis undertaken with the 99 *Pectobacterium* isolates by digesting the 434-bp amplified fragment with the four restriction enzymes revealed the presence of 12 different profiles (RFLP groups: 1, 3, 4, 8, 9, 10, 12, 14, 22, 25, 26 and 27) demonstrating polymorphism among the various isolates. Three of these RFLP groups have not been previously described and were assigned numbers 25–27. The RFLP groups 1, 8, 9 and 22 were the most frequent in the collection.

Pathogenicity of *Pectobacterium* isolates

Among the number of *P. carotovorum* tested, *most of them* were pathogenic to both cultivars Bintje and Désirée at (2.10^7 cel. ml^{-1}). However, some of the *Pcc* strains failed to induce symptoms on the two cultivar, others caused rotting on cv. Bintje but not on cv. Désirée.

The visual aspect of symptoms in both cultivars was different between *Pca* and *Pcc*: the former group produced a dark brown slimy necrosis surrounded by a black ring, while soft rot caused by the latter group was light brown and drier. These symptoms were similar to those caused by the *Pca* and *Pcc* reference isolates, respectively. Moreover, *P. c.* subsp *odorifera* (*Pco*) strain showed a high level of aggressiveness; rot aspect was intermediate between *Pca* and *Pcc* isolates. In all cases,

control plants did not exhibit any symptoms. Twenty five of the 85 typical and atypical *Pcc* isolates entirely failed to cause rotting At 2.10^5 cel. ml^{-1} on both Bintje and Désirée, and the volume of rotted tissue produced by the remaining isolates remained small (0.1 - 0.9 ml). This result is summarized in a Table 1.

Pathogenicity differences between groups

As expected, aggressiveness was significantly high at the higher inoculum concentration (2.10^{7}cel. ml^{-1}) for all isolates and on both cultivars (Figure 1). Overall, *Pca* and*Pco* isolates were significantly more aggressive than *Pcc* isolates: the volumes of rotted tissue caused by *Pca* isolates at an intermediate inoculum concentration (2.10^5 cel. ml^{-1}) were nearly the same as those caused by *Pcc* isolates at 2.10^7 cel. ml^{-1}.

Significant aggressiveness differences between and within RFLP groups were also observed (Figure 2). Interestingly, the four isolates belonging to RFLP group 25 recently described were more aggressive than the reference *Pcc* strain CFPB 20-46 at the same concentration (two fold difference in the mean rot volume at 2.10^7 cel. ml^{-1}. Two of these isolates were obtained from seed of cvs Désirée and Timate multiplied in Algeria, and the other two were isolated from Dutch seed lots of cvs Désirée and Diamant. *Eca* isolates obtained from Algerian seed lots also were generally more aggressive than the reference strains CFPB 15.26 and 86.20 on the two cultivars. The aggressiveness of isolates collected from Dutch seed lots was similar to that of the corresponding type strains.

Cultivar Bintje proved more susceptible than cv Désirée to *Pcc,* whereas the opposite was true for *Pca* and *Pco*. This effect was more pronounced at 20°C than at 25°C for the last two subspecies (Table 2).

ANOVA analysis of the rot volumes after inoculation of the two cultivars with 40 isolates representative of the collection at 20 and 25°C showed highly significant effects of temperature on disease severity, as well as a highly significant temperature pathogen interaction (Table 3). Analyses conducted separately for each bacterial group revealed that *Pca* isolates were the most aggressive, particularly at 20°C. Inoculation at 25°C of the two cultivars with an inoculum of 2.10^7 cel. ml^{-1} resulted in similar amounts of rot with *Pcc* and *Pca* isolates.

DISCUSSION

Our results confirmed that the half-tuber inoculation method is suitable for assessing the aggressiveness variability among *Pectobacterium* isolates. The differences in disease produced by the isolates tested here were consistent with those observed by other workers (e.g. Jones, 1910; Johnson et al., 1989; Smith and Bartz, 1990). They were also related to the taxonomic diversity

Table 1. Half tubers inoculation of cvs Bintje and Désirée at 20 °C with *Pectobacterium* isolates with 2.10^6 and 2.108 cel. ml^{-1}.

N°	Ssp	MIBb	MIDb	MIBa	MIDa	N°	Ssp	MIBb	MIDb	MIBa	MIDb
*1	Pcc	0.75	0.60	0.10	0.10	54	Pcc[nt2]	1.27	1.20	0.40	0.80
*2	Pcc	1.04	1.08	0.30	0.10	55	Pcc[nt2]	1.00	1.10	0.30	0.05
*3	Pcc	0.05	0.05	0.10	0.10	56	Pcc[nt2]	0.90	1.09	0.05	0.05
4	Pcc	1.05	0.52	0.05	0.05	57	Pcc[nt2]	1.00	0.05	0.15	0.05
5	Pcc	0.40	0.25	0.05	0.05	58	Pcc[nt3]	1.10	0.90	0.05	0.05
6	Pcc	0.90	0.84	0.05	0.05	59	Pcc[nt]	1.02	1.55	0.60	0.80
7	Pcc	0.05	0.05	0.05	0.05	60	Pcc[nt3]	1.10	0.91	0.30	0.05
8	Pcc	0.05	0.05	0.05	0.05	*61	Pcc[nt]	1.09	0.90	0.05	0.05
9	Pcc	0.60	0.20	0.05	0.05	*62	Pcc[nt3]	0.94	0.30	0.15	0.05
10	Pcc	1.52	1.00	0.75	0.05	*63	Pcc[nt]	0.05	0.05	0.05	0.05
11	Pcc	1.40	0.32	0.70	0.05	64	Pcc[nt3]	1.00	0.40	0.60	0.05
12	Pcc	1.04	0.20	0.10	0.05	65	Pcc[nt]	0.40	0.05	0.05	0.05
13	Pcc	0.88	0.48	0.05	0.05	66	Pcc[nt3]	1.30	1.00	0.90	0.05
*14	Pcc	1.60	0.60	0.25	0.05	67	Pcc[nt3]	0.90	0.05	0.05	0.05
*15	Pcc	0.50	0.40	0.05	0.10	68	Pcc[nt]	0.05	0.05	0.05	0.05
*16	Pcc	1.22	0.90	0.05	0.05	69	Pcc[nt3]	0.20	0.20	0.05	0.05
*17	Pcc	1.30	1.10	0.20	0.05	70	Pcc[nt]	0.22	0.05	0.05	0.05
*18	Pcc	1.10	0.90	0.05	0.05	71	Pcc[nt3]	0.40	0.40	0.05	0.05
19	Pcc	1.10	1.60	0.43	0.50	72	Pcc[nt]	1.30	0.71	0.45	0.05
20	Pcc	0.90	0.40	0.05	0.05	73	Pcc[nt3]	1.40	0.72	0.05	0.05
*21	Pcc	1.65	1.00	0.50	0.05	74	Pcc[nt3]	1.20	0.91	0.80	0.05
*22	Pcc	1.50	0.60	0.30	0.05	75	Pcc[nt]	1.39	0.20	0.80	0.25
*23	Pcc	1.20	0.70	0.20	0.12	76	Pcc[nt3]	1.32	1.07	0.20	0.40
24	Pcc	1.10	0.51	0.40	0.30	77	Pcc[nt]	0.71	0.35	0.40	0.05
25	Pcc	1.10	0.60	0.30	0.20	78	Pcc[nt3]	0.71	0.05	0.05	0.05
26	Pcc	0.90	0.68	0.10	0.30	*79	Pcc[nt4]	1.20	1.28	0.71	0.60
27	Pcc	0.90	0.30	0.10	0.10	80	Pcc[nt4]	0.90	1.56	0.70	0.80
28	Pcc	1.10	0.90	0.30	0.05	81	Pcc[nt4]	1.39	1.30	0.60	0.70
*29	Pcc	1.08	0.80	0.10	0.10	82	Pcc[nt4]	0.89	0.05	1.00	0.05
*30	Pcc	1.30	0.80	0.75	0.25	*83	Pcc[nt']	0.05	0.05	0.05	0.05
*31	Pcc	0.80	0.60	0.10	0.10	*84	Pcc[nt5]	0.77	0.52	0.05	0.05
*32	Pcc	0.90	0.30	0.05	0.05	*85	Pcc[nt6]	0.89	1.30	0.60	0.50
*33	Pcc	1.19	0.70	0.10	0.10	*86	Pco[n7]	1.69	2.29	0.70	0.70
*34	Pcc	1.30	0.80	0.15	0.05	*87	Pca	2.00	3.42	0.80	1.90
*35	Pcc	1.00	0.50	0.40	0.25	*88	Pca	2.24	3.11	1.09	1.50
*36	Pcc	1.10	1.10	0.05	0.15	89	Pca	2.81	3.28	1.40	1.60
37	Pcc	1.00	0.90	0.15	0.15	*90	Pca	2.56	3.29	1.23	1.90
38	Pcc	0.80	0.71	0.45	0.05	*91	Pca	2.90	4.10	1.72	200
*39	Pcc	0.80	0.40	0.30	0.05	*92	Pca	3.62	4.20	2.39	2.20
40	Pcc	0.75	0.60	0.30	0.15	*93	Pca	3.72	4.20	2.10	2.50
41	Pcc[nt1]	0.60	0.20	0.10	0.05	94	Pca	3.05	3.62	1.10	1.60
*42	Pcc[nt2]	0.84	0.60	0.20	0.05	95	Pca	2.40	2.91	2.09	0.90
*43	Pcc[nt2]	1.50	1.10	0.60	0.50	96	Pca	2.70	2.89	1.95	1.00
44	Pcc[nt2]	1.50	1.01	0.80	0.10	97	Pca	2.45	2.96	1.00	1.10
45	Pcc[nt2]	1.20	0.90	0.50	0.05	98	Pca	2.20	2.52	1.60	1.20
*46	Pcc[nt2]	0.90	0.80	0.05	0.30	99	Pca	2.40	2.60	1.50	1.00
47	Pcc[nt2]	1.30	1.30	0.15	0.80	*100	Pca	2.38	2.91	1.00	0.80
*48	Pcc[nt2]	1.10	0.72	0.40	0.40	86 20	Pca*	2.31	3.39	1.30	1.50
*49	Pcc[nt2]	0.60	0.52	1.60	0.80	15 26	Pca[T]	2.82	3.01	1.50	1.30
50	Pcc[nt2]	2.00	1.60	0.70	0.60	91 8	Pcc*	0.90	1.19	0.05	0.05

Table 1. Contd.

51	Pcc^{nt2}	1.10	0.29	0.35	0.05	20 46	Pca^{T}	1.20	1.15	0.43	0.40
52	Pcc^{nt2}	1.30	1.21	0.10	0.50	0.48	Pch^{T}	0.05	0.05	0.05	0.05
53	Pcc^{nt2}	0.90	1.09	0.40	0.50						

MIB: mean of 10 inoculations of cultivar Bintje; MID: mean of 10 inoculation of cultivar Desiree. (a) at 2.10^{6} and(b) at 2.10^{8} cel. $ml^{-1.}$ Pcc^{nt} :non typical (1 to 6) and Pco^{nt7} Pectobacterium. carotovoum. subsp. odorifera inable to grew at $37°C$. Pca* 86.20: P. c. subsp. atrosepticum; Pcc* 91.8 : P. c. subsp. carotovorum; Pch: P. chrysathemi. * Bernard Jouan, National Institute of Agronomic Research, Rennes, France, personal collection. Pca^{T} 15.26, Pcc^{T} 20.46 and PchT 20.48 (Strains type from CFBP: French Collection of Phytopatogenic Bacteria, INRA, Angers, France).

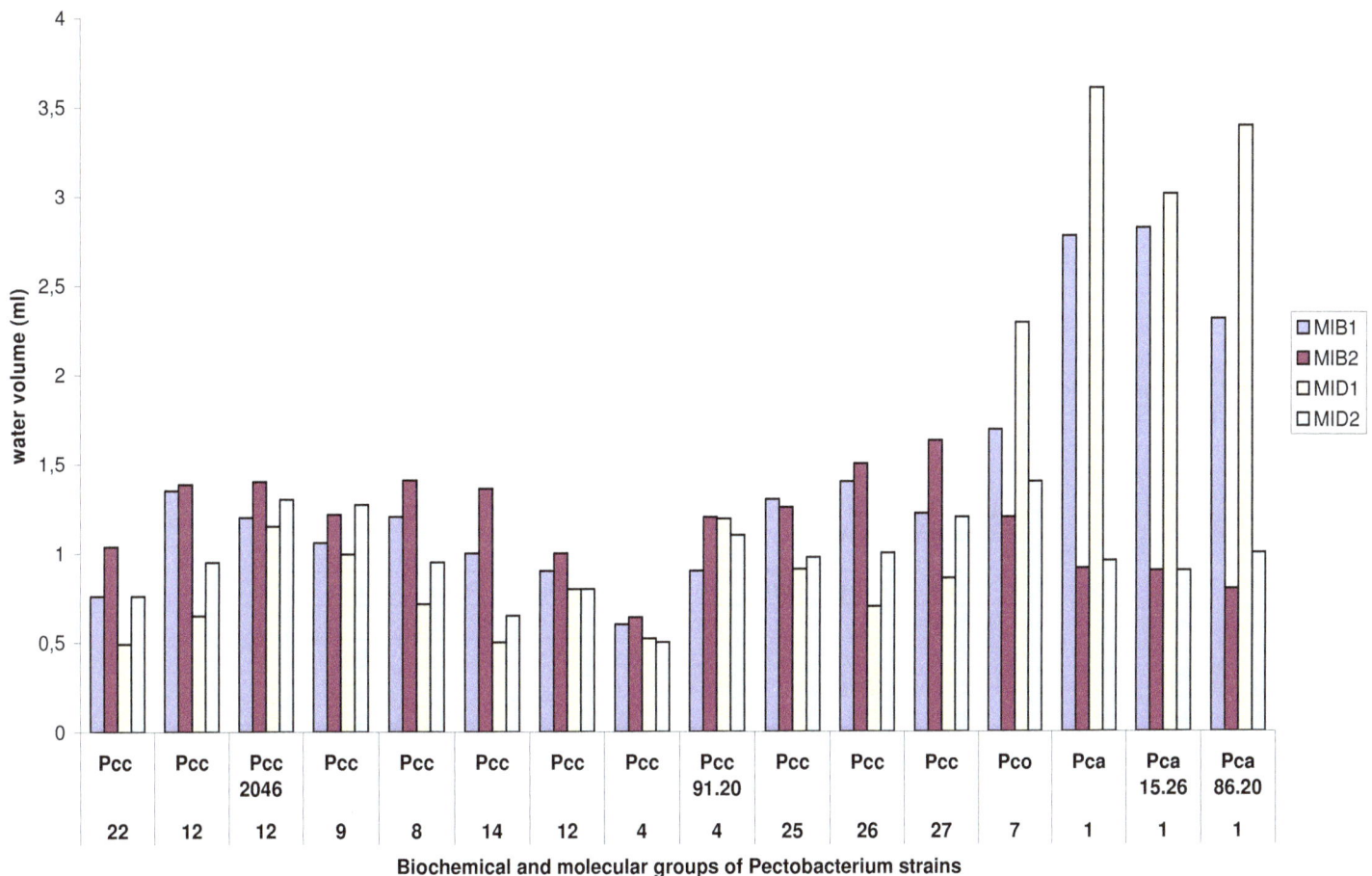

Figure 1. Pathogenic variability within biochemical and molecular groups of *Pectobacterium carotovorum* sp. on two cultivars Bintje and Désirée inoculated with (2.10^{7}cel. ml^{-1}), and incubated at 20 and 25 °C.

existing within this collection isolated from various cultivars and derived from populations of different geographic origins.

At relatively moderate inoculum concentrations (2.10^{5} cel. ml^{-1}), many *Pcc* isolates from our collection failed to induce rotting, whereas all the *Pca* isolates proved pathogenic (Table 1). The higher aggressiveness to potato tubers of *Pca* was confirmed at higher inoculum concentrations, confirming previous observations about aggressiveness differences between *P. carotovorum*

subspecies (Lapwood et al., 1984). Interestingly, the unique *Pco* isolates from our collection were also highly pathogenic on potato tubers. This observation confirmed the diversity existing within the *Pcc* isolates. The pathogenic variability observed within the *Pcc* isolates on the two cultivars extends earlier observations (Jones, 1910; Johnson et al., 1989; Smith and Bartz, 1990). Disease failed to develop in both cultivars Désirée and Bintje inoculated with some *Pcc* isolates. The failure of some *Pcc* isolates to cause tuber rot could be explained

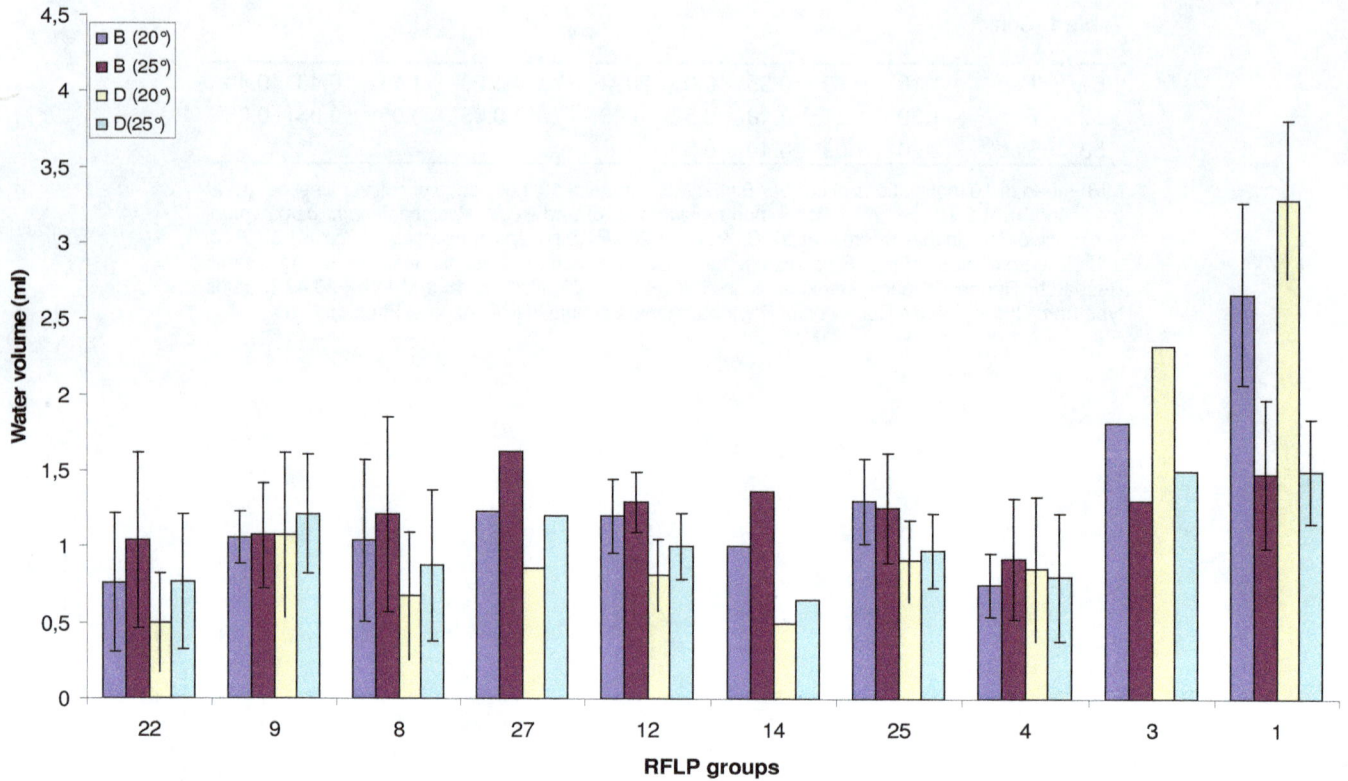

Figure 2. Pathogenic differences between 10 RFLP groups (1, 3, 4, 8, 9, 12, 14, 22, 25 and 27) of *Pectobacterium carotovorum* on two potato cultivars Bintje (B) and Désirée (D).

Table 2. Results of potato tubers inoculation, with 40 strains of *P. carotovorum* from Algeria.

Strain Number	Subsp	MIB$_1$	MIB$_2$ (ml)	MID$_1$	MID$_2$	RFLP groups	Years
1	*Pcc*	0.75	1.25	0.60	1.20	22	1995
2	*Pcc*	1.04	1.18	1.08	1.30	9	1995
3	*Pcc*	0.05	0.05	0.05	0.05	22	1995
14	*Pcc*	1.60	2.00	0.60	1.00	8	1995
15	*Pcc*	0.50	0.85	0.40	0.60	22	1995
16	*Pcc*	1.22	1.63	0.86	1.20	27	1995
17	*Pcc*	1.30	1.70	1.10	1.40	22	1995
18	*Pcc*	1.10	1.20	0.90	0.95	8	1995
21	*Pcc*	1.65	1.80	1.00	1.40	8	1995
22	*Pcc*	1.50	1.42	0.60	0.90	12	1995
23	*Pcc*	1.20	1.35	0.70	1.00	12	1995
29	*Pcc*	1.08	1.41	0.80	1.00	9	1994
30	*Pcc*	1.30	1.70	0.80	0.90	22	1994
31	*Pcc*	0.80	1.06	0.60	0.75	9	1994
32	*Pcc*	0.90	1.08	0.30	0.80	22	1994
33	*Pcc*	1.19	1.22	0.70	0.95	22	1994
34	*Pcc*	1.30	1.40	0.80	1.05	9	1995
35	*Pcc*	1.00	1.36	0.50	0.65	14	1995
36	*Pcc*	1.10	1.16	1.10	1.30	9	1995
39	*Pcc*	0.80	1.28	0.40	0.60	22	1996
42	*Pcc*[at]	0.84	0.90	0.60	0.65	8	1996

Table 2. Contd.

43	Pcc^{at}	1.50	1.51	1.10	1.15	25	1996
46	Pcc^{at}	0.90	1.00	0.80	0.80	12	1996
48	Pcc^{at}	1.10	1.00	0.72	0.80	25	1996
49	Pcc^{at}	0.60	0.64	0.52	0.50	4	1996
61	Pcc^{at}	1.09	1.40	0.90	1.09	8	1996
62	Pcc^{at}	0.94	1.14	0.30	0.60	8	1996
63	Pcc^{at}	0.05	0.05	0.05	0.05	8	1996
79	Pcc^{at}	1.20	1.40	1.28	1.50	9	1995
83	Pcc^{at}	0.05	0.05	0.05	0.05	22	1995
84	Pcc^{at}	0.77	1.20	0.52	1.05	22	1995
85	Pcc^{at}	0.89	0.90	1.30	2.00	9	1995
86	Pco^{at}	1.69	1.20	2.29	1.40	7	1996
87	Pca	2.00	0.60	3.42	1.00	1	1995
88	Pca	2.24	0.85	3.11	0.80	1	1996
90	Pca	2.56	0.90	3.29	0.90	1	1996
91	Pca	2.90	1.00	4.10	1.00	1	1996
92	Pca	3.62	1.30	4.20	1.20	1	1995
93	Pca	3.72	1.10	4.20	1.00	1	1995
100	Pca	2.38	0.65	2.91	0.80	1	1995
1526^{TS}	Pca	2.82	0.90	3.01	0.90	1	TS
8620	Pca	2.31	0.80	3.39	1.00	1	RS
2046^{TS}	Pcc	1. 40	1.15	1.30	1. 20	12	TS
918	Pcc	0.90	1.20	1.19	1.10	4	RS

MIB_1 and MIB_2: mean (ml) of ten inoculations on cultivar Bintje at 20 and 25 °C respectively with 2.10^8 cel. ml^{-1}. MID_1 and MID_2: mean (ml) of ten inoculations on cultivar Desiree at 20 and 25 °C respectively.[TS] with 2.10^8 cel. ml^{-1}: Type strain of French Collection of hytopathogenic Bacteria, Angers, France. RS: Reference strain. The forty representative isolates (1 - 100) were previously described by Yahiaoui-Zaidi et al., 2003.

Table 3. ANOVA analysis of the rot volumes after inoculation of the two cultivars/ Effects of temperature on disease severity.

Cultivars	Temperature	Subsp	Temp/ Subsp	Var/Temp	
F values	24.58	26.94	198.83	222.34	7.58
P	0.0001	0.0001	0.0001	0.0001	0.006

by the lost of the pathogenicity due to the successive subculturing as reported by Priou (1992). However, in our tests, the pathogenicity of certain isolates differed between cultivars, and varied among isolates that were isolated at the same time. Thus, the loss of aggressiveness or failure to cause disease did not appear to have resulted from subculturing. Bartz (1980) obtained similar results, and suggested that some isolates collected from symptoms would not be likely to initiate disease in nature. The low aggressiveness of several isolates towards their cultivar of origin is not easily explained. Dickey (1981) observed that certain isolates of *P. chrysanthemi* were not pathogenic on their host of origin and suggested that such isolates may not have initiated the lesions from

which they were obtained.Our results also suggest a differential susceptibility of potato cultivars Bintje and Désirée to the various *P. carotovorum* subspecies. Differences in the susceptibility of the two cultivars to rotting were evident when comparing their behaviour towards *Pca,* more pathogenic on Desirée than on Bintje, and, more pathogenic on Bintje than on Désirée. The behaviour of Algerian *Pca* isolates was intriguing, because according to Hélias et al. (2000) Bintje was known to be more susceptible to soft rot than Désirée. The fact that most Algerian isolates originated from Désirée, a popular cultivar in northern Africa, may explain their higher aggressiveness towards this cultivar.

The resistance of tuber tissue of cultivars Désirée to

soft rot caused by *Pca* isolates could be related to the presence of some phenolics compounds like anthocyanins as described by Wegener and Jansen (2007), in coloured potato.

As expected, we observed a significant effect of temperature on soft rot development (Table 2; Figure1), as well as a highly significant temperature pathogen interaction, consistent with previous data indicating that *Pcc* isolates have a higher optimum temperature for growth than *Pca* isolates (Pérombelon and Kelman, 1980). Certain differences in pathogenicity and aggressiveness among the taxons of the soft rot *Pectobacterium* have been related to the optimum growth temperature. Indeed, besides the genetic characteristics of the host cultivar and pathogen isolate, environmental conditions also condition largely the influences of the quantum of disease (Bain et al., 1990). Temperature exerts a differential effect on the aggressiveness of *Pectobacterium* taxa by regulating of their enzyme production (Kotoujansky, 1987, Pérombelon and Kelman, 1980; Pérombelon 1982). The inability of some isolates to induce disease can be explained by the absence of enzyme production, particularly pectate lyase (PL). In *P. chrysanthemi* for example the low production of PL causes a reduction of pathogenicity (Hugouvieux-Cotte-Pattat et al, 1992).

Overall, our data further strengthen the recommendation of Priou (1992) to use Désirée and Bintje as control cultivars to allow comparisons of pathogenicity tests between laboratories. The differential reaction of these two cultivars to Algerian *Pectobacterium* isolates, coupled with the genetic characteristics of the bacterial populations involved, provide a starting point for a more comprehensive study of pathogenic adaptation mechanisms in soft rot *Pectobacterium*. Yishay et al. (2008) reported that a clear decline in virulence of the monocot isolates towards the dicot hosts suggests these isolates are better adapted to monocot hosts.

ACKNOWLEDGEMENT

We thank B. Jouan, D. Anrivon, (INRA LeRheu /Rennes for their collaboration.

REFERENCES

Bain RA, Pérombelon MCM, Tsror L, Nachmias A (1990). Blackleg development and tuber yield in relation to numbers of *Erwinia carotovora* subsp. *atroseptica* on seed potatoes. Plant Pathol. 39: 125-133.

Bourne WF, McCalmont DC, Wastie RL (1981). Assessing potato tubers for susceptibility to bacterial soft rot (*Erwinia carotovora* subsp. *atroseptica*). Potato Res. 24: 409-415.

Bartz JA (1980). Causes of post harvested losses in Florida tomato shipment. Plant Disease 64: 934-937.

Carputo D, Basile B, Cardi T, Frusciante L (2000). *Erwinia* resistance in backcross progenies of *Solanum tuberosum* X *S. tarijense* and *tuberosum* (+) *S. commersonii* hybrids. Potato Res. 43: 135-142.

Cazelles O, Ruchi A, Schwärzel R (1995). Pourriture bactérienne des tubercules de pomme de terre: progrès dans la détection des infections latentes d'*Erwinia chrysanthemi* et dans la prévision des attaques au champ. *Revue Suisse d'Agriculture* 27: 17-22.

De Boer SH, Kelman A (1978). Influence of oxygen concentration and storage factors on susceptibility of potato tubers to bacterial soft (*Erwinia carotovora*). Potato Res. 21: 65-80.

Dickey RS (1981). *Erwinia chrysanthemi*: Reaction of eight plant species to strains from several hosts and to strains of other *Erwinia* species. Phytopathol. 71: 23-29.

Farrar JJ, Nunez JJ, Davis RM (2000). Influence of soil saturation and temperature on *Erwinia chrysanthemi* soft rot of carrot. Plant Disease 84: 665-668.

Garibaldi A, Batman DF (1971). Pectic enzymes produced by *Erwinia chrysanthemi* and their effect on plant tissue. Physiol. Plant Pathol. 1: 25-40.

Gregg M (1952). Study in physiology of parasitism. XVII. Enzyme secretion by strain of *Bacterium carotovorum* and other pathogens in relation to parasitic vigor. Annales de Botanique 16: 235-250.

Hélias V, Andrivon D, Jouan B (2000). Development of symptoms caused by *Erwinia carotovora* ssp *atroseptica* under field conditions and influence of their effects on the yield of individual potato plants. Plant Pathol. 49: 23-32.

Hugouvieux-Cotte-Pattat N, Domenguez H, Robert-Baudouy J (1992). Environmental conditions affect the transcription of the pectinase genes of *Erwinia chrysanthemi* 3937. J. Bacteriol. 174: 7807-7818.

Ibrahim M, Jouan B, Samson R, Poutier F, Sailly M (1978). Prospect of a pathogenicity test concerning *Erwinia carotovora* subsp *atroseptica* and *Erwinia carotovora* subsp. *carotovora* on half tubers. Proceedings of the 4th International Conference on Plant Pathogenic Bacteria, Angers pp. 591-602.

Johnson DA, Regner KM, Lunden JD (1989). Yeast soft rot of onion in the Walla Walla Valley of Washington and Oregon. Plant Disease 73: 686-688.

Jones LR (1910). A soft rot of carrot and other vegetables caused by *Bacillus carotovorus* Jones. Vermont Agricultural Experimental Station 13th Annual Report, 1889-1900.

Kotoujansky A (1987). Molecular genetics of pathogenesis by soft rot erwinias. Annual Rev. Phytopathol. 25: 405-430.

Lapwood DH, Read PJ, Spokes J (1984). Methods for assessing the susceptibility of potato tubers of different cultivars to rotting by *Erwinia carotovora* subsp. *atroseptica* and *carotovora*. Plant Pathol. 33: 13-20.

Lumb VM, Pérombelon MCM, Zutra D (1986). Studies of a wilt disease of the potato plant in Israel caused by *Erwinia chrysanthemi*. Plant Pathol. 25: 196-202.

Pérombelon MCM (1979). Factors affecting accuracy of the tuber incubation test for the detection of contamination of potato stocks by *Erwinia carotovora*. Potato Res. 22: 63-68.

Pérombelon MCM (1982). The impaired and soft rot *Erwinia* In: Lacy GN,Mount MS, eds *Phytopathogenic procaryotes*, 2. New York, USA: Academic Press pp. 55-68.

Pérombelon MCM (2002). Potato disease caused by soft rot erwinias: an overview of pathogenesis. Plant Pathol. 51: 1-12.

Pérombelon MCM, Kelman A (1980). Ecology of soft rot *erwinias*. Annual Rev. Phytopathol. 18: 361-387.

Pérombelon MCM, Kelman A (1987). Blackleg and other potato diseases caused by soft rot erwinias: proposal for revision of terminology. Plant Disease 71: 283-285.

Pérombelon MCM, Salmond GPC (1995). Bacterial soft rots. In: Singh US, Singh RP, Kohmoto K, eds. Pathogenesis and Host Specificity in Plant Diseases, 1. Oxford, UK: Pergamon Press pp. 1-20.

Priou S (1992). Variabilité phénotypique et génétique et caractérisation des sous espèces d'Erwinia carotovora en relation avec leur pouvoir pathogène sur pomme de terre. PhD. Thesis, Ecole Nationale Supérieure Agronomique de Rennes, Rennes, France.

Rabot B, Pasco C, Schmidt J (1994). Assessing six Austrian cultivars for resistance to *Erwinia carotovora* subsp. *atroseptica*. Potato Res. 37: 197-203.

Sambrook J, Fritsch EF, Maniatis T (1989). Molecular Cloning: a Laboratory Manual, 2nd ed. Cold Spring, NY, USA: Cold Spring Harbor Laboratory.

Smith C, Bartz JA (1990). Variation in the pathogenicity and aggressiveness of strains of *Erwinia carotovora* subsp. *carotovora* isolated from different hosts. Plant Disease 74: 505-509.

Stead D (1999). Bacterial diseases of potato: relevance to *in vitro* potato seed production. Potato Res. 42: 449-456.

Waterer DR, Pritchard MK (1984). Monitoring of volatiles: a technique for detection soft rot (*Erwinia carotovora*) in potato tubers. *Canadian* J. Plant Pathol. 6: 165-175.

Wegener CB, Jansen G (2007). Soft-rot Resistance of Coloured Potato Cultivars (*Solanum tuberosum* L.): The Role of Anthocyanins. Potato Res. 50: 31-44

Yahiaoui-Zaidi R, Jouan B, Andrivon D (2003). Biochemical and molecular variability among *Erwinia* isolates from potato in Algeria. Plant Pathol. 52: 28-40.

Yishay MI, Burdman S, Valverde A, Luzzatto T, Ophir R, Yedidia I, (2008). Differential pathogenicity and genetic diversity among *Pectobacterium carotovorum* s sp *carotovorum* isolates from monocot and dicot hosts support early genomic divergence within this taxon. Environ. Microbiol. 10: 2746–2759.

9

Urinary schistosomiasis: Efficacy of praziquantel and association of the ABO blood grouping in disease epidemiology

M. O. Oniya and O. Jeje

Department of Biology, Federal University of Technology, P. M. B. 704 Akure, Ondo State, Nigeria.

Schistosomiasis, one of the neglected tropical diseases continues to plague communities with little or no access to potable water and with high water contact activities. Ipogun village in Ondo State is one of such communities in Nigeria. This study assessed the efficacy of praziquantel and the association of the ABO blood grouping in disease epidemiology in the only private primary school in the village. Ten milliliters of urine and 3 ml of blood samples were collected for urinalysis and blood grouping test respectively from a total of 113 pupils. Results showed that, 60 (53.1%) were infected with *Schistosoma haematobium*. Infected pupils were treated with praziquantel (40 mg/kg body weight) and subsequently re-screened 5 months after the administration of the chemotherapeutic. Results also showed that a single dose of praziquantel conferred a 94.44% cure rate. ABO blood grouping was also observed not to be associated with the epidemiology of the disease as frequency or severity of infection was not significant (p > 0.05) among the three represented blood groups (*A*, *B* and *O*).

Key words: Urinary schistosomiasis, praziquantel, ABO blood grouping, epidemiology.

INTRODUCTION

Schistosomiasis continues to threaten millions of people, particularly the poor in rural settlements in the developing countries (Chitsulo et al., 2000; Engels et al., 2002). Of all the estimated 200 million infected people globally, more than half are asymptomatic and 20 million exhibit severe disease manifestations (WHO, 1993). Historically, the disease has always been with man. Ruffer (1910) reported calcified eggs in the kidneys of two Egyptian mummies of the twentieth century dynasty.

There are five species of *schistosomes* that can infect humans, of which *Schistosoma mansoni*, *Schistosoma japonicum*, and *Schistosoma haematobium* are the most prevalent ones. While infection with the former two species is associated with chronic hepatic and intestinal fibrosis, infection with *S. haematobium* can lead to ureteric and bladder fibrosis and calcification of the urinary tract (Ross et al., 2002; Utzinger et al., 2001). The other two species, *Schistosoma mekongi* and *Schistosoma intercalatum* are not too common. In Nigeria

and other tropical African countries, two species of these causative organisms have been reported. These are *S. mansoni* and *S. haematobium*, causing intestinal and urinary schistosomiasis respectively with the latter more widely spread (Ejezie et al., 1989). Estimates suggest that 85% of all schistosomiasis current burden is concentrated in sub-Saharan Africa (Chitsulo et al., 2000).

Chemotherapy still remains the principal tool in the global battle against the scourge with Praziquantel being the current drug of choice. Recently, there have been concerns on drug resistance in praziquantel-induced therapy in schistosomiasis. praziquantel resistance has been reported in *S. mansoni*. Ernould et al. (1999) and Doenhoff et al. (2002) discussed the emergence of resistance by *S. mansoni* to praziquantel. Lawn et al. (2003), while expressing the concern on heavy reliance upon praziquantel and the potential development of drug resistance, described a British traveller who acquired *S. mansoni* infection in East Africa and in whom repeated standard 40 mg/kg doses of praziquantel failed to clear the infection despite no opportunity for reinfection. Similarly in *S. haematobium,* King et al. (2001) analysed

---

*Corresponding author. E-mail: onixous@yahoo.co.uk.

that, attempts to increase community treatment coverage to 100% would accelerate the emergence of clinically significant resistance, and thus emphasized that, targeted treatment has the potential advantage to prolong the useful lifespan of praziquantel.

The blood group frequencies in small inbred populations reflect the influences of genetic drift. In a small community, an allele can be lost from the genetic pool, if persons carrying it happen to be infertile, while it can increase in frequency, if advantage exists (Encyclopædia Britannica, 2007). Though, there is a dearth of information on immunity and/or susceptibility to schistosomiasis vis-à-vis blood group in Nigeria, blood tests are occasionally useful in supporting the diagnosis or assessing the severity of *schistosomiasis* infection. Kassim and Ejezie (1982) reported no significant association between the ABO blood group and *S. haematobium* from two hundred and sixty nine individuals in Epe, Lagos. However, information on this subject remains essential particularly in disease epidemiology, as very scanty data is available. The present study was conducted to assess the efficacy of praziquantel, and the association of the ABO blood grouping in *schistosomiasis* epidemiology, in an untreated population in an endemic community in south west Nigeria.

MATERIALS AND METHODS

The study area

The study was carried out in Ipogun, a village in Ifedore local government area of Ondo State, south west Nigeria. Ipogun (7°19'N; 5°05'E), is about 14 km away from Akure, the capital city. In Ipogun, there is a wet season (April-October) characterized by heavy rains with occasional flooding of river banks and a dry season distinguished by increased temperature, very little or no rainfall and consequently, the river dries up with a few stagnant pockets of water along its course. (November- March). The primary source of water for agrarian and most domestic activities is the 'Aponmu' river, flowing through the village. The inhabitants are mainly farmers who use water from the river in carrying out their daily and recreational activities (bathing and washing).

Ethical considerations, study subjects and collection of samples

Ethical clearance was sought from, and provided by the Ondo State ministry of health. Written informed consent was sought from the parents and guardians of the children before the study began. Results were made known to the parents and all infected children were treated with praziquantel. The pupils of Morohunkeji nursery and primary school, a private establishment in Ipogun were the subjects of the research. The pupils from the school had hitherto not enjoyed any intervention programme in the past from the state's chemotherapeutic control measures. Pupils from primary one to six in the school were all screened. The survey was conducted between November and December 2006, while sample collection to determine the efficacy of the administered drug, praziquantel was subsequently carried out on May, 2007. Urine samples were collected between 09:00 am and 12 noon. Each pupil was given a

clean, dry and labelled screw-capped urine bottle. Five months after initial chemotherapy, urine samples were collected and analysed from the treated pupils to assess the cure rate. Other Information collected from the subjects included name, age, weight, class and gender.

For the purpose of the research, 3 ml of blood was collected from each pupil by venipuncture, with the aid of hypodermic needle and syringe by qualified health workers. This was done after sterilizing the skin surface by dabbing with methylated spirit. The samples were collected into EDTA bottles after which the bottles were shaken to ensure homogeneity, to prevent coagulation. Samples were then taken to the laboratory in cooler boxes for analysis.

Laboratory analyses

Urine

Urine samples were analysed using the centrifugation method as described by Chugh et al., (1986). The samples were left to stand on the bench for about 45 min. Following this, excess samples were decanted before subsequent agitation and pipetting of 10 ml of urine centrifuged 1,500 rpm for 3 min. The supernatant was discarded and the residue was put on a clean glass slide and examined under x10 objective lens of the microscope. Intensity was recorded as geometric mean egg count/10 ml urine. The mean of three separate counts were used for each subject.

Blood

Blood grouping test was done using anti- sera A, B and D reagents (Biotech laboratories Ltd., United Kingdom). The reagents were normally stored at 8°C, and allowed to attain ambient temperature prior to use. Three drops from each blood sample was placed on a clean white tile. Anti sera were then added and the reactions read to determine the blood group.

Treatment

Chemotherapy: Infected subjects following urinalysis were treated with the standard dose of praziquantel (40mg/kg body weight). The pupils were made to take the drugs in the presence of the village health supervisor, before going back to their classrooms. The drugs were supplied by the Ondo State ministry of health's *Schistosomiasis* control programme.

Statistical analysis

Prevalence rate was determined as the percentage of the infected subjects of the total number of the examined population, and results was further tested using chi square analysis. The geometric mean given as GM= antilog $[\{\sum \log(x+1)\}/n]-1$ (Sturrock, 2001) was calculated using Microsoft Excel 2007 in the sex and blood group categories. The test of significance for the blood group relative to the prevalence rate was done with the chi square analysis (x^2) using SPSS version 11 for windows.

RESULTS

A total of 113 pupils were screened, out of which, 60 were found to be infected with *S. haematobium*. This gave 53.1% (Table 1) prevalence in the examined population. About 46.67% (n=25) of this number showed

Table 1. Prevalence and intensity of infection for the urinary *schistosomiasis* among school pupils, in the examined population.

Gender	No. examined	No. infected	Haematuria (%)	Prevalence (%)	Intensity of infection
Male	59	32	45.16	54.23	16.16
Female	54	28	54.84	51.85	18.64
Total	113	60	100.00	53.10	17.60

x^2, df = 1, p = 0.001.
There was a significant difference in prevalence of infection within the genders (P<0.05).

Table 2. Prevalence of urinary *schistosomiasis* within age groups among school pupils, in the examined population.

Age (years)	No. examined	No. infected	Haematuria (%)	Prevalence (%)
5	1	0	0.00	0.00
6 -10	94	49	64.52	52.13
11-15	18	11	35.48	61.11
Total	113	60	100.00	53.10

x^2, df = 1, p = 0.001.
There was a significant difference in prevalence among the age groups (P<0.05).

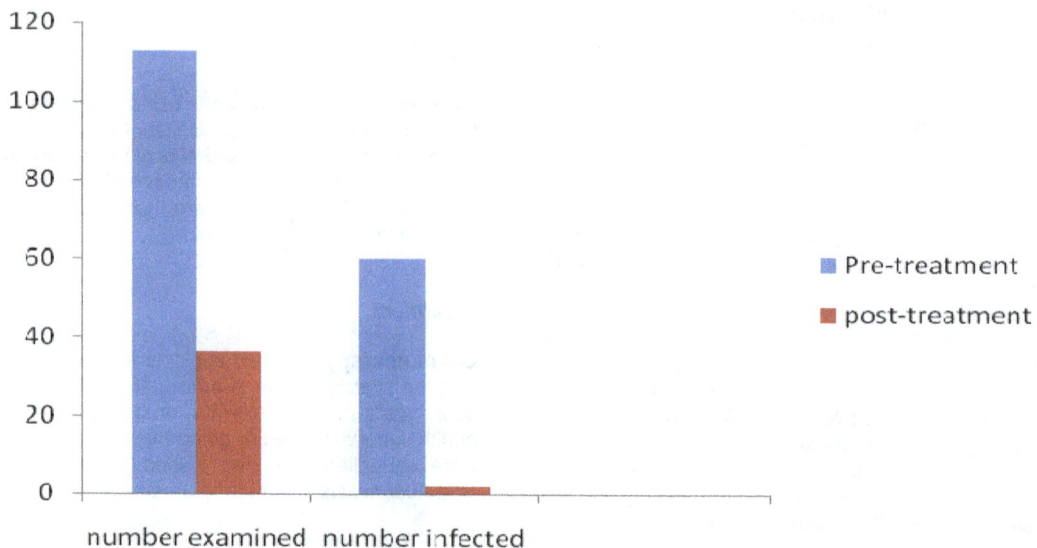

Figure 1. Prevalence of infection in the examined population post-treatment with praziquantel.

positive *haematuria* before centrifugation while 10% (n = 6) showed only after centrifugation. The geometric mean intensity of infection was 18.64 (Table 1) among the female pupils, while 16.16 was recorded for male.

Disease prevalence was highest (61.11%) within the age group 11 - 15 years and this was significant (p < 0.05) (Table 2). In this age group is also recorded the highest cases of *haematuria* (64.52%). Chemotherapy, using praziquantel was employed in treatment and control of infection. This anti-*schistosomal* drug was administered on a total of 36 pupils who agreed to continue with the programme from the initial 60 positive

cases. The outstanding 24 pupils were either absent from school or refused to be treated. The drug was administrated according to subject's body weight (40 m g/kg). Only 2 of this number tested positive after 5 months of the treatment giving 94.44% cure rate for praziquantel (Figure 1).

A total of 64 blood samples was also collected out of which 17 were 'A', 11 were 'B' while group 'O' recorded 36. Group 'AB' on the other hand was absent from the entire sampled population. Intensity of infection among infected pupils in the blood groups was 12.95, 15.18 and 14.45 (Table 3) for blood groups 'A', 'B' and 'O'

Table 3. Blood group specific prevalence of urinary schistosomiasis among school pupils in the examined population.

Blood group	No. examined	No. infected	Percentage prevalence (%)	Intensity of infection
A	17	10	58.82	12.95
B	11	6	54.55	15.18
O	36	24	66.67	14.45

χ^2, df = 2, p = 9.15.
There was no significant difference in prevalence within the blood groups (p>0.05).

respectively. There was no significant difference (p > 0.05) in relation to prevalence of infection.

DISCUSSION

The prevalence of urinary *schistosomiasis* was considerably high in the screened population, since more than half (53.10%) suffered from infection with 56.65% of the infected having visible *haematuria*. The high proportion in prevalence, observed in the screened population may be an indication that most villagers still frequent the stream for their daily activities, occupational or recreational purposes.

Gender specific prevalence revealed that the number of male pupils infected was slightly higher (54.23%), though significantly, than females (51.85%). This observation may be due to the fact that males are more involved in activities that have to do with water contact e.g. swimming, washing, and irrigation.

Pupils in the village have been reported to have more frequent contact with fresh water from 'Aponmu' stream during recreational activities such as swimming (Oniya, 2007). The observed prevalence along age group showed that age group 11 - 15 years had the highest prevalence of infection (61.11%), while 52.10% was recorded for age group 6 - 10 years and no infection was found in the 5 years old. Those in age group 11 - 15 years were obviously mostly affected probably because they frequently indulge in activities that bring them in contact with the source of infection. Frequent water contact activities by this age group may also promote transmission potential in the community. Furthermore, disease prevalence was higher in male pupils (54.23%) than the female with 51.85%. Intensity, however, was higher in the female population (18.64) and lower in the male pupils (16.16). Prevalence generally seems not to determine the worm load in the screened population as also shown in the blood group specific prevalence compared with the intensity.

From the results (Table 3), blood group O was predominant in the population with 56.25% of the pupils in this blood group. Blood group A was 26.56% while 17.19% individuals belonged to blood group *B*. There was higher prevalence of infection among individuals of blood group O, where 66.67% prevalence was recorded, this was followed by 58.82% prevalence from blood

group *A* and 54.55% from blood group *B*. *AB* blood group was absent from the screened population. Similar results were observed by Kassim and Ejezie (1982) and Ndamba et al. (1997). They later reported highest prevalence rates among individuals of blood group O (61.30%), 60.80% for blood group A, and 53.80% for blood group B, while blood group AB was altogether absent in a population screened for *schistosomiasis* in East Africa. Though we report the highest intensity and infection rates in the blood group O, our findings also showed that, there was no significant association between disease prevalence and ABO blood grouping in the examined population.

The efficacy of praziquantel treatment was very high in the studied population as the prevalence of S. *haematobium* dropped to 5.56%, without any case of *haematuria*, in the treated population after 5 months of chemotherapeutic administration (Figure 1). This gave 94.44% efficacy or cure rate for praziquantel. The examined population had not enjoyed any control coverage programme from the government as it was a privately owned school as opposed to the other public schools in the village, which had benefited from previous chemotherapeutic intervention programmes (Oniya and Odaibo, 2006; Oniya, 2007). The likelihood of reinfection is also high in the treated population as abstinence from the transmission sites cannot be guaranteed.

Presently, disease control is principally centred on chemotherapy, this alone cannot solve the problem (Sturrock, 2001). The rate of reinfection following parasitological cure is another concern for a multi pronged approach.

Chemotherapy alone may not be sufficient for the eradication of the disease as long as the intermediate hosts persist. Even in the previously untreated population, 100% cure rate was not recorded and with the growing concerns of the predicted failure of praziquantel (King et al., 2000 and 2001), an integrated control approach is desirable, to prolong the efficient use of praziquantel. There is also an urgent need for state and local governments in endemic countries to show genuine political commitment in order to halt transmission in their endemic communities. Such designs should however be implemented over a five years uninterrupted period.

ACKNOWLEDGEMENTS

We like to thank the Ondo State Ministry of Health, the members of Ipogun community, Miss Bola Alabi, and

Mrs. M. B. Oniya for their assistance.

REFERENCES

Chitsulo L, Engels D, Montresor A, Savioli L (2000). The global status of schistosomiasis and its control. Acta. Trop. 77(1): 41-51.

Chugh KS, Harries AD, Dahniya MH, Nwosu AC, Gashau A, Thomas J, Thaliza TD, Hogger S, Ajewski Z, Onwuchekwa AC (1986). Urinary Schistosomiasis in Maiduguri, Northeast Nigeria. Annals Trop. Med. Parasitol. 80(6): 593-599.

Doenhoff MJ, Kusel JR, Coles GC, Cioli D (2002). Resistance of Schistosoma mansoni to Praziquantel: is there a problem? Trans. Roy. Soc. Trop. Med. Hyg. 96(5): 465-469.

Ejezie GC, Gemade EI, Utsalo SJ (1989). The Schistosomiasis problem in Nigeria. J. Hyg. Epidemiol. Microbiol. Immunol. 33(2): 169-179.

Encyclopaedia Britannica (2007). Encylopaedia Britannica online 23 Apr. http://www.britannica.com/eb/article-33511.

Engels D, Chitsulo A, Montresor A, Savioli L (2002). The global epidemiological situation of schistosomiasis and new approaches to control and research. Acta Trop. 82: 139-146.

Ernould JC, Ba K, Sellin B (1999). Increase of intestinal Schistosomiasis after Praziquantel treatment in a Schistosoma haematobium and Schistosoma mansoni mixed focus. Acta Trop. 73(2): 143-152.

Kassim OO, Ejezie GC (1982). ABO blood groups in malaria and Schistosomiasis haematobium. Acta Trop. 39(2): 174-184.

King CH, Muchiri EM, Ouma JH (2000). Evidence against rapid emergence of praziquantel resistance in Schistosoma haematobium, Kenya. Emerg. Infect. Dis. 6(6): 585-594.

King CH, Muchiri EM, Ouma JH (2001). Evidence against rapid emergence of praziquantel resistance in Schistosoma haematobium, Kenya. Letters to the editor. Emerg. Infect. Dis. 7(6): 1069-1070.

Lawn SD, Lucas SB, Chiodini PL (2003). Case report: Schistosoma mansoni infection: failure of standard treatment with Praziquantel in a returned traveller. Trans. Roy. Soc. Trop. Med. Hyg. 97(1): 100-101.

Ndamba J, Gomo E, Nyazema N, Makaza N, Kaodera KC (1997). Schistosomiasis infection in relation to the ABO blood groups among school children in Zimbabwe. Acta Trop. 65(3): 181-190.

Oniya MO, Odaibo AB (2006). Reinfection pattern and predictors of urinary schistosomiasis among school pupils from a Southwestern village in Nigeria. Inter. J. Trop. Med. 1(4): 173-176.

Oniya MO (2007). Socio-cultural practices promoting the transmission of urinary schistosomiasis among school aged pupils in a South western village in Nigeria. Res. J. Biol. Sci. 2(1): 1-4.

Ross AGP, Bartley PB, Sleigh AC, Olds GR, Li YS, Williams GM, McManus DP (2002). Schistosomiasis. N. Engl. J. Med. 346(16): 1212-1220.

Ruffer M (1910). Note on the presence of Bilharzia haematobium in Egyptian mummies of the twentieth dynasty (1250-1000 B.C.). Br. Med. J. 1: 16.

Sturrock RF (2001). Schistosomiasis epidemiology and control: How did we get here and where should we go? Memórias do Instituto Oswaldo Cruz..Suppl. 96: 17-27.

Utzinger J, N'Goran EK, Bergquist R, Tanner M (2001). The potential of artemether for the control of Schistosomiasis. Inter. J. Parasitol. 31(14): 1549-1562.

World Health Organization (1993). The control of Schistosomiasis. Second Report of a WHO expert committee. World Health Organization Technical report Series 830.

10

Optimization of carbon and nitrogen sources of submerged culture process for the production of mycelial biomass and exopolysaccharides by *Trametes versicolor*

Krishna Bolla, B. V. Gopinath, Syed Zeenat Shaheen and M. A. Singara Charya

Department of Microbiology, Kakatiya University, Warangal - 506 009, A.P, India.

Medicinal mushrooms have profound health-promoting benefits. Polysaccharides constitute an important percentage of fungal biomass, where the hyphal wall frequently contains more than 75% of polysaccharide. *Trametes versicolor* is a medicinal fungus producing exopolysaccharides (EPS). The media were tested with different carbon and nitrogen sources which maximize the production of EPS by *T. versicolor*. The media were optimized with different carbon (glucose, fructose, sucrose, maltose, lactose, raffinose, mannitol and xylose) and nitrogen sources (peptone, glycine, gelatin, casein, yeast extract, ammonium sulphate, KNO_3 and $NaNO_2$) for the higher yield of polysaccharides. Biomass, pH changes along with the EPS production of the broth were followed during fermentations lasting 7 and 14 days. Fructose (8 g. dr. w/l) was shown to have yielded the highest production of EPS for 7 days, and gelatin (11 g. dr. w/l) to have produced the highest biomass. An experimental design to do this was adopted, in which the effects of pH were considered.

Key words: Basidiomycetes, exopolysaccharide, biomass, submerged culture.

INTRODUCTION

Many fungi are able to produce extracellular polysaccharides. They fulfill different tasks during the growth on natural substrates, such as adhesion to surfaces, immobilization of secreted enzymes, prevention of hyphae from dehydration and increased residence time of nutrients inside the mucilage (Rau, 1999). Many of them contain α- (Pullulan) or β-linked (e.g., Scleroglucan, Schizophyllan) glucose units. The alignment and disposition of linkage and branching affect the three-dimensional structure and determine the physicochemical characteristics of the gum. The branched β-glucans are biologically active and consequently are used in medicine and biotechnology, as well as additives in food and cosmetics (Manzoni and Rollini, 2001).

Trametes versicolor, belonging to the basidiomycetes class, can produce both extracellular and intracellular polysaccharides that have received special attention due to their physiological and biological activity. These fungi are well known as a medicinal mushroom in traditional therapeutic practice in Japan, China, Korea and other Asian countries (Cui and Chisti, 2003). Their polysaccharides have shown antitumour activity and include protein-bound polysaccharides extracted from the fungal mycelium like the Krestin (PSK) and Polysaccharopeptide (PSP) (Sugiura et al., 1980; Ng, 1998) and the extracellular polysaccharide Coriolan (Miyazaki et al., 1974).

Most of the reported studies have focused on polysaccharides isolated from the mycelium. However, a few studies on EPS from *T. versicolor* in submerged culture have been reported (Kim et al., 2002). Although a number of works have attempted to obtain the best culture conditions and EPS characterization from different fungi, the effect of medium composition on fermentations

---

*Corresponding author. E-mail: bollakrishna@gmail.com.

Table 1. Effect of carbon and nitrogen sources on biomass and exopolysaccharides production by *T. versicolor*.

Carbon source	Days	*Trametes versicolor*		
		pH	B	EPS
Glucose	7	3.88	4.0	3.0
	14	3.5	6.2	6.4
Sucrose	7	3.92	4.6	5.6
	14	3.70	8.8	3.4
Maltose	7	2.92	7.6	7.4
	14	4.39	9.2	6.2*
Lactose	7	5.70	2.6	2.8
	14	5.90	3.0	3.0
Fructose	7	3.53	4.0	8.0
	14	5.70	2.6	2.8
Xylose	7	3.40	5.0	6.0
	14	4.00	6.2	2.8
Mannitol	7	3.78	5.8	6.6
	14	3.69	6.8	4.6
Raffinose	7	3.64	6.8	3.8
	14	4.42	8.4	7.0
Nitrogen source				
$(NH_4)_2SO_4$	7	3.88	4.0	3.0
	14	3.5	6.2	6.4
Casein	7	4.48	9.2	5.2
	14	5.83	9.0	5.6
Gelatin	7	4.13	10.4	5.0
	14	5.06	11.0	3.2
Glycine	7	3.85	7.4	4.6
	14	5.00	10.2	6.4
KNO_3	7	5.22	7.2	10.2*
	14	5.82	7.8	4.4
$NaNO_2$	7	6.23	4.6	4.0
	14	6.17	6.2	7.0
Peptone	7	4.51	4.2	3.2
	14	6.03	7.6	6.4
Yeast extract	7	5.75	8.6	4.6
	14	4.96	8.0	5.4

* Precipitate, B - biomass (g. dr. w/ L), EPS - exopolysaccharides (g. dr. w/l).

and cultivation kinetics, which are important parameters to EPS production, remain relatively unexplored.

Response surfaces, based on experimental designs, have been used in several fields of bioprocesses and it has been demonstrated to be an adequate tool to evaluate the effects and interactions of the different parameters that rule a biochemical system (Box et al., 1978). However, so far experimental designs have not been applied to EPS production by *T. versicolor*.

The aim of this study was to define experimental conditions to optimize EPS production by *T. versicolor*. Firstly a fermentation broth was selected and after that an experimental design and different carbon and nitrogen sources was used to optimize the medium and the culture conditions.

MATERIALS AND METHODS

Trametes versicolor were collected from Mushroom Culture Collection Lab, Kakatiya University, Warangal, Andhra Pradesh, India, which was maintained on the malt extract medium.

Liquid culture medium (g/l): Peptone 1.0; yeast extract 2.0; K_2HPO_4 1.0; $MgSO_4.7H_2O$ 0.2; $(NH_4)_2SO_4$ 5.0; glucose 20.0; pH 6.0. This medium was selected in preliminary studies as adequate for exopolysaccharide production. Erlenmeyer flasks containing 100 ml of sterilized culture medium were inoculated with the suspension in sterile water of fungal mycelium grown on malt extract agar slants. Incubation was done at 27°C.

The incubation times were 7 and 14 days. The culture was filtered to separate fungal biomass, which was washed twice with distilled water and quantified as dry weight (105°C to constant weight). Isopropanol was added to the culture filtrate (1:1 v/v) and after 24 h at 4°C the precipitated biopolymer was separated by centrifugation (8,000 rpm for 15 min) and also quantified as dry weight.

T. versicolor was tested with different carbon and nitrogen sources (Wasser et al., 2003; Hsieh et al., 2005). The media is optimized with different carbon (glucose, fructose, sucrose, maltose, lactose, raffinose, mannitol and xylose) and nitrogen source (peptone, glycine, gelatin, casein, yeast extract, ammonium sulphate, KNO_3 and $NaNO_2$) for the higher yield of EPS (Jonathan et al., 2006).

RESULTS AND DISCUSSION

The results were calculated with standard deviation and the graph was plotted with error bar by using SPSS software. The results were significant and shown for pH, biomass and EPS are 0.89, 2.15 and 1.87 respectively for carbon sources and for the nitrogen source, the results were significant and shown for pH, biomass and EPS are 0.89, 2.15 and 1.26 (Figure 1).

Fructose (8 g. dr. w/l) and raffinose (7 g. dr. w/l) are effective carbon sources (Table 1) and Casein (5.2 g. dr. w/l) and $NaNo_2$ (7 g. dr. w/l) are effective nitrogen sources (Tables 2a and 2b) in the production of exopolysaccharides after 7 and 14 days of incubation, respectively.

Maltose (7.6 and 9.2 g. dr. w/l) and gelatin (10.4 and 11 g. dr. w/l) are effective carbon and nitrogen sources

Optimization of carbon and nitrogen sources of submerged culture process for the production of mycelial biomass...

67

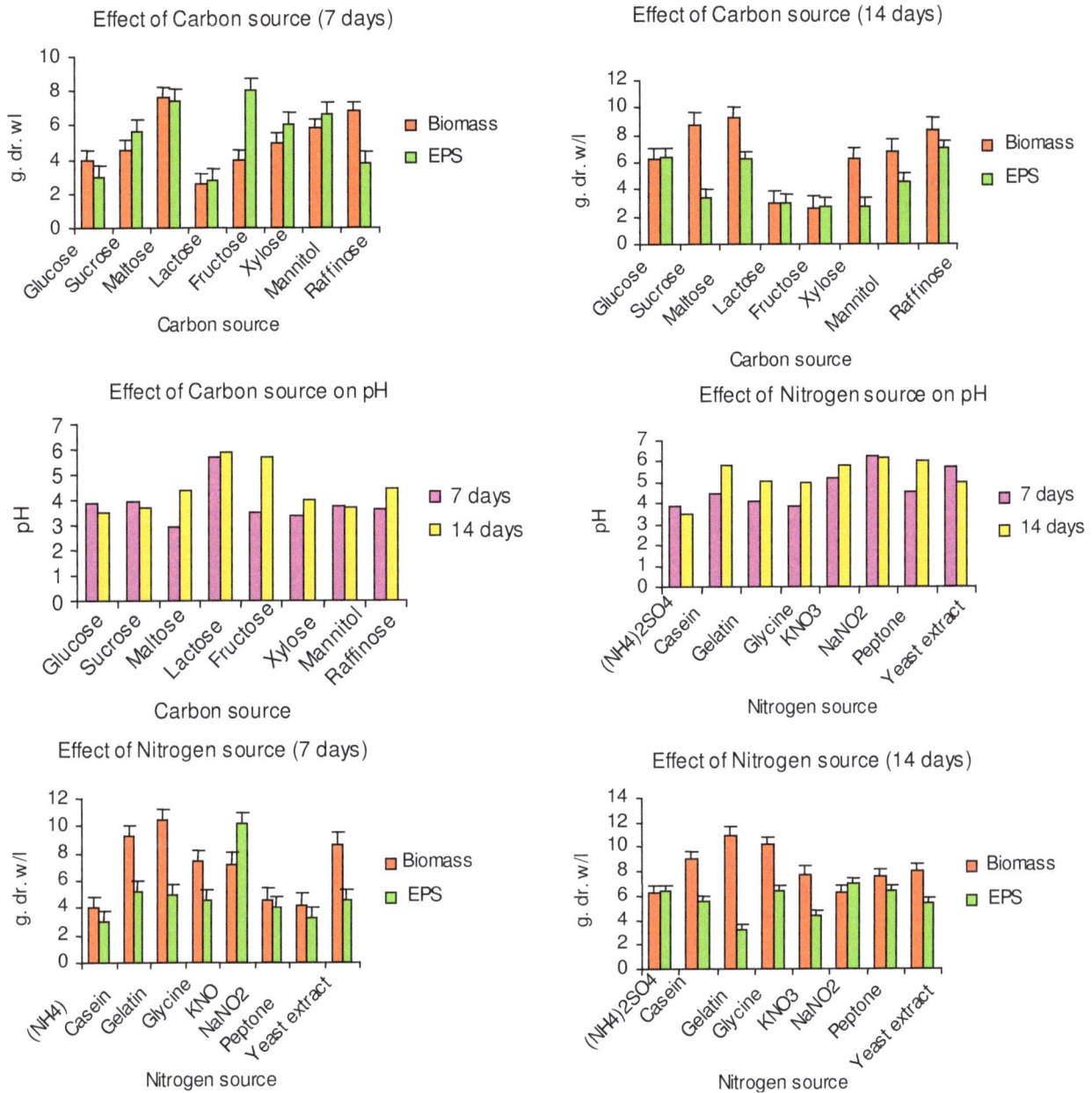

Figure 1. Effect of carbon and nitrogen source on biomass and exopolysaccharides production by *T. versicolor*.

(Tables 1, 2a and 2b) in the production of biomass after 7 and 14 days of incubation respectively.

An interesting observation was made concerning the formation of an insoluble gel when the culture filtrate was frozen prior to polysaccharide precipitation. In the Tables these strains are marked. This peculiar characteristic could aid polymer separation, since there is no need of an organic solvent such as isopropanol, ethanol or acetone for the precipitation of the polymer, thus increasing the process viability. Moreover, it is important to observe that the product obtained by solvent

precipitation cannot be considered pure polysaccharide because proteins and salts present in the medium co-precipitate. The strain studied here were submitted to a lignin degradation activity test (Capelari and Zadrazil, 1997; Hou and Chen, 2008) was proved to be effective.

The conditions used for the submerged culture could be considered adequate for biomass production. Data presented in literature (Manachini, 1979; Compere et al., 1980; Masaphy and Levanon, 1992; Burns et al., 1994) showed lower production for *Pleurotus* sp. with other culture parameters. During estimation of polymer and

Table 2a. Effect of carbon source-statistics (VAR00002).

		Frequency	Percent	Valid percent	Cumulative percent	Std. deviation
Valid	2.92	1	6.3	6.3	6.3	.88818
	3.40	1	6.3	6.3	12.5	
	3.50	1	6.3	6.3	18.8	
	3.53	1	6.3	6.3	25.0	
	3.64	1	6.3	6.3	31.3	
	3.69	1	6.3	6.3	37.5	
	3.70	1	6.3	6.3	43.8	
	3.78	1	6.3	6.3	50.0	
	3.88	1	6.3	6.3	56.3	
	3.92	1	6.3	6.3	62.5	
	4.00	1	6.3	6.3	68.8	
	4.39	1	6.3	6.3	75.0	
	4.42	1	6.3	6.3	81.3	
	5.70	2	12.5	12.5	93.8	
	5.90	1	6.3	6.3	100.0	

N = 16   Missing = 0   Total = 100.0   Total = 100.0

SD for biomass-statistics (VAR00002).

		Frequency	Percent	Valid percent	Cumulative percent	Std. deviation
Valid	2.60	2	12.5	12.5	12.5	2.15515
	3.00	1	6.3	6.3	18.8	
	4.00	2	12.5	12.5	31.3	
	4.60	1	6.3	6.3	37.5	
	5.00	1	6.3	6.3	43.8	
	5.80	1	6.3	6.3	50.0	
	6.20	2	12.5	12.5	62.5	
	6.80	2	12.5	12.5	75.0	
	7.60	1	6.3	6.3	81.3	
	8.40	1	6.3	6.3	87.5	
	8.80	1	6.3	6.3	93.8	
	9.20	1	6.3	6.3	100.0	

N = 16   Missing = 0   Total = 100.0   Total = 100.0

For EPS- statistics (VAR00002).

		Frequency	Percent	Valid percent	Cumulative percent	Std. deviation
Valid	2.80	3	18.8	18.8	18.8	1.87506
	3.00	2	12.5	12.5	31.3	
	3.40	1	6.3	6.3	37.5	
	3.80	1	6.3	6.3	43.8	
	4.60	1	6.3	6.3	50.0	
	5.60	1	6.3	6.3	56.3	
	6.00	1	6.3	6.3	62.5	
	6.20	1	6.3	6.3	68.8	

Table 2a. Contd.

6.40	1	6.3	6.3	75.0
6.60	1	6.3	6.3	81.3
7.00	1	6.3	6.3	87.5
7.40	1	6.3	6.3	93.8
8.00	1	6.3	6.3	100.0
	N = 16 Missing = 0	Total = 100.0	Total = 100.0	

Table 2b. Effect of nitrogen- For pH: Statistics (VAR00002).

		Frequency	Percent	Valid percent	Cumulative percent	Std. deviation
Valid	3.50	1	6.3	6.3	6.3	.89455
	3.85	1	6.3	6.3	12.5	
	3.88	1	6.3	6.3	18.8	
	4.13	1	6.3	6.3	25.0	
	4.48	1	6.3	6.3	31.3	
	4.51	1	6.3	6.3	37.5	
	4.96	1	6.3	6.3	43.8	
	5.00	1	6.3	6.3	50.0	
	5.06	1	6.3	6.3	56.3	
	5.22	1	6.3	6.3	62.5	
	5.75	1	6.3	6.3	68.8	
	5.82	1	6.3	6.3	75.0	
	5.83	1	6.3	6.3	81.3	
	6.03	1	6.3	6.3	87.5	
	6.17	1	6.3	6.3	93.8	
	6.23	1	6.3	6.3	100.0	
	N = 16 Missing = 0	Total = 100.0	Total = 100.0		

For biomass- statistics (VAR00002).

		Frequency	Percent	Valid percent	Cumulative percent	Std. deviation
Valid	4.00	1	6.3	6.3	6.3	2.15283
	4.20	1	6.3	6.3	12.5	
	4.60	1	6.3	6.3	18.8	
	6.20	2	12.5	12.5	31.3	
	7.20	1	6.3	6.3	37.5	
	7.40	1	6.3	6.3	43.8	
	7.60	1	6.3	6.3	50.0	
	7.80	1	6.3	6.3	56.3	
	8.00	1	6.3	6.3	62.5	
	8.60	1	6.3	6.3	68.8	
	9.00	1	6.3	6.3	75.0	
	9.20	1	6.3	6.3	81.3	
	10.20	1	6.3	6.3	87.5	
	10.40	1	6.3	6.3	93.8	
	11.00	1	6.3	6.3	100.0	
	N = 16 Missing = 0	Total = 100.0	Total = 100.0		

Table 2b. Contd (For EPS- statistics [VAR00002]).

		Frequency	Percent	Valid percent	Cumulative percent	Std. deviation
Valid	3.00	1	6.3	6.3	6.3	1.26537
	3.20	2	12.5	12.5	18.8	
	4.00	1	6.3	6.3	25.0	
	4.40	1	6.3	6.3	31.3	
	4.60	2	12.5	12.5	43.8	
	5.00	1	6.3	6.3	50.0	
	5.20	1	6.3	6.3	56.3	
	5.40	1	6.3	6.3	62.5	
	5.60	1	6.3	6.3	68.8	
	6.20	1	6.3	6.3	75.0	
	6.40	3	18.8	18.8	93.8	
	7.00	1	6.3	6.3	100.0	
	N = 16 Missing = 0		Total = 100.0	Total = 100.0		

biomass produced it is important to consider that EPS adherent to the hyphae are also entrapped into the pellets formed during the submerged culture which means that the dry weight of biopolymer which precipitated from the culture filtrate does not correspond to the total EPS and that the biomass can be overestimated. To minimize this problem biomass was washed twice with distilled water.

The pellets formed can be regular or irregular in form and size. The form varies from spherical to cylindrical and the size from 1 to 20 mm. In some cases the formation of pellets was not observed, but rather a mycelial agglomeration without a defined form (Maziero, 1996). The pellets were smooth, hairy (with looser outer zones) or with fringes of aggregated hyphae that give the pellet a star form. The color and consistency were also different, as well as the flavour. Sometimes the culture filtrate was very clear, other times was turbid and very viscous. In most of the cultures the presence of crystals with different forms was observed, which could indicate, in some cases, the presence of excreted metabolites. When there is a depletion of glucose in the medium it was observed that pellets begin to become darker and break up. The dead hyphae are decomposed and the resulting substances are reabsorbed by the mycelium.

ACKNOWLEDGEMENT

The authors are thankful to Prof. M. A. Singara Charya Head, Department of Microbiology for providing, necessary facilities.

REFERENCES

Box GEP, Hunter WG, Hunter JS (1978). Statistics for Experiments: An Introduction to Design, Data Analysis, and Model Building, New York: John Wiley & Sons, Inc 306-3170-471- 09315-7.

Burns PJ, Yeo P, Keshavarz T, Roller S, Evans CS (1994). Physiological studies of exopolysaccharide production from the Basidiomycete Pleurotus florida. Enzyme Microb. Technol. 16: 566-572.

Capelari M, Zadrazil F (1997). Lignin degradation and in vitro digestibility of wheat straw treated with Brazilian Tropical species of white rot fungi. Folia Microbiol. 42: 481-487.

Compere AL, Griffith WL, Greene SV (1980). Polymer production by Pleurotus. Dev. Ind. Microbiol. 21: 461-469.

Cui J, Chisti Y (2003). Polysccharopeptides of Coriolus versicolor: physiological activity, uses and production. Biotechnol. Adv. 21: 109-122.

Hou X, Chen W (2008). Optimization of extraction process of crude polysaccharides from wild edible BaChu mushroom by response surface methodology. Carbohydrate Polymers 72: 67-74.

Hsieh C, Tsai MJ, Hsu TH, Chang DM, Lo CT (2005). Medium optimization for polysaccharide production of Cordyceps sinensis. Appl. Biochem. Biotech. 120:145-157.

Jonathan G, Ayodele S, Damilola A (2006). Optimization of sub-merged culture conditions for biomass production in Pleurotus florida (mont.) Singer, a Nigerian edible fungus. Afr. J. Biotech. 5: 1464-1469.

Kim SW, Hwang HJ, Park JP, Cho YJ, Song CH, Yun JW (2002). Mycelial growth and exo-biopolymer production by submerged culture of various edible mushrooms under different media. Lett. Appl. Microbiol. 34: 56-61.

Manachini PL (1979). Screening di funghi superiori commestibili per la produzione di biomasse e di metabolici esocellulari in coltura sommersa. Tecnol. Alim. (9):17-24.

Manzoni M, Rollini M (2001) Isolation and characterization of the exopolysaccharide produced by Daedalea quercina. Biotechnol Lett. 23:1491–1497.

Masaphy S, Levanon D (1992). The effect of lignocellulose on lignocellulolytic activity of Pleurotus pulmonarius in submerged culture. Appl. Microbiol. Biotechnol. 36: 828-832.

Maziero R (1996). Produção de exopolissacarídeos por basidiomicetos em cultura submersa: screening., caracterização química preliminar e estudo de produção utilizando Irpex lacteus (Fr.:Fr.) Fr. São Paulo, Ph.D. Thesis. Instituto de Biociências, USP p. 181.

Miyazaki T, Yadomae T, Sgiura M, Ito H, Fujii K, Naruse S, Kunihisa, M (1974). Chemical structure of antitumor polysaccharide, coriolan, produced by Coriolus versicolor. Chem. Pharm. Bull. 22: 1739-1742.

Ng TB (1998). A review of research on the protein-bound polysaccharide (polysaccharopeptide, PSP) from the mushroom Coriolus *versicolor* (Basidiomycetes: Polyporaceae). Gen. Pharmacology 30: 1–4.

Rau U (1999). Production of Schizophyllan. In: Bucke C (ed) Methods in biotechnology, Vol. 10: Carbohydrate biotechnology protocols. Humana, Totowa pp. 43-57.

Sugiura M, Ohno H, Kunihisa M, Hirata F, Ito H (1980). Studies on antitumor polysaccharides, especially D-II, from mycelium of *C. versicolor*. Japanese. J. Pharmacol. 30: 503–513.

Wasser SP, Vladimir I, Elisashvili, Kok KT (2003). Effects of Carbon and Nitrogen Sources in the Medium on *Tremella mesenterica* Retz: Fr. (Heterobasidiomycetes) Growth and Polysaccharide Production. Int. J. Med. Mushr. 5:49-56.

11

Large scale recovery of tetanus toxin and toxoid from fermentation broth by microporous tangential flow filtration

Chellamani Muniandi[1]*, **Kavaratty Raju Mani**[1] **and Rathinasamy Subashkumar**[2]

[1]Pasteur Institute of India, Coonoor-643 103, The Nilgiris, Tamil Nadu, India.
[2]PG and Research Department of Biotechnology, Kongunadu Arts and Science College, Coimbatore – 641 029, Tamil Nadu, India

The commercial production of purified tetanus toxoid mainly depends on the effective separation of the bacterial toxin and toxoid from large volumes of fermentation broth of *Clostridium tetani* (Harvard 49205) vaccine strain. Tangential flow or cross-flow filtration system was used as rapid drive in the processing of immunobiological assays of tetanus toxin. Tetanus toxoid was prepared by detoxifying the culture filtrates of *C. tetani* and further purified by ultrafiltration, salt fractionation and adsorption onto aluminium phosphate. Present study deals with the separation of tetanus toxins using a microporous membrane (0.22 µm) and concentration of tetanus toxoids using an ultrafiltration membrane (30 kDa, NMWL pore size) with operational variables like average trans-membrane pressure (ATP), cross flow rate, flux. Under the best conditions, >96% recovery was achieved. Additionally, potency control of 10 batches of tetanus toxoid, prepared from the filtered toxins/toxoid lots by microporous tangential flow filtration system, was evaluated by *in vitro* passive haemagglutination (PHA) assay and the results obtained in the *in vitro* PHA were compared with *in vivo* toxin neutralization (TN) test. An excellent correlation between *in vitro* test and *in vivo* TN test was observed by Spearman's correlation coefficient. It reveals that the process development in which employing available equipment and the *in vitro* PHA is a promising alternative to the toxic TN test in the potency assay of tetanus vaccine.

Key words: Tetanus toxin, Seitz filtration, tangential flow filtration, microfiltration module, flux, tetanus toxoid, immunogenicity test.

INTRODUCTION

Tetanus is an infectious bacterial disease caused by a highly toxinogenic strain of bacillus *Clostridium tetani*. Under favorable anaerobic conditions in necrotic wounds, and with dirt, this ubiquitous bacillus may produce tetanospasmin, a highly potent neurotoxin. This bacterial toxin is the second most potent toxin, known after botulinus toxins with a minimal lethal dose of less than 2.5 ng kg^{-1} of human body weight and blocks inhibitory neurotransmitters in the central nervous system and causes the muscular stiffness and spasms typical of generalized tetanus (WHO, 2006). Tetanus toxin is synthesized intracellularly by *C. tetani* as a single polypeptide chain with a molecular weight of 150 kDa during the logarithmic phage. After cell lysis, the toxin is released into the surrounding medium and cleaved by endogenous protease enzyme to give an NH$_2$-terminal light chain of 50 kDa (toxic moiety) and COOH-terminal heavy chain of 100 kDa (binding). The light and heavy chains are held together by disulfide bridge (Plotkin and Orenstein, 1999).

Now-a-days, basic immunization of infants against teta-

*Corresponding author. E-mail: drcmunish@yahoo.co.in or rasubash@rediffmail.com.

tetanus is usually given with the combined adsorbed Diphtheria-Tetanus-Pertussis (DTP) vaccine, which results in formation of protective tetanus antitoxin in serum. Presently, many manufacturer produce DTP group of vaccines for mass immunization program using large volume capacity bioreactors. In these bioreactors, after the end of their incubation period, the concentration of toxin in supernatant usually reaches 80 to 120 flocculation units. The separation of bacterial cells, detoxification, concentration and sterile filtration of toxoids are some of the major unit operations in the processing of large volumes of fermentation media in the production of bacterial vaccines.

The separation of cells is achieved traditionally by centrifugation or by the dead-ended depth filtration method. Such solid-liquid separation methods are time consuming and toxin loss may exceed 25%. Furthermore, membrane filters separate components and suspensions on the basis of molecule size and it is hard to validate a depth filtration system, which is mandatory under current Goods Manufacturing Practices (cGMP). In the conventional method, the dead-end filtration, incorporating filter pads of suitable pore sizes, the separation must provide complete retention of cells and maximal passage and recovery of soluble end-product. When tetanus fermentation broth is clarified, the large bacterial mass blocks most of the dead ended filtration pads requiring still higher inlet pressures or change of pads altogether. The clarified toxin solution is then polished by passing it through pad filters to produce sterile tetanus toxin. Additionally, the pads also adsorb the active material, resulting in reduced yields. For these reasons, an alternative bio-separation technique was looked at seriously, that could be validated, and the trials were conducted. Membrane-based methods have been widely suggested for the processing of bioactive species (Reid and Adlam, 1974; Tanny et al., 1980; Brown and Kavanogh, 1987). In modern filtration techniques, microporous tangential flow filtration (TFF) system, the particles or solutes retained by membrane are continuously removed in the retentate flowing tangentially across the membrane surface. The clarified solution flows through the membrane into permeate, also called the cross flow filtration. The TFF system provides a practical and economical alternative to the dead ended filtration. The entire operation conforms to the GMP norms and can be validated as and when required (Rao et al., 1992; Levine and Castillo, 1999).

In this study, the conventional depth filtration method normally used for the clarification of tetanus culture broth has been successfully replaced with that of microporous TFF system for large scale recovering of tetanus toxin at a commercial level and further use of a similar system for concentration of crude tetanus toxoid. The study establishes operation parameters like ATP, cross flow rate, flux that optimizes overall filtration, concentration and result in improvement in the preparation of purified tetanus toxoid. The TFF system was completed and validated for GMP

point of view. This paper also describes immunogenicity tests on the tetanus vaccines, prepared from the filtered toxin/toxoid lots by TFF system, the *in vivo* toxin neutralization (TN) test and *in vitro* passive haemagglutination assay (PHA), and results were compared by Spearman's correlation coefficient.

MATERIALS AND METHODS

Strain and culture medium for production of toxin

The Harvard strain of *C. tetani* (obtained in lyophilized state from Central Research Institute, Kasauli, India, A National Control Authority) was used for the production of tetanus toxin in fermenter vessels (500 L). Modified Mueller Miller (MMM) medium was used for growing *C. tetani* (WHO, 1977). The fermentation vessel had a working volume of 400 L. The pH of medium was adjusted to 7.4 with 40% sodium hydroxide solution. Sterilization of the medium was carried out by steam at 115°C for 20 min and after cooling it down to 35°C. it was inoculated with one day seed and prolonged for about one week at 35°C under continuous mild agitation and aeration. At the end of the incubation period, samples were drawn for purity checks and estimation of antigen concentration.

Clarification of fermentation broth

The fermented broth showed a pure and satisfactory concentration of the toxin (Lf mL^{-1}), it was then clarified by (i) centrifugation (ii) Seitz filtration (iii) tangential flow filtration to study the toxin recovery. Tetanus cells were centrifuged (Model-Sorvall RC 3B plus, Sorvall Instruments, New Town, Germany) at a speed of 4000 rpm. The load (tetanus culture broth filled in polypropylene containers 6 x 1 l) was distributed symmetrically around the rotating assembly and centrifuged for 60 min using the timer mode. The supernatant, which contained the toxin, was collected.

Conventional method

Using the conventional method, the alternative was Seitz filtration, which is a dead-ended depth filtration method. With this method, the fermented culture fluid was then first clarified by filtration through T500 Seitz filter pads, 20 x 20 cm, 0.45 μ pore size (Seitz Werke Bad Kreuznach, West Germany) and subsequently sterilized by filtration through EKS Seitz filter pads, 20 x 20 cm, 0.22 μ pore size using a filter press assembly (Straussburger, West Germany) which hold 14 pads. This operation was carried out at 35°C under 12 *psi* pressure. After collecting 80 to 100 L of filtrate, physiological saline was pumped into the body to filter the balance toxin.

Tangential flow filtration method

This method was carried out using a Millipore's Prostak system and Pellicon system with relevant open-channel modules for rapid clarification of tetanus culture broth and concentration of crude toxoid, respectively. These open-channel modules are tangential flow stacked plate membrane devices with open feed channels and are available with both microporous membranes (Prostak MF modules) and ultrafiltration membranes (Pellicon, UF modules). The separation between ultrafiltration and microfiltration is based on the pore size of the membranes. The Prostak system (Sys No. TFF05P178; Cat.No.BM5AN8834; Millipore) includes prostak holder, Rotary Lobe pump (Jhonson Pumpen AG, Wadenswil, Switzerland) coupled with a motor and the speed of the same is controlled through a variable speed controller and a feed tank. The insulation consists of temperature and pressure gauges, flow sensors, air flow meter and sanitary piping. A sanitary type diaphragm valve was also installed on the retentate line, which is used in controlling the trans-membrane pressure, across the membrane. The Prostak holder is designed to accommodate modules with serpentine flow path from feed channel to retentate channel.

The Prostak MF modules (Catalogue No PSGV AG201, 0.22 μm pore size) are made of void free composite microporous hydrophilic polyvinylidene diflouride (PVDF) membrane with the Millipore (Molshein, France) trade name Durapore. The material is autoclavable, a very low protein-binding characteristic and has broad chemical compatibility. The system with open-channel modules is fully on-line steam sterilizable. The filter area can be varied from 10 to 50 ft^2. In the Prostak system, five 20 ft^2 Prostak MF modules were installed in series and the system was operated at a cross flow of 2000 L h^{-1} at an operating temperature of 23°C.

Toxoid preparation

After the clarification of fermented broth in the Prostak system, the filtrate containing the toxin was detoxified by adding 0.5% (v/v) formaldehyde (35-40% AR grade). The containers were incubated at 36°C for 4 weeks to obtain the crude toxoid and then the specific toxicity was checked. Once the toxoid passed the specific toxicity test, it was then concentrated using the Pellicon system, described subsequently (WHO, 1977; Pasteur Institute of India, 1991).

Concentration of toxoid using Pellicon TFF system

Concentration is a pressure driven membrane process used to concentrate, separate or purify macromolecules. The separation is based on molecular weight of the macromolecule. The membrane is a thin semi-permeable polymeric material that will retain macromolecules and allow smaller dissolved solutes to pass through the membrane. The pore sizes of ultrafiltration membranes are in the range of 0.001 to 0.1 μm and the retentative abilities of ultrafiltration membranes are described by the nominal weight limit (NMWL) (Zeman and Zydney, 1996).

Concentration of crude tetanus toxoid was determined using a Pellicon TFF system with ultrafiltration membranes. The ultrafiltration membranes are available in a range of molecular weight cutoffs of 500 Daltons (Da) to 1000 KDa and offer filtration areas of 0.5 (5.5 ft^2) and 2.5 m^2 (27.5 ft^2). The Pellicon TFF system (Sys No. TFF05C184; Cat.No.BM5CN9473; Millipore) accommodates an acrylic filter holder and other insulations as in the case of Prostak TFF system. The basic material of construction is modified polyethersulfone with a pore size of 30 kDa (NMWL) with the trade name Biomax-30. The material has a higher flux, excellent chemical resistant, integrity testable, void-free structure for higher yield and reliability. The batch size at this stage of concentration of tetanus toxoid was in the range of 800 to 1000 L. The molecular weight of tetanus toxoid is about 150 kDa, which includes the monomer, dimers and oligomers of various sizes. The area of the device Biomax 30K NMWL cassettes was 50 ft^2. Two cassettes were installed in the acrylic holder of the Pellicon system and the system was operated at a cross flow of 1000 L h^{-1} at an operating temperature of 25°C.

Both the TFF systems were flushed with sterile distilled water as recommended by the manufacturer. The clean water flux was measured to assess the cleaning efficiency after each trial. After optimization of the operating parameters based on flow and pressure excursion experiments, the volume of permeate (clarification of broth by Prostak system) and retentate (concentration of toxoid by Pellicon system) was measured over time until the known initial volume of the broth/toxoid had been reduced 10-fold (10x level). At this point, the cell and module wash started to recover completely

the balance toxin/toxoid retained.

To clean the modules between experiments, the TFF systems were flushed with sterile normal saline followed by warm distilled water (40°C). During cleaning procedure, the pump was operated at a higher velocity to flush out all material. Then the prostak and pellicon systems were sanitized with sodium hypochlorite (600 ppm) and 0.1 N sodium hydroxide solutions, respectively. A 2% formaldehyde solution was used as storage agent for both the modules to ensure membranes still wet and to prevent microbiological growth without damage to the filter. Prior to the formaldehyde step, the clean water flux was measured to assess the cleaning effectiveness.

Validation of the TFF system

The selection of a membrane, as specified earlier, was mainly guided by the possibility of validation of the system. Validation of the system included: (i) determination of the pre-and post-operation integrity testing of the membrane device, which ensured repeated use, (ii) removal of the storage agent, i.e., formalin from the modules. Formalin was removed by flushing the system with distilled water. Samples were collected at different intervals. To 3 ml sample, 0.5 ml chromotropic acid was added and the mixture was placed in boiling water bath for 5 min. The presence of formalin was confirmed by the development of pink colour, (iii) checking absence of *C. tetani* – as the membrane used was of pore size 0.22 μ, the absence of *C. tetani* from the clarified toxin needed to be validated. This was carried out by passing a representative sample of 250 ml through the GVWP disc (Durapore PVDF 0.22μ disc), which was placed on the surface of MMM agar. On another MMM agar plate, overnight grown culture of *C. tetani* in fluid thioglycollate medium was streaked as a positive control, (iv) sanitization of the system – for sanitizing the Prostak TFF and Pellicon TFF systems, sodium hypochlorite and NaOH were used, respectively. The cleaning procedure ensured removal of the cleaning agent. Removal of sodium hypochlorite (below 10 ppm)/NaOH was confirmed by comparing serially diluted sample of sodium hypochlorite/NaOH with a universal indicator.

TT vaccine lot preparation

The crude tetanus toxoid was concentrated, purified by fractionation with ammonium sulphate and adsorbed onto aluminium phosphate gel. It is supplied as a sterile solution in physiological buffer usually with preservative. Ten licensed lots of adsorbed tetanus toxoid vaccine, containing 10 Lf mL^{-1} of tetanus toxoid adsorbed onto 3.0 mg mL^{-1} of aluminium phosphate were used for immunogenicity study.

Analytical methods

Flocculation test (Lf)

The immunological precipitation test was carried out to determine the concentration of toxin or toxoid, which is useful for calculating the antigen content. The antigen content is a good indicator of consistency of production. The tetanus antitoxin (100 Lf ml^{-1}) supplied by the quality control division of this institute was standardized against the international reference reagent of tetanus toxoid. Lf per ml of toxin/toxoid was determined by the method of Ramon as explained in the WHO manual (WHO, 1994a).

Minimum lethal dose (MLD)

The MLD test was performed on mice with a weight range of 17 to 20 g. The purpose of the test is to confirm that the strain used really produces toxin. The MLD of tetanus toxin was determined by the method described in the WHO manual (WHO, 1994b).

Protein nitrogen assay

The total protein nitrogen content of the toxin/toxoid is determined by the method described by Kjeldahl, which is the standard method included in pharmacopeias (WHO, 1994c). The ratio of toxin/toxoid concentration (Lf ml^{-1}) to protein nitrogen concentration (mg PN ml^{-1}) denoted the antigenic purity of toxin/toxoid and was expressed as Lf per mg of protein nitrogen.

Immunogenicity/potency test

The potency test was done on the tetanus vaccines to predict the effectiveness of vaccine in humans by using animal models (guinea pigs and mice). The classical *in vivo* TN test and serological *in vitro* PHA have been employed to determine the end point of tetanus antitoxin content (WHO, 1990; Galazka, 1993). The TN test was performed in mice and it measured the amount of antitoxin, which could neutralize the toxic activity of the toxin. The TN test was the reference assay for the PHA in use. It was standardized in order to correlate them with a known level of immune response seen in the *in vivo* test.

In vivo toxin neutralization test

The *in vivo* TN test on pooled sera was performed in mice at Lp/200 dose level of tetanus toxin according to the Indian Pharmacopoeia potency method (Indian Pharmacopoeia, 1985). The Lp/200 toxin dose level is defined as the minimum amount of tetanus toxin which when mixed with 0.005 (1/200) IU tetanus antitoxin causes a defined degree of tetanic paralysis in mouse of a defined weight in four days. The level of antitoxin was calculated against the standard and expressed in IU/ml. The Lp/200 dose of the toxin was 0.00008 Lf. The tetanus antitoxin levels of most serum pools were tested at end points level.

The total protein nitrogen content of the toxin/toxoid is determined using Kjeldahl Method.

In vitro passive haemagglutination assay

The *in vitro* PHA was performed by the method described in the WHO manual (WHO, 1997). According to which, the purified antigen was covalently coupled to tannin-treated turkey erythrocytes. Briefly, pooled serum samples for individual TT vaccine lots were prediluted at 1/20 in PBS, pH 7.2. Each serum samples and standard tetanus antitoxin were tested in duplicate. For 2-fold dilution, all the wells were filled with 50 µl of PBS, pH 7.2 except for wells A12-D12, which were filled with 25 µl of PBS, pH 7.2. All the wells of row 1A-10A were filled with 50 µl of 1/20 diluted serum and the well of row 11A was filed with 50 µl of PHA reference serum (1 IU/ml) and mixed well. The 50 µl of the mixture in the wells of the row 1A-11A were transferred to the next wells of the row 1B-11B and mixed well and the transferring and mixing process was continued through all the rows from B-H. To equalize the volumes, 50 µl from each of the wells of the row 1H-11H were ultimately discarded. A volume of 25 µl of the 1/20 prediluted positive control serum was added in the wells A12-D12, which served as positive control and the wells E12-H12, used as a negative control with PBS only. To all the wells was then added 50 µl of the diluted red blood cell suspension. Microtitre plates were incubated at room temperature for 20 min. The tetanus antibody titre in the serum samples was estimated by optical reading of agglutination and expressed in IU/ml by multiplying the reciprocal of the specific titre of the reference serum with the sensitivity of the test serum. The titration of the tetanus reference serum was found to be 0.0078 IU/ml. The PHA titres were converted into antitoxin units (AU or IU) by running in parallel with the test sera and a reference serum with known antitoxin content, determined and calibrated against the WHO standard serum in IU/ml.

Statistical analysis

The Spearman's correlation coefficient was used to compare the relationship between *in vitro* PHA method and *in vivo* TNT. These statistical analyses were done at KMCH College of Health Sciences, Coimbatore, India using the Statistical Package of Social Sciences – SPSS software, version 11.0 (SPSS Inc.2000-01).

RESULTS AND DISCUSSION

Clarification by traditional methods

A high biomass was produced by stationary pot culture method and fermentor cultivation (Pasteur Institute of India 1991). Applying the methods of centrifugation and Seitz filtration, a volume of 100 L culture was processed as a single batch. Two and four trails were carried out by centrifugation and Seitz filtration, respectively and the output of each batch was 80 to 100 L filtrate. The Lf titre was estimated and the total titre was calculated. An average toxin recovery of about 95% was obtained. Even though the recovery is good, there is an inherent danger of potentially hazardous aerosol formation, spillage and the process is maximum exposure and labor intensive. Table 1 shows the result of four clarification trials by Seitz filtration method. From Table 1, it is clear that the maximum toxin recovery obtainable was about 88 to 90%. This was mainly because of the high hold-up volume, which resulted in dilution. Further, there must be some toxin loss due to adsorption by the filter pads. The initial feed pressure (4-8 psi) increased to 12 psi due to the build-up of cell debris from fermentation broth and the flow through the filter was reduced considerably. Moreover, the process seriously lacked scale-up criteria. This method required 150 to 200 filter pads to filter 400 L broth. The output thus obtained was subjected to another depth filtration using EKS pads and 28 pads were required to obtain a sterile product. The procedure detailed required change of pads during the process. The process time for a 400 L batch was usually about 8 to 9 h including post operation sanitation. A slight variation in inlet pressure results in a turbid output, needing yet another clarification cycle. The problem was encountered in 14 out of 40 batches processed in a year. Cost of each filter pads is very expensive.

Clarification and concentration by tangential flow filtration method

In the recent past, the demand for tetanus toxoid has been very high. To enhance production, our organization introduced a pilot scale bioreactor to produce 400 L of culture in the batch mode. Before this study, the standard method for clarification was Seitz filtration. The tetanus culture broth was divided into 4 lots of 100 L. Considerable loss in the yield of toxin was noticed apart from the process being time consuming and labor intensive. In order to find a suitable solution to the above problems, a preliminary trial

Table 1. Tetanus toxin recovery in production batches clarified by Seitz filtration.

Trail number	Volume of culture broth (L)	Flocculation unit in culture broth (Lf/ml)	pH	MLD (in million)	Volume of toxin collected (L)	Flocculation unit in toxin (Lf/ml)	Recovery (%)[a]
SF-1	100	120	7.60	8	106	90	88.8
SF-2	100	100	7.53	4	102	80	89.1
SF-3	100	110	7.55	8	104	85	90.0
SF-4	100	120	7.58	8	107	90	89.2

MLD- Minimum lethal dose (WHO requirements); [a]Recovery is calculated based on the total Lf titre in the broth after reducing 10% of the initial volume (which is equivalent to the volume of the biomass).

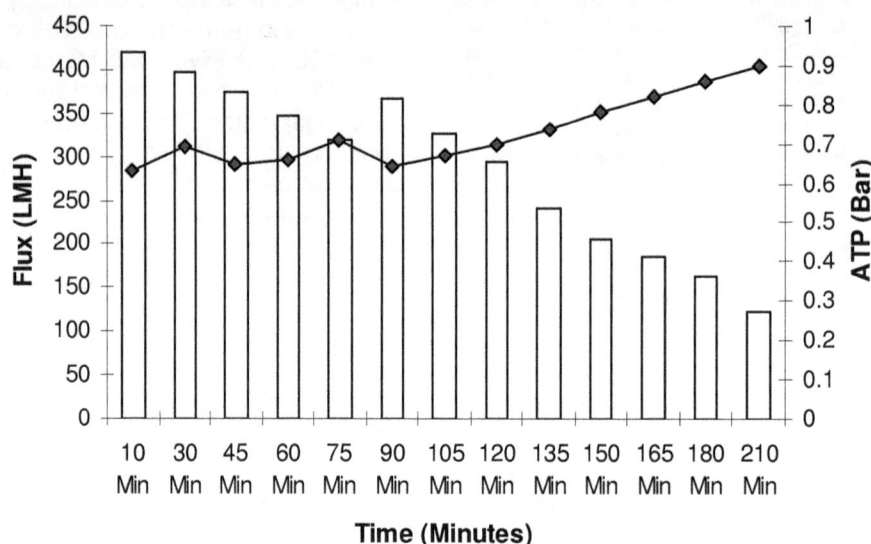

Figure 1. Product flux profile of tetanus toxin clarified through Prostak modules. Conditions: 0.22-μm pore size; 100-ft^2 filter area; process volume 400.

was conducted on a pilot unit of TFF system using two open-channel 2 ft^2 modules with hydrophilic PVDF membrane of 0.22 μm pore size (Type GVPP; Durapore) by the Millipore India team under the initiative of UNICEF. The trials showed better recovery and also effective and economical clarification of the tetanus toxin fermentation due to good flux. The data were scaled up for processing a 400 L batch and also processed with the commercial-scale data on the other unit operation. According to Ravetkar et al. (2001) the tetanus toxin recovery was >95% for a batch volume of 100 L.

Unlike "dead-end" filtration, the product fluid flows tangentially across the surface of the membrane. Briefly, the product fluid present in the feed tank enters the system through the pump and flows into the feed manifold of Prostak/Pellicon holder through the strainer in order to separate coarse particles on the fluid then flows into the Prostak/Pellicon modules and tangentially across the membrane of the different plates. Materials smaller than the pore size are able to pass through the membrane as filtrate (or permeate), which can exist from the module at its outlet and flows out of the filtrate tubing and collected in a collection tank. The membranes retain large particles (mammalian and bacterial cells, particles and high molecular weight organics). The retained material is called retentate (or concentrate), which is returned to the feed tank due to the action of the pump. This sweeping action helps to keep retained material from settling and eventually restricting the flow.

The experimental trails with Prostak module at maximum pump speed (pump speed 400) and diaphragm valve (60 to 80% open) showed an initial permeate flow rate of 9 L/min at a cross-flow rate of 1800 L/h. The permeate flow rate gradually reduced to 4.5 L/min at the end. The flux is expressed in liters per square meter per hour (LMH) and the decay profile is given in Figure 1. An average trans-membrane pressure (ATP) was initially 8.7 psi. There was a gradual increase of ATP as the cell concentration increased and stabilized at 0.9 bar during the cell wash step (Table 2).

An average trans-membrane pressure (ATP) of 0.8 bar was kept by adjusting the pump speed (300 rpm) and diaphragm valve (30% open) in the retentate line in the case of the Pellicon module. The ATP was increased, as the concentration of low molecular weight species was so high and stabilized at 1.1 bar. Comparing both the modules.

Table 2. Average trans-membrane pressure (ATP) during tetanus toxin clarification using Prostak modules.

Volume in feed tank (L)	Permeate volume (L)	Concentration factor (X)	Pressure inlet (P_{IN}) bar	Pressure outlet (P_{OUT}) bar	Pressure permeate (P_{PER}) bar	ATP[a] bar (P_{in} + P_{OUT} - P_{PER})
400	0	1	0.8	0.4	0.0	0.60
200	200	2	0.8	0.4	0.0	0.60
150	250	3	0.8	0.5	0.0	0.65
125	275	4	0.9	0.5	0.0	0.70
110	290	5	1.0	0.5	0.0	0.75
100	300	6	1.0	0.6	0.0	0.80
88	312	8	1.0	0.6	0.0	0.80
80	320	10	1.2	0.6	0.0	0.90

[a]Average value in five trails; Conditions: 0.22-μm pore size; 20- ft^2 filter area.

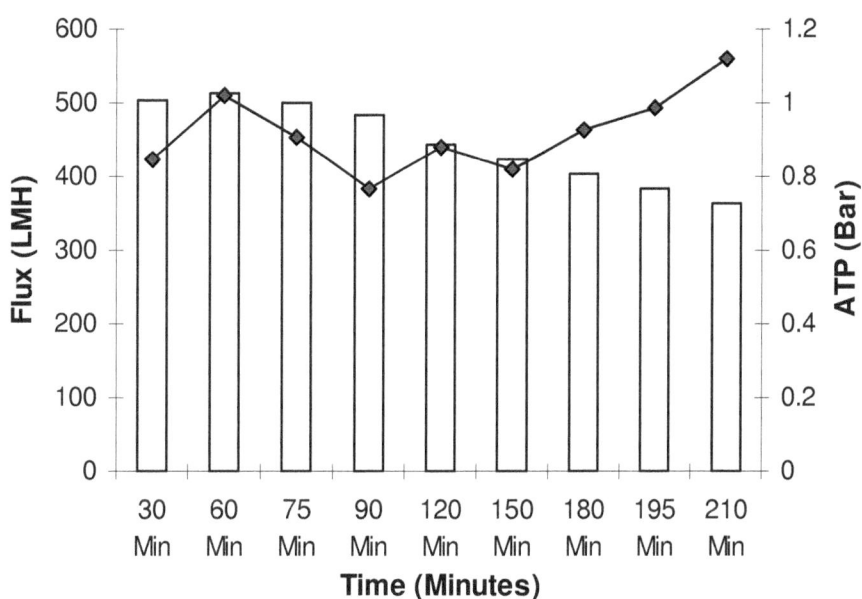

Figure 2. Flux decay of tetanus toxoid concentrated through Pellicon modules. Conditions: 30 kD pore size; 50-ft^2 filter area; process volume 950.

modules, there was no great change in ATP, denoting that the cross flow was optimal. The concentration polarization layer (gel layer) formed by the bacterial cells and low molecular weight species was acting to retard the flow of toxin and toxoid either through the polarization layer or the membrane itself as evidenced by the rapid reduction in product flux at the start (Figure 2). The product flux obtained is similar to the pattern observed in the Prostak system. The higher flux obtained in the Prostak TFF system is mainly due to the large filtration area (100 ft^2) and higher capacity pump, which gives an enhanced cross flow rate. The TFF permitted the more rapid removal of macromolecules such as formaldehyde than the traditional dialysis process; it reduces the level of residual formaldehyde by about 50 fold (Vancetto et al., 1997).

The recovery rate for tetanus toxin and toxoid using the Prostak modules (5 trails) and Pellicon modules (5 trails), respectively are given in Tables 3 and 4. Consistently, >96% yield was obtained in all the trials. Additional trials were conducted in the Prostak modules to find out the effect of pH and MLD on toxin recovery. The batches from routine production were selected in such a way that the pH and MLD at the time of harvesting ranged from 7.0 to 7.6 and 1 to 8 million, respectively. The recovery was identical (>96%) in all the experiments and required 4.5 h including post operation sanitation for the filtration and concentration of 400 L batch fermentation broth and 950 L crude toxoid, respectively. The benefits of TFF system for clarification are (i) minimal exposure of the working personnel as it is a closed operation and no spillage, (ii) no recurring cost of pads (iii) Savings on autoclaving charges for sterilization of filter assemblies, (iv) savings on disposal expenses like decontamination and incineration of used pads, (v) no problem of final sterile filtration, as 0.2 μ filtered output was

Table 3. Tetanus toxin recovery by tangential flow filtration.

Trail number[a]	Volume of culture broth (L)	Flocculation unit in culture broth (Lf/ml)	Volume of saline wash (L)	Volume of toxin collected (L)	Flocculation unit in culture broth (Lf/ml)	Recovery (%)[b]
PRO -1	400	110	120	480	80	97.0
PRO-2	400	95	120	470	70	96.2
PRO -3	400	80	120	470	60	98.0
PRO-4	400	120	120	490	85	96.4
PRO-5	400	90	120	480	65	96. 3

[a]Toxin recovery using Prostak modules (PRO-1to5). [b]Recovery is calculated based on the total Lf titre in the culture broth after reducing 10% of the initial volume (which is equivalent to the volume of the biomass).

Table 4. Tetanus toxoid recovery by tangential flow filtration.

Trail no.[a]	Volume of toxoid (L)	Flocculation unit in toxoid (Lf/ml)	Volume of saline wash (L)	Volume of toxoid collected (L)	Flocculation unit in toxoid (Lf/ml)	Recovery (%)[b]
CON-1	970	70	30	29	2300	98.2
CON -2	960	60	30	35	1600	97.2
CON -3	970	90	30	33	2600	98.2
CON -4	950	85	30	31	2600	99.8
CON -5	980	65	30	30	2100	98.7

[a]Toxoid recovery using Pellicon modules (CON-1to 5); [b]Recovery is calculated based on the total Lf titre in the crude toxoid after concentration of the initial volume.

used.

The decline in flux with time in both the modules suggested a dynamic gel layer formed on the membrane surface as evidenced by the reduction in Lf titre. As the cell concentration increased, so too did the instantaneous concentration of toxin in the permeate (Figure 3). The presence of a significant concentration of toxin in permeates was also indicative of toxin retention by the above layer. The optimal cross flow and membrane wash procedure enhanced the recovery of toxin/toxoid to a great extent.

The establishment of an effective cleaning strategy was our next goal. The same Prostak / Pellicon modules were used for 5 successive clarification/concentration and cleaning cycles. After sanitation with sodium hypochlorite /0.1 N NaOH and cleaning, the normalized water permeability (NWP) was calculated and compared with the original value. The NWP of Prostak module after each cleaning cycle was over 90% of the original value (742.87 /h/m²/bar at 25°C) indicative of correct cleaning protocols. Further, the same Prostak and Pellicon modules were employed for the clarification of toxin and concentration of toxoid, respectively from regular production batches. Over a span of 6 years, about 60,000 L of culture broth and about 40,000 L of crude toxoid were clarified and concentrated, respectively before irreversible clogging set in.

The toxin clarified by Seitz filtration or centrifugation needed polishing filtration prior to sterile filtration through 0.22 μm filter pads. The quality of the toxin filtrate obtained by TFF was excellent as judged by its bright, shiny appearance and the ease with which it passed through a sterile cartridge. The toxin obtained from 5 trials in the Prostak module was detoxified to toxoid by treatment with 0.5% (v/v) formaldehyde. The toxoid obtained from 5 trials in the Pellicon module was ammonium sulphate fractionated, dialyzed and finally sterile filtered to yield a purified toxoid ready for formulation into a vaccine. The potency control of these formulated tetanus vaccine was evaluated by the in vivo TNT and in vitro PHA according to the methods described in this study (Indian Pharmacopoeia, 1985; WHO, 1997). Table 5 summarizes the antigenic purity and tetanus antitoxin levels determined by in vivo TNT and in vitro PHA in pooled sera from guinea pigs immunized with different TT vaccine. The antigenic purity ranged from 1500 to 2000 Lf per mg of protein nitrogen. According to the WHO requirements, the purity of TT for the preparation of vaccine should not be less than 1000 Lf per mg/PN (WHO, 1990). All the preparations had anti-genic purities higher than the minimum purity required by the WHO regulations.

An in vitro PHA method for titration of tetanus antitoxin in sera of guinea pigs immunized with tetanus vaccines is shown in Figure 4. The results obtained in the in vitro PHA were compared with the in vivo TNT used routinely to perform the potency control of tetanus component in adsorbed vaccines. The correlation coefficient was between the in vitro and in vivo tests by Spearman's correlation coefficient (Figure 5). According to this coefficient, a perfect correlation, positive or negative, was determined

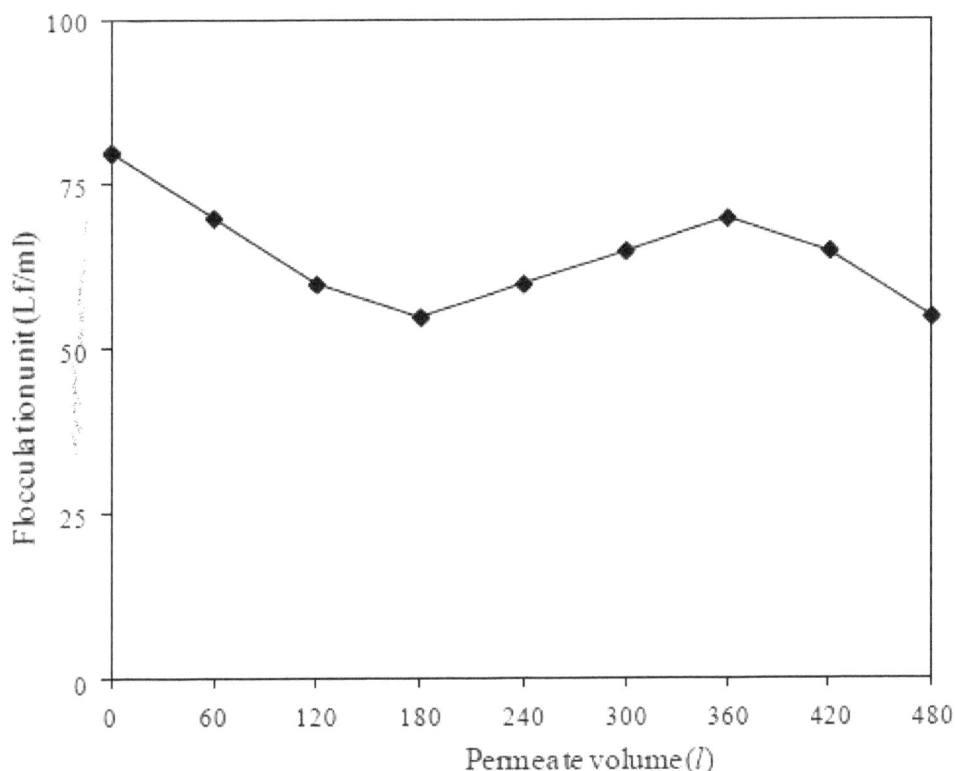

Figure 3. Lf titre of tetanus toxin clarified through Prostak modules. Conditions: 0.22-µm pore size; 100-ft^2 filter area; process volume 400.

Table 5. Potency of tetanus toxoid vaccines estimated by *in vivo* TNT and *in vitro* PHA.

Vaccine lots[a]	Antigenic purity Lf/mg PN	Tetanus antitoxin levels (IU/ml) determined by	
		In vivo TNT	*In vitro* PHA
TFF-TT-1	1970	6	5
TFF-TT-2	1984	7	10
TFF-TT-3	1388	4	2.5
TFF-TT-4	1870	5	5
TFF-TT-5	1913	5	5
TFF-TT-6	1639	4	2.5
TFF-TT-7	1373	2	2.5
TFF-TT-8	1724	4	2.5
TFF-TT-9	1785	4	5
TFF-TT-10	1477	3	5

[a]TT vaccine lots from clarified toxin (Prostak modules), detoxified, concentrated (Biomax modules), purified and coded as TFF-TT (TT vaccine).

by an approximated value of r = 1.0 or r = -1.0, respectively. In this work, correlation coefficient between *in vitro* PHA method and *in vivo* TNT was r = 0.655 and significant at the 0.05 level (2 tailed). This set of results confirms an excellent degree of correlation among the results obtained by the *in vitro* and *in vivo* tests. The TNT is expensive, time consuming, requires well-trained personnel,

a large number of animals and a relatively large amount of serum. Introduction of the PHA procedure in potency testing of tetanus vaccines contributes in two ways to improvement of animal welfare. First, replacement of the traumatic lethal paralytic technique by serum titration results in a considerable refinement of the *in vivo* TNT. Secondly, a substantial saving in number of animals is

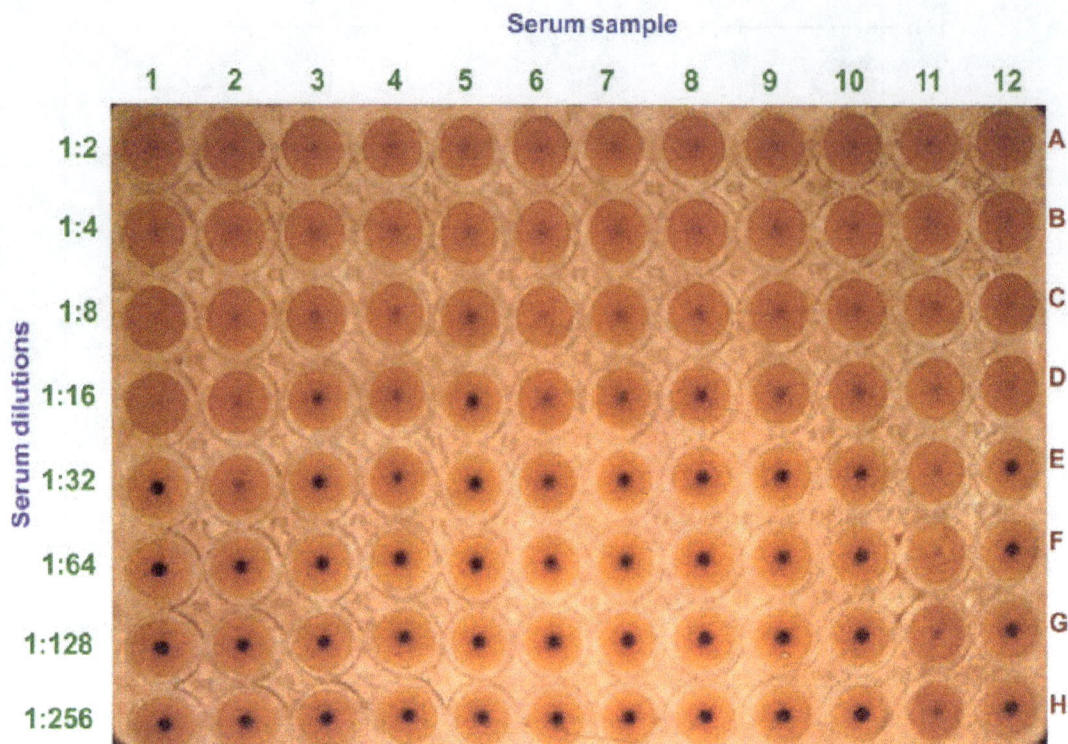

Figure 4. Passive haemagglutination test using TRBC.

feasible as quantitative data (antitoxin units, IU mL^{-1}) are generated instead of qualitative data (paralysis or lethality). Comparing the results, it was evident that the PHA is simple, specific, inexpensive, more sensitive, reproducible and the results of PHA are determined within 20 min as observed by Relyveld et al. (1996). The PHA is therefore a useful test for monitoring antitetanus antibodies in epidemiological surveys and in high-risk groups and also widely used to compare the effectiveness of different vaccines and immunization schedules. The PHA standardized in our laboratory is sufficient for determining the potency of the antitoxin content of the serum of the immunized guinea pigs. The minimum protective level of tetanus antitoxin in serum is 0.01 IU/ml (Galazka, 1993).

The introduction of a fermentor is a prerequisite to enhance toxin production. The fermentation broth must be clarified quickly otherwise deterioration will take place due to the presence of proteolysis enzymes (Pasteur Institute of India, 1991). Conventional methods like centrifugations or Seitz filtration are either cumbersome or time consuming. The other drawbacks are a lack of scale-up criteria or environmental problems like aerosol formation. Also, the depth filtration is not a closed process (due to the need to change pads during the process) and the working personnel are exposed to the toxin. Another GMP aspect observed is the inability to test integrity of pads and possible leaching of extractable matter. A modern filtration method, like what we described in this study, is the need of the hour. Operational parameters like cross flow, ATP and

filtration areas are based on the pressure excursion test (unpublished data). The Millipore systems used in this study combine gentle but high-velocity cross flow, allowing a higher recovery of toxin/toxoid without the loss of biological activity associated with other techniques. An effective protocol suitable for bench-scale and pilot-scale production units is presented in this article. The higher recovery is mainly due to the low hold-up volume and weak protein binding nature of the membrane materials. The reusable modules maintain integrity over a large number of runs. The modules are completely sealed from the environment; generate no aerosols and both the TFF systems can be validated as per the GMP requirement. In the present report, we showed that it is a versatile system for processing hundreds of liters in hours, rather than days and we also evaluated the suitability of the PHA as a serological method for the titration of tetanus antitoxin in sera of immunized guinea pigs which gives comparable results to the *in vivo* TN method. When consistency in production and quality control is well established, the PHA method is superior to the TNT both in precision and in reducing the number of animals.

ACKNOWLEDGEMENTS

The authors wish to thank Mr. S. D. Ravetkar (Serum Institute of India, Pune, India), Dr. M. LujimMala Muniandi (Medical Officer, Govt. Primary Health Centre, Tirunelveli, India), Mr. Roshan (M/S Millipore Corp., India) and UNICEF

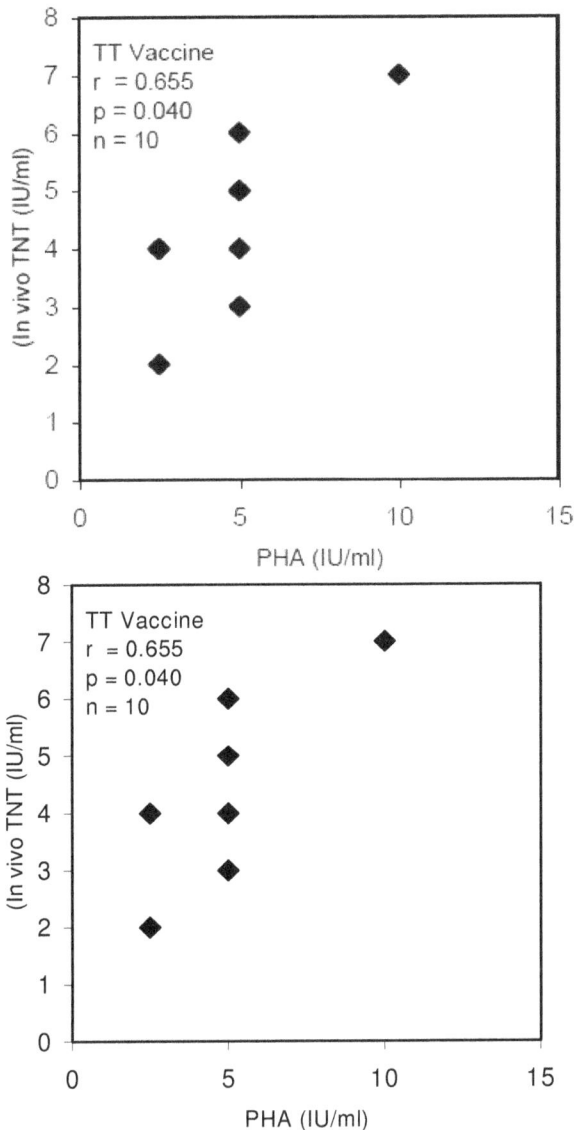

Figure 5. Spearman's correlation coefficient was applied to estimate the correlation between *in vitro* PHA and *in vivo* neutralization test in TT vaccine lots. n, lots of vaccine.

for their initiative, scientific discussion and valuable guidance. The technical assistance of Mr. C. Palaniappan, Mrs. Chandra Charles and staff of the Tetanus Production Laboratory are greatly acknowledged.

REFERENCES

Brown DE, Kavanogh PR (1987). Cross-flow separation of cells. Process Biochem. 22:96-101.

Galazka AM (1993). The immunologic basis for immunization series: Tetanus document. Geneva, World Health Organization. WHO/EPI/GEN/93.13.

Indian Pharmacopoeia (1985). Monograph on DTP vaccine (Adsorbed), DT vaccine (Adsorbed) and Tetanus vaccine (Adsorbed) Vol. I, 3rd Edition.

Levine HL, Castillo FJ (1999). In Biotechnology, Quality assurance and validation, drug manufacturing technology Series, In: Avis KE, Wagner CM and Wu VL, editors. Inter Pharma Press Inc., Baffalo Grove: Illinoisp. p. 154.

Pasteur Institute of India (1991).Manual for the production and standardization of Diphtheria-Tetanus-Pertussis vaccine (DTP). Pasteur Institute of India, Coonoor, TamilNadu, India.

Plotkin SA, Orenstein WA (1999). Tetanus toxoid, 3rd edition, Vaccines. p441-474.

Rao YUB, Mahadevan MS, Michaels LS (1992). Evaluation of microporous tangential-flow filtration in the production of Diphtheria and Pertussis vaccines. Pharm. Technol. 16:102-110.

Ravetkar SD, Rahalkar SB, Kulkarni CG (2001). Large scale processing of tetanus toxin from fermentation broth. J. Sci. Ind. Res. 60:773-778.

Reid DE, Adlam C (1974). Large-scale harvesting and concentration of bacteria by tangential flow filtration. J. Appl. Bacteriol. 41:321-324.

Relyveld EH, Huet M, Lery L (1996). Passive hemagglutination tests using purified antigens covalently coupled to turkey erythrocytes. Dev. Biol. Standardization 86:225-241.

Tanny GB, Mireman D, Pistole T (1980). Improved filtration technique for concentration and harvesting bacteria. Appl. Environ. Microbiol. 40:269-273.

Vancetto MDC, de Oliveira JM, Prado SMA, Fratelli F, Higash HG (1997). Tetanus toxoid purification: A case study. Nature Biotechnol. 15:807 – 808.

WHO (1990). Requirements for diphtheria, tetanus, pertussis and combined vaccines. (Revised 1989) Technical Report Series 800

WHO (1977). Manual for the production and control of vaccines. Tetanus toxoid. BLG/UNDP/77.2 Rev.1.

WHO (1990). Requirements for tetanus vaccine (adsorbed). Tech. Rep. Series 800, WHO, Geneva, pp.109-126.

WHO (1994a). Production and control of tetanus vaccine. A training curriculum, World Health Organization. WHO/VSQ/GEN/94.5.

WHO (1994b). Production and control of tetanus vaccine. A training curriculum, World Health Organization. WHO/VSQ/GEN/94.7.

WHO (1994c). Production and control of tetanus vaccine. A training curriculum, World Health Organization. WHO/VSQ/GEN/94.6.

WHO (1997). Manual of laboratory methods for testing of vaccines used in the WHO Expanded Programme on Immunization, World Health Organization, Geneva. WHO/VSQ/97.04.

WHO (2006). Weekly Epidemiological Record. 20(81):197-208.

Zeman LJ, Zydney AL (1996). Microfiltration and Ultrafiltration. Dekker, New York.

12

Temperature as a factor in the elaboration of mycotoxins by two fungi in groundnut fodder

Kiran Saini*, S. Kalyani, M. Surekha and S. M. Reddy

Toxicology Laboratory, Department of Botany, Kakatiya University, Warangal-506009 (AP), India.

Production of patulin and terreic acid by *Aspergillus terreus* and zearalenone by *Fusarium oxysporum* in groundnut fodder in relation to temperature was investigated. Biodeterioration activity of *F. oxysporum* and *A. terreus* was more at incubation temperature of 20 and 30°C respectively. *F. oxysporum* elaborated maximum amount of zearalenone at 20 to 25°C, while *A. terreus* could produce maximum terreic acid and patulin at 25°C.

Key words: Groundnut fodder, temperature, *Aspergillus terreus*, *Fusarium oxysporum*, terreic acid, zearalenone and patulin.

INTRODUCTION

The nutritive value of groundnut fodder is comparable with most of the pulse straws like cowpea, green gram, etc. Storing of groundnut fodder in open atmosphere is an unusual practice at least in Andhra Pradesh, India and practically no attention is given to the storage of fodder. Under fluctuating environmental conditions, the fodder is exposed to variety of stress conditions such as moisture and temperature creating congenial conditions for proliferation of storage moulds (Lacey, 1991; Visconti and Girolamo, 2002). Of these factors, temperature is of major concern and responsible for changing the metabolic activity of the mycotoxigenic fungi. Though the influence of temperature on biodeterioration and elaboration of mycotoxins in food grains by some fungi has been studied (Girisham et al., 1987; Schneweis et al., 2000; Cairns-Fuller et al., 2005; Hope et al., 2005), practically there is limited information is available for feeds and fodder (Dos et al., 2003; El-Shanawany et al., 2005). The present investigations were aimed to study the effect of temperature on biodeterioration and mycotoxin production in groundnut fodder by two mycotoxigenic fungi.

MATERIALS AND METHODS

The freshly harvested groundnut fodder from fields of Warangal

*Corresponding author. Email: sainik244@gmail.com.

was collected and brought to the laboratory. It is air dried at laboratory conditions. From this bulk sample thirty gram of fodder is randomly collected and chopped into pieces and inoculated separately with 7 day old fungal mat of *Aspergillus terreus* and *Fusarium oxysporum* which were grown on cellophane implanted malt agar medium. The groundnut fodder thus prepared was incubated at different temperatures (15, 20, 25, 30, 35, and 40°C) for 30 days. Moisture and other parameters were kept constant in all the experiments. Uninoculated groundnut fodder treated in a similar manner was served as control. At the end of incubation period, protein (Lowry et al., 1951), crude fibre, lignin, cellulose, total nitrogen and starch (Chopra and Kanwar, 1982), ash content (AOAC, 1990) and loss of weight were determined by using standard methods. The experiments were conducted in triplicate and repeated atleast twice. The results obtained are statistically analysed using SPSS software (Version 12.0). Extraction and estimation of patulin (Subramanian, 1982), terreic acid (Subramanian et al., 1978) and zearalenone (Kamimura et al., 1981) were carried out by employing standard methods.

RESULTS AND DISCUSSION

Groundnut fodder inoculated with both the fungi suffered maximum weight loss at incubation temperature of 25°C (Table 1). The deteriorating activity of *F. oxysporum* was comparatively high at low incubation temperature (15°C), while *A. terreus* was more active at 25°C and above. Biodeteriorating activity of both fungi under investigation was only marginal at incubation temperature of 40°C.

The protein and ash content increased due to fungal inoculation. The increase in protein and ash content was more significant at 25°C which decreased both under low

Table 1. Statistical analysis of effect of temperature on biodeterioration and toxin production in groundnut fodder by two fungi.

Temperature (°C)	Name of the fungus	Statistical parameter	Protein (mg/g)	Crude fibre (mg/g)	Cellulose (mg/g)	Lignin (mg/g)	Starch (mg/g)	Total nitrogen (mg/g)	Ash content (mg/g)	Loss of weight (%)	Mycotoxin	Amount
15	A. terreus	Mean±St.d	5.40±0.56	274.0±2.05	309.0±1.41	37.2±1.34	104.5±0.70	19.3±0.98	109.4±3.39	9.10±1.83	-	Nil
		S.E	0.40	1.45	1.00	0.95	0.50	0.70	2.40	1.30	TA/Pat	18/Nill
	F. oxysporum	Mean±St.d	5.50±0.70	273.7±2.47	308.2±2.47	37.6±084	104.5±0.70	19.2±1.60	109.5±3.53	10.0±3.18	Zearalenone	-
		S.E	0.5	1.75	1.75	1.75	0.5	0.75	2.5	2.25		
20	A. terreus	Mean±St.d	5.75±1.60	272.1±4.73	307.1±4.03	36.2±2.82	102.6±3.32	17.6±3.32	112.3±7.99	12.8±6.43	-	Nil
		S.E	0.75	3.35	2.85	2	2.35	2.35	6.5	3.5	TA/Pat	14
	F. oxysporum	Mean±St.d	5.65±0.91	271.4±5.79	306.9±4.38	34.7±4.94	103.6±1.90	17.6±3.32	113.5±9.19	12.8±6.43	Zearalenone	3
		S.E	0.65	4.1	3.1	3.5	1.35	2.35	6.5	4.55		
25	A. terreus	Mean±St.d	6.10±1.55	271.9±5.19	305.2±6.78	35.0±4.52	101.9±4.38	17.0±4.29	114.9±11.1	33.3±21.2	-	Nil
		S.E	1.10	3.60	4.80	3.20	3.10	3.00	7.90	15.0	TA/Pat	2
	F. oxysporum	Mean±St.d	6.15±1.62	270.5±7.00	305.0±7.07	33.4±6.78	101.1±5.51	17.0±4.24	115.8±11.3	33.9±23.2	Zearalenone	3
		S.E	1.15	4.95	5.00	4.80	3.90	3.00	8.00	15.6		
30	A. terreus	Mean±St.d	5.65±0.91	272.9±3.67	307.0±4.24	36.2±2.82	102.8±3.11	18.1±2.61	112.2±7.42	27.7±13.5	-	Nil
		S.E	0.65	2.60	3.00	2.00	2.20	1.85	5.25	9.55	TA/Pat	4
	F. oxysporum	Mean±St.d	5.65±0.91	272.6±4.03	307.0±4.17	35.1±4.38	103.9±1.55	17.8±3.11	112.9±8.34	24.5±8.90	Zearalenone	+
		S.E	0.65	2.85	2.95	3.10	1.10	2.20	5.90	6.30		
35	A. terreus	Mean±St.d	5.50±0.70	273.6±2.61	308.4±2.26	37.2±1.34	103.9±1.55	18.9±1.55	108.6±2.33	9.45±5.86	-	Nil
		S.E	0.50	1.85	1.60	0.95	1.10	1.10	1.65	4.15	TA/Pat	2
	F. oxysporum	Mean±St.d	5.50±0.70	272.4±4.38	309.3±0.98	37.2±1.34	104.2±1.06	19.4±0.84	108.9±2.66	9.65±6.15	Zearalenone	Nil
		S.E	0.50	3.10	0.70	0.95	0.75	0.60	1.90	4.35		
40	A. terreus	Mean±St.d	5.40±0.56	274.4±1.48	309.4±0.84	38.1±0.14	104.8±0.28	19.6±0.56	108.5±2.12	7.30±3.25	-	Nil
		S.E	0.40	1.05	0.60	0.10	0.20	0.40	1.50	2.30	TA/Pat	10/Nil
	F. oxysporum	Mean±St.d	5.05±0.70	274.5±1.34	310.0±0.00	38.1±0.14	104.7±0.35	20.0±0.00	108.5±2.12	6.60±2.26	Zearalenone	Nil
		S.E	0.50	0.95	0.00	0.10	0.25	0.00	0.50	1.60		

S.E= Standard error, St.d= standard deviation. By comparing means, standard deviation and standard error of triplicates of fungi at different temperature we observe that loss of weight is maximum at 25°C and minimum 40°C. Similarly protein and ash content is maximum at 25°C where as in all other cases like crude fibre, cellulose, lignin, starch and total nitrogen is minimum at 40°C.

and higher incubation temperature. Crude fibre, cellulose, lignin, starch and total nitrogen contents of groundnut fodder decreased at 25°C due to inoculation of fungi and the decline of these compounds with increase or decrease in incubation temperature was observed. *A. terreus* elaborated terreic acid at all incubation temperatures tried, and it was maximum at 25°C. On the other hand, patulin could be detected only between incubation temperature of 20 and 30°C. Patulin production was maximum at 25°C.

Zearalenone production by *F. oxysporum* was maximum at an incubation temperature of 20 to 25°C. No mycotoxins could be detected in groundnut fodder when incubated at 30°C and above. Mirocha et al, (1979) reported that low temperature or alternating moderate and low temperature will be favorable for mycotoxins production by species of *Fusarium*. It is reported that the enzymes responsible for the biosynthesis of zearalenone are active at 12-14°C. Similarly Milano and Lopez (1991) have also reported that zearalenone production was inhibited at higher temperature.

From the present investigations it is clear that the temperature plays an important role in infestation of groundnut fodder by *A. terreus* and *F. oxysporum*. Storing of groundnut fodder above 30°C will be more safer and can be made free from fungal infestation by specially *F. oxysporum* and *A. terreus* and mycotoxins contamination. However, more detailed studies dealing with interaction of other environmental factors have also to be investigated before reaching any decesive conclusions.

ACKNOWLEDGMENTS

Thanks are due to the Head, Department of Botany, Kakatiya University for providing laboratory facilities and University Grants Commission for financial assistance. We also thank Dr. J. Srinivas Department of Statistics for the statistical analysis.

REFERENCES

AOAC (1990). Official method of analysis of the Association of Official Analytical Chemists, 15th ed. Association of Official Analytical Chemists. Washington, DC.

Cairns-Fuller V, Aldred D, Magan (2005). Water, temperature and gas composition interactions affect growth and ochratoxin a production by isolates of *Penicillium verrucosum* on wheat grain. J Appl. Microbiol., 99: 1215-1221.

Chopra SL, Kanwar JS (1982). In analytical agricultural chemistry, Kalyani Publishers, New Delhi, p. 337.

Dos Santos VM, Dorner JW, Carrira F (2003). Isolation and toxigenicity of *Aspergillus fumigatus* from moldy silage. Mycopathologia, 156(2): 133-138.

El-Shanawany AA, Eman Mostafa M, Barakat A (2005). Fungal populations and mycotoxins on silage in Assiut and Sohag governorates in Egypt, with a special reference to characteristic Aspergilli toxin. Mycopathologia, 159: 281-289.

Girisham S, Reddy SM, Krishna RV (1987). Factors influencing the growth and patulin production. J. Indian Bot. Soc., p. 67.

Hope R, Aldred D, Magan N (2005). Comparison of environmental profiles for growth and deoxynivalenol production by *Fusarium culmorum* and *F. graminearum* on wheat grain. Lett. Appl. Microbiol., 40: 295-300.

Kamimura H, Nishijima H, Yasuda K, Saito K, Ibe A, Nagayama T, Ushiyama H, Naoi Y (1981). Simultaneous detection of several Fusarium mycotoxins in Cereals, grains and food stuffs. J. Assoc. Anal. Chem., 64: 1067.

Lacey J (1991). Natural occurrence of mycotoxins in growing and conserved forage crops. Mycotoxins and Animal Foods. CRC Press, Boca Raton. In: J. E. Smith and R. E. Henderson (Eds.), pp. 363-397.

Lowry CH, Rose BNJ, Farr L, Randall (1951). Protein measurement with Folic- phenol reagent. J. Biochem. Chem., 193: 265.

Milano GD, TA Lopez (1991). Influence of temperature on zearalenone production by regional strains of *Fusarium graminearum* and *Fusarium oxysporum* in culture. Int. J. Food Microbiol., 13: 329-333.

Mirocha CJ, Sachuerhamer B, Christensen CH, Kommedahl T (1979). Zearalenone, deoxynivalenol and T-2 toxin associated with stalk rot in Corn. Appl. Environ. Microbiol., 38: 557-558.

Schneweis I, Meyer KH, Rmansdorfer S, Bauer J (2000). Mycophenolic acid in silage. Appl. Environ. Microbiol., 66(8): 3639-3641.

Subramanian T (1982). Colorimetric determination of patulin production by *Penicillium patulin*. J. Assoc. Anal. Chem., 65: 5-7.

Subramanian T, Kuppouswamy MN, Shanmuga SERB (1978). Colorimetric determination of terreic acid produced by Aspergillus terreus. J. Assoc. Anal. Chem., 61: 581-583.

Visconti A, De girolamo A (2002). *Fusarium* mycotoxins in cereals storage, processing and decontamination. In: Scholten, O.E., Ruckenbauer, P., Visconti, A, van Osenbrugger, W.A, den Nijs, A.P.M. (Eds.), Food Safety of cereals: A chain-wide approach to reduce Fusarium Mycotoxins, EU Commission, FAIR-CT98-4094.

13

Incidence of mycotoxigenic penicillia in feeds of Andhra Pradesh, India

V. Koteswara Rao*, P. Shilpa, S. Girisham and S. M. Reddy

Department of Microbiology, Kakatiya University, Warangal-506 009, India.

Incidence of different species of *Penicillium* in poultry feeds (starter, breeder, boiler and layer) and cattle feeds was analyzed. In all twenty three species of *Penicillium*, *Penicillium aethiopicum*, *Penicillium alli*, *Penicillium aurantiogriseum*, *Penicillium brevicampactum*, *Penicillium camemberti*, *Penicillium caseifulvum*, *Pchrysogenum*, *Penicillium citrinum*, *Penicillium commune*, *Penicillium crustosum*, *Penicillium digitatum*, *Penicillium dipodomyis*, *Penicillium discolor*, *Penicillium expansum*, *Penicillium flavigenum*, *Penicillium griseofulvum*, *Penicillium italicum*, *Penicillium nalgiovense*, *Penicillium nordicum*, *Penicillium olsonii*, *Penicillium roqueforti*, *Penicillium rubrum*, *Penicillium tricolor* and *Penicillium verrucosum* could be recorded in 400 samples of feed samples, poultry feed (280) and cattle feed (120) was analyzed for fungus isolation by dilution plate method, screening of 483 *Penicillium* species by thin layer chromatography (TLC) and spray reagents for mycotoxin production, 299 strains were positive to be mycotoxigenic which elaborates a variety of mycotoxins such as Citrinin, Cyclopiazonic acid, Mycophenolic acid, Ochratoxin A, Patulin, Penitrems, PR toxin, Roquefortine C,Rubratoxin and Terric acid etc.

Key words: *Penicillium* species, *mycotoxins*, cattle feed, poultry feed.

INTRODUCTION

In recent times poultry and dairy farming became an important agro-business, millions of small and marginal farmers use crop residues and natural herbage to feed their live stock. The availability and type of feed depends on local resources, climatic and the socioeconomic condition of the people, lack of scientific knowledge on improper processing during harvest, unseasonal rains and high moisture content provides an ideal condition for the proliferation of moulds and mycotoxins production in foods and feeds (Frisvad, 1995; Fazekas et al., 1996; Trucksess, 2001; Ana et al., 2009). These mycotoxins can be very stable to food processing (Molinie et al., 2005) can be present in fungal product. Penicillia from moldy feeds may cause infections, provoke allergic responses in sensitized objects or poison with toxic metabolites (Answorth and Austick, 1973; Lacey, 1975; Abramson, 1997). Hence, the present investigation was aimed to undertake an extensive and intensive survey of different feeds and feed ingredients for the incidence

of *Penicillium* species.

MATERIALS AND METHODS

An extensive survey of different feeds (cattle and poultry) of different geographical regions of Andhra Pradesh State (A.P.), India was undertaken. The samples were collected randomly, and analyzed for the presence of *Penicillium* species by dilution plate technique (Waksmann, 1922). Specific medium such as Czapek Yeast Autolysate (CYA) agar (Pitt, 1979) medium was employed for isolation of *Penicillium* species. In addition macro morphology of structure and branching of the conidiophores, the shape and ornamentation of conidia, colony characters that including growth rate, conidium color and reverse color of the colony, diffusing pigment characteristics for few species were observed and documented. Most of the *Penicillium* isolates inoculated on four enriched media such CYA agar (Pitt, 1979) Blakeslee Malt extract Autolysate (MEA) agar (Raper and Thom, 1949) Yeast extract sucrose (YES) agar (Frisvad et al., 1992) and Creatine sucrose(CREA) agar (Frisvad, 1985) for their identification, and these media gave characteristics aereal and reverse colour on type of media. The sub genus *Penicillium* species were identified by with the help of standard manuals and protocols. (Hyde, 1990; Filtenberg et al., 1992; Svendsen and Frisvad, 1994; Pitt et al., 2000; Samson et al., 2002; Frisvad and Samson, 2004).

*Corresponding author. E-mail: Koti_micro08@yahoo.co.in.

The percentage of incidence, frequency and abundance of each fungus with special emphasis on *Penicillium* was calculated by the following formulae:

% of incidence = (No. of colonies of a species in all plates / Total no. of colonies of the all the species in all plates) × 100

% of frequency = (No of observations in which a species appeared / Total no. of observations) × 100

% of abundance = (No. of colonies of species in all observations / Total no. of colonies in all observations) × 100

Penicillium mycotoxins (Extrolites) were analyzed by employing thin layer chromatography (TLC) (Frisvad and Filtenberg, 1989; Filtenberg et al., 1983, 1992; Lund, 1995). The TLC plates (Silica Gel GF 254) were impregnated in 10% solution of oxalic acid in methanol solution for 10 min, after heating at 110°C for two minutes the plates were kept for cooling and immediately the mycotoxin extract (20 µl) was spotted on activated and cooled TLC plates (Smedsgaard, 1997). The spotted plates were developed in suitable solvents system (Samson and Pitt, 2000) by ascending chromatography. The compounds thus separated were identified either by the color of the fluoresce under (U.V.333 nm) or by the Rf value, they were further confirmed by chemical tests using different spray reagents (Pitt and Hocking, 1996, and Frisvad et al., 2004), and U.V Spectrum (U.V-10 VIS). The R_f value was calculated by the following formulae:

$R_f =$ Distance traveled by the compound / Distance traveled by the solvent.

RESULTS AND DISCUSSION

The identification of subgenus *Penicillium* species is difficult (Thom, 1930; Raper and Thom, 1949; Smith 1960; Ciegler et al., 1969; Frisvad, 1981, Samson and Pitt, 1990; Larsen and Frisvad, 1995) because the micro morphology of the strains is very similar. In total twenty three *Penicillium* species were associated with both poultry and cattle feed (Table 1) collected from different geographical region of Andhra Pradesh with the employment of CYA agar media. *Penicillium rubrum*, *Penicillium citrinum* and *Penicillium olsonii* occurred with highest percentage of incidence followed by *Penicillium chrysogenum*, *Penicillium aethiopicum Penicillium alli* and *Penicillium aurantiogriseum* could be isolated only feed collected from Adilabad Districts. The incidence of *Penicillium* species was dominated and followed by *Aspergillus* species. *Penicillium brevicampactum* was associated with all the samples except in Nalgonda and Guntur. *Penicillium flavigerum was* associated with all the samples except the sample collected from Khammam, Warangal, and Nalgonda Districts. *Penicillium nalgiovense* could be detected in all the samples except those collected from Adilabad. *Penicillium roqueforti* and *P. rubrum* were associated with the samples of both poultry and cattle feed. *Penicillium verrrucosum* was detected in all poultry feed samples of Warangal, similarly. *Penicillium caseifulvum* could not be detected

in poultry feed samples of Adilabad. On the other hand, *P.chrysogenum*, *P.citrinum* and *Penicillium commune* were common to all samples. The incidence of *Penicillium crustosum* could not be recorded in cattle feed samples of Warangal, Khammam, Nalgonda, Krishna, Guntur and Adilabad .The incidence of *Penicillium* species more in poultry feeds than in cattle feeds.

Screening for toxigenic potential of different species of *Penicillium* revealed (Table 3) that, large numbers of *Penicillium* isolates were mycotoxigenic. However, the percentage of mycotoxigenic strains varied with the species and place of collection. All the isolates of *Penicillium* sub genus were mycotoxigenic and many cases more than one mycotoxin was detected thus so called OSMAC (one strains many compounds), (Bode et al., 2002).

Table 2 reveled that out of 483 strains of *Penicillium* species isolated from cattle and poultry feed 299 strains were mycotoxigenic and 65 toxigenic out of 95 strains of *P. citrinum*, *P. expansum*, *P. nordicum* and *P. verrucosum* were screened for citrinin production. More strains of *P. citrinium* were toxigenic from the cattle feed samples collected from Khammam and poultry feed samples of Warangal, Nalgonda and Krishna Districts. Out of 25 strains of *P. camemberti* and *P. commune* were screened, 17 strains produced Cyclopiazonic acid. Out of 40 strains of *P. verrucosum* and *P. nordicum,* 28 strains produced Ochratoxin A. Comparatively more numbers of strains were toxigenic in cattle feeds of Khammam, Guntur, Adilabad and poultry feed of Krishna districts. Out of 23 strains of *P.expansum* and *P.dipodomyis* screened, 7 strains were positive for Patulin elaboration. The incidence of mycotoxigenic strains were more in cattle feed samples collected from all districts of Khammam and Nalgonda.Out of 16 strains of *P. flavigenum*, 10 strains produced penitrems (penitrems A). Contamination of penitrems was comparatively more in poultry feeds of Guntur and Adilabad District. When 19 strains of *P.aurantiogriseum* and *P.alli* from Adilabad districts were screened, 8 strains were positive for Penicillic acid, Similarly 93 strains of *P. alli*, *P. chrysogenum*, *P. crustosum*, *P. flavigenum*, *P. expansum* and *P. roqueforti* when screened for their mycotoxigenic potential, 57 strains were found to produce roquefortin C toxin, However, *P. alli* failed to produce the toxin. When 50 strains of *P. rubrum* were screened, 36 strains were positive for rubratoxin B.The incidence of this mycotoxin was comparatively more in poultry feeds of Khammam, and cattle feed of Adilabad. When 18 strains of *P. crustosum* and 53 strains of *P. chrysogenum* and *P. roqueforti* were screened, 5 strains were terric acid positive and 30 produced PR toxin. Similarly out of 41 strains of *P. brevicampactum* and *P. roqueforti,* 33 strains elaborated mycophenolic acid.

The order of percentage of contamination of different secondary metabolites by *Penicillium* species were

Table 1. Incidence of mycotoxigenic *Penicillia* in feed samples.

Name of fungus	Incidence												Frequency		Abundance	
	Khammam		Warangal		Nalgonda		Krishna		Guntur		Adilabad					
	A	B	A	B	A	B	A	B	A	B	A	B	A	B	A	B
P. aethiopicum	--	--	--	--	--	--	--	--	--	--	2.1	1.3	16.6	16.6	0.85	0.46
P. alli	--	--	--	--	--	--	--	--	--	--	0.6	1.4	16.6	16.6	0.24	0.48
P. aurantiogriseum	--	--	--	--	--	--	--	--	--	--	1.3	0.8	16.6	16.6	0.53	0.28
P. brevicampactum	5.2	1.4	2.2	1.6	--	--	--	1.5	2.9	--	1.4	3.7	50	66.6	4.77	2.93
P. camemberti	0.9	2.1	--	3.7	3.1	1.4	2.5	6.7	--	1.3	2.2	--	50	83.3	3.55	5.44
P. caseifulvum	1.5	1.3	2.6	4.4	1.2	6.2	1.6	4.4	2.4	2.8	--	1.1	83.3	100	3.7	7.24
P. chrysogenum	4.6	3.4	3.1	4.8	1.6	2.1	5.1	1.8	1.9	5.3	3.9	4.3	100	100	8.24	7.77
P. citrinum	2.7	5.9	6.2	1.8	5.8	1.8	4.6	5.3	2.3	4.6	4.2	5.1	100	100	10.2	8.78
P. commune	2.2	1.5	1.6	0.6	2.4	1.2	1.8	0.6	3.8	2.3	3.8	1.3	100	100	6.36	2.68
P. crustosum	1.6	--	--	2.8	--	--	--	2.1	7.2	--	1.9	2.7	50	50	4.36	2.72
P. digitatum	--	2.8	--	2.6	--	--	2.6	2.9	--	2.7	2.1	1.6	33.3	83.3	1.91	4.51
P. dipodomyis	--	--	--	--	--	--	0.6	--	--	--	1.9	2.2	33.3	16.3	1.02	0.78
P. discolor	2.1	1.8	--	--	--	1.9	--	1.3	3.1	2.8	1.6	2.7	50	83.3	2.77	3.76
P. expansum	2.3	4.5	2.7	0.8	2.7	5.9	3.8	0.8	--	1.1	3.3	2.3	83.3	100	6.04	5.51
P. flavigenum	0.9	--	1.9	--	1.9	2.5	0.7	2.9	0.5	2.1	4.5	1.8	66.6	100	4.24	3.33
P. italicum	--	1.2	--	1.5	--	4.2	--	1.4	2.6	1.5	0.7	3.1	33.3	100	1.34	4.62
P. nalgiovense	3.2	3.6	4.2	--	4.2	2.1	2.5	3.3	0.2	2.9	--	--	83.3	66.6	5.83	4.26
P. nordicum	1.5	2.6	3.1	3.7	3.1	0.9	5.4	4.7	1.2	2.2	--	0.9	83.3	100	5.83	5.37
P. olsonii	2.4	3.8	2.8	5.4	2.8	2.6	6.5	4.2	1.7	5.4	2.1	2.7	100	100	7.4	8.63
P. roqueforti	2.3	2.8	1.4	4.5	1.4	2.2	2.3	4.9	2.5	2.3	--	0.3	83.3	100	4.04	6.09
P. rubrum	6.4	4.9	2.9	3.2	2.9	3.1	2.9	1.4	1.2	2.4	2.3	4.1	100	100	7.59	6.84
P. tricolor	0.2	0.9	1.2	--	--	--	0.4	1.1	0.5	--	--	--	66.6	33.3	0.93	0.71
P. verrucosum	2.9	4.1	3.4	--	3.9	3.7	2.7	2.2	2.3	4.6	3.7	4.2	100	83.3	7.71	6.73
Otherfungi =	57.1	51.4	60.7	58.6	63.0	58.2	54.1	46.5	63.7	53.7	56.4	52.4	100	100		

Otherfungi: *Aspergillus Fusarium, Mucor Rhizopus, Neurospora, Cladosporium* species.

Rubratoxin, Citrinin, Ochratoxin Roquefortine followed by Patulin, PR, Mycophenolic acid, Cyclopiazonic acid, Penicillic acid and Terric acid respectively.

The critical perusal of Table 1 reveals that the cattle feed was most ideal substratum for the proliferation of *Penicillia* and mycotoxin elaboration. The mycotoxigenic potential of *Penicillium* species isolated from poultry feed was intermediate. Hyde (1990) has isolated *Cladosporium herbarum, Alternaria tennuissima* and *Aspergillus fumigatus* from feeds, which are responsible for various allergic diseases in cattle and farm workers. Fungal growth in feeds may also deplete the nutritive value and can alter the availability of micronutrients (Zohri et al., 1993; Broster, 1998). In all Cyclopiazonic acid followed by Patulin, Citrinin, Ochratoxins

Table 2. Contamination of Mycotoxins produced by *Penicillium* spp.

Name of fungus	TS	PS	Ts (%)	Name of the toxin
P. citrinum	40	28	70	Citrinin
P. verrucosum	30	22	73	
P. nordicum	10	6	60	
P. camemberi	10	5	50	Cyclopiazonic acid
P. expansum	15	9	60	
P. commune	15	12	80	
P. verrucosum	30	22	70	Ochratoxin A
P. nordicum	10	6	60	
P. dipodomyis	8	2	25	Patulin
P. expansum	15	5	45	
P. flavigenum	16	10	62	Penitrem A
P. aurantiogriseum	14	6	42	Penicillic acid
P. alli	5	2	40	
P.chrysogenum	35	24	68	Roquefortine C
P. crustosum	9	4	44	
P. expansum	15	9	60	
P. flavigenum	16	8	50	
P. roqueforti	18	12	66	
P. rubrum	50	36	72	Rubra toxin B
P. crustosum	18	5	27	Terric acid
P. chrysogenum	35	24	68	PRtoxin
P. roqueforti	18	6	33	
P. brevicampactum	33	25	75	Mycophenolic acid
P. roqueforti	18	11	66	
	483	299		

TS= Total strains; PS= Positive strains; %Ts= Percentage of toxigenic strains.

Table 3. Detection Penicillium producing Mycotoxin by different spray reagents.

Name of toxin	TEF	U.V	Spray reagents				
	Rf		CesO$_4$	2.4.DNP	FeCl3	*P.anisaldehyde*	AlCl3
Citrinin	0.52	y	Y	Bry	Br	--	LY
Cyclopiazonic acid	0.52	Y	Br	Y	Br	--	Lb
Ochratoxin A	0.32	B	B	Lo	Pbr	--	Lbr
Patulin	0.22	P	G	Y	--	--	Gr
Penitrem A	0.4	LP	Lo	P	G	--	LY
Penicillic acid	0.16	B	O	--	Lo	--	--
Roquefortine C	0.3	P	--	Gr	--	--	--
Rubra toxin B	0.35	Y	--	--	--	--	--
Terric acid	0.23	B	--	--	--	--	--
PRtoxin	0.19	PB	--	--	Br	--	--
Mycophenolic acid	0.36	Y	--	--	--	--	--

Detection color: Y=Yellow, LP=Light purple, B= blue, Pb= Purple blue, Br= brown, Ybr=Yellow brown, G= green, Lo= Light orange, O= orange, Bry= brown yellow, RBr= Red brown, Gr= grey, vo= violet, LY= Light yellow, Lb= Light brown, Ly= Light yellow, Lbr= Light brown, Pbr=Purple brown.
Spray reagents: 1= CeSO$_4$ 1% IN 6N H2SO$_4$, 2 = 2,4 DNP,3= FeCl3 3% in Ethanol, 4=p-anisaldehyde,5=50% H2SO4, 6 = 1%FeCl3 in Ethanol,Iodine ,AlCl3.Solvent system: TEF= Toluene, Ethyl acetate, Formic acid (6;3;1).

Rubratoxins, Roquefortine, Mycophenolic acid and Penicillin etc. could be spotted in feed samples analyzed from different geographical places of Andhra Pradesh.

ACKNOWLEDGEMENT

Thanks to the Head of Department of Microbiology, KakatiyaUniversity, Warangal, and UGC.F.No.36-129-2008(SR) MRP for necessary facilities and financial assistance respectively.

REFERENCES

Abramson D (1997). Toxicants of the genus *Penicillium*. In Flex: Flex D'Mello, J.P, (eds). Handbook of Plant and fungal toxicants. Florida: CRC, pp. 303-317.

Answorth GC, Austwick PKC (1973). Fungal diseases of animals. Second edition.Fairham Royal: common Wealth Agricultural Bureau.

Ana MP, Emilia CB, Hector HL, Gonzalez EM, White C, Elena JM, SilivaLR (2009). Fungal and fumonisins contamination and Argentinemaize (*Zea mays L.*) silico bags. J. Agric. Food Chem., 57: 2778-2781.

Bode HB, Bethe B, Hof R, Zeeck A (2002). Effects from small changes: Possible ways to explore nature's chemical diversity. Chem. Bio. Chem., 3: 619-627.

Broster WH, Broster VJ (1998). Body score of dairy cows. J. Dairy Res., 65: 155-173.

Ciegler A. (1969). A tremorgenic mycotoxin from *Penicillium*. Appl. Microbiol., 18: 128-129.

Fazekas B, Kis M, Haidu ET (1996). Data on the contamination of maize with fumonisins B1 and other fusarial toxins in Hungary. Acta Vet. Hung, 44: 25-37.

Filtenberg O, Frisvad JC, Lund F, Thrane U (1992). Simple identification procedure for association of spoilage and toxigenic mycocoflora in foods .In modern method in food mycology ed.Samson R.A., Hoking, AD., Pitt, J.I. and King, A.D. Amsterdam: Elsevier, pp. 248-258

Frisvad JC (1981). Physiological criteria and mycotoxin production as aids in identification of common asymmetric *Penicillia*. .Appl. Environ. Microbiol., 41: 568-579.

Frisvad JC (1985). Creatine sucrose agar, a differential medium for mycotoxin producing terverticillate *Penicillium* species. Lett. Appl. Microbiol., 1: 109-113.

Frisvad JC (1995). Mycotoxins and Mycotoxigenic fungi in storage In: Stored-grain Ecosystems (Jayas, D.S., White, N.D.G.and Muir, W.E., Eds.), Marcel Dekker, New York, pp. 251-288.

Frisvad JC, Filtenborg O (1989). Terverticillate *Penicillia* chemotaxonomy and mycotoxin production. Mycol., 81: 837-861.

Frisvad JC, Filtenborg O, Lund F, Thrane U (1992). New selective media for the detection of toxigenic fungi in cereal products, meat and cheese. In modern methods in food mycology (eds) Samson, R.A., Hoking, A.D., Pitt, J.I and King, A.D. Amsterdam: Elsevier, pp. 259-268.

Frisvad JC, Samson RA (2004). Polyphasic taxonomy of *Penicillium* subgenus *Penicillium*. A guide to identification of food and air-borne terverticillate Penicillia and their mycotoxins. Stud. Mycol., 49: 1-173.

Frisvad JC, Smedsgaard J, Larsen TO, Samson RA (2004). Mycotoxins, drugs and other extrolites produced by species in *Penicillium* subgenus *Penicillium*. Stud. Mycol. (Utrecht), 49: 201–242.

Hyde KD (1990). Intertidal mycota of five mangrove tree species. Asian Mar. Biol., 7: 93-107.

Lacey (1975) Potential hazards to animals and man from microorganisms in fodders and grain. Transactions of the British Mycological Society.

Larsen TO, Frisvad JC (1995). Chemosystematics of *Penicillium* based on profiles of volatile metabolites. Mycol. Res., 99: 1167-1174.

Lund (1995). Differentiating Penicillium species by detection of indole metabolites using a filter paper method. Lett. Appl. Microbiol,, 20: 228-231.

Molini'e A, Faucet V, Castegnaro M, Pfohl-Leszkonicz A (2005). Analysis of some breakfast cereals on the fresh market for their contents of ochratoxin a, citrinin and fumonisins B1: development of method for simultaneous extraction of ochratoxin and citrinin. Food Chem., 92: 391-400.

Pitt JI, Hocking AD (1996). Current knowledge of fungi and mycotoxins associated with food Commodities in Southeast Asia in Highley E, Johnson GI. (Eds) Mycotoxin Contamination in Grains. Canberra. Australian Centre for International Agricultural Research. ACIAR Tech. Reports, 37: 5-10.

Pitt JI, Samson RA, Frisvad JC (2000). List of accepted species and their synonyms in the family Trichocomaceae In: Integration of Modern Taxonomic Methods for *Penicillium* and *Aspergillus* Classification (Samson, R.A. and Pitt, J.I., Eds.), Harwood Academic Publishers, Amsterdam, pp. 9-47.

Pitt JI (1979). The genus *Penicillium* and its teleomorphic states *Eupenicillium* and *Talaromyces*. Academic Press Ed., London.

Raper KB, Thom C (1949). Manual of the *Penicillia*. Williams and Wilkins, Baltimore, p. 875.

Samson RA, Pitt JI (1990). Modern Concepts in *Penicillium* and Aspergillus Classification (eds). Plenum Press, New York, USA.

Samson RA, Pitt JI (2000). Integration of Modern Methods for *Penicillium* and *Aspergillus*. Harwood Academic Publishers, Amsterdam, the Netherlands, p. 510.

Samson RA, Hoekstra ES, Frisvad JC, Filtenborg O (2002). 6th Edn Introduction to Food- and Airborne Fungi, 202. Centraalbureau voor Schimmelcultures, Utrecht, pp. 379-381.

Smedsgaard J (1997). Micro-scale extraction procedure for standardized screening of fungal metabolite production in cultures. J. Chromatogr., 760: 264-270.

Smith G (1960). An Introduction to Industrial Mycology. 5th ed. Edward Arnold Ltd., London, p. 399.

Svendsen A, Frisvad JC (1994). A chemotaxonomic study of the terverticillate *Penicillia* based on high performance liquid chromatography of secondary metabolites. Mycol. Res., 98: 1317-1328.

Thom C (1930). The *Penicillia*. Williams and Wilkins, Baltimore, p. 643.

Truckess MW, Tang Y (2001). Solid phase extraction for Patulin in apple juice and unfiltered apple juice. In M.W.Truckees and A.F.pohland(ed.),Mycotoxin protocols. Human press, Totowa, N. J., pp: 205-213.

Waksman SA (1922). A method for counting the number of fungi in the soil. J. Bot., 7: 339-341.

Zohri AA, Abdei-Gawad KM (1993). Survey of mycoflora and mycotoxins of some dried fruits in Egypt, pp. 279-288.

14

Ethnomedicinal plants and other natural products with anti-HIV active compounds and their putative modes of action

Kazhila C. Chinsembu[1] and Marius Hedimbi[1,2]*

[1]Department of Biological Sciences, Faculty of Science, University of Namibia, Windhoek, Namibia.
[2]Multidisciplinary Research Centre, Science and Technology Division, University of Namibia, P/Bag 13301, Windhoek, Namibia.

The use of ethnomedicines to manage HIV/AIDS has recently gained public interest, although harmonization with official HIV/AIDS policy remains a contentious issue in many countries. Plants and other natural products present a large repertoire from which to isolate novel anti-HIV active compounds. In this literature survey, 55 plant families containing 95 plant species, and other natural products, were found to contain anti-HIV active compounds that included diterpenes, triterpenes, biflavonoids, coumarins, caffeic acid tetramers, hypericin, gallotannins, galloylquinic acids, curcumins, michellamines, and limonoids. These active compounds inhibited various steps in the HIV life cycle. However, further studies are needed to determine their interactions with current regimes of antiretroviral drugs. More clinical trials of candidate drugs developed from these novel compounds are also encouraged.

Key words: Anti-HIV active compounds, other natural products, plants.

INTRODUCTION

In 2008, an estimated 2.7 million new HIV infections occurred worldwide; this was 30% lower than the 3.5 million new infections at the peak of the epidemic in 1996 (UNAIDS, 2009). Sub-Saharan Africa remains the most heavily affected region, accounting for about 71% of all new HIV infections in 2008. There are two related but distinct types of HIV: HIV-1 and HIV-2 (Fletcher et al., 2002). HIV-1 is the most pathogenic and causes over 99% of HIV infections (Cos et al., 2004). HIV-2 is also known to cause AIDS but is much less prevalent, being present in fewer and isolated geographic locations such as West Africa. Therefore, most research is done on HIV-1 (Klos et al., 2009).

AIDS-related diseases remain one of the leading causes of death globally. According to UNAIDS, the number of people living with HIV/AIDS worldwide was estimated at 33.4 million in 2008; >20% higher than the number in 2000 (UNAIDS, 2009). It was estimated that 2million deaths due to AIDS-related illnesses occurred worldwide in 2008; this was ~10% lower than in 2004 (UNAIDS, 2009). The declines in new infections and AIDS-deaths may be attributed to the scale-up of anti-retroviral therapy (ART) programmes, especially in the developing world. As of December 2008, approximately 4 million people in low- and middle-income countries were on ART, representing a 10-fold increase over five years (UNAIDS, 2009). In eastern and southern Africa, ART coverage rose from 7% in 2003 to 48% in 2008 (UNAIDS, 2009).

Despite this impressive progress, Chinsembu (2009) reports that poverty in southern Africa still plays a major role in the dynamics of the HIV/AIDS. There are concerns that free public sector ART programmes are not sustainable due to their heavy reliance on donor funding. Besides funding, access to treatment still has many shortcomings, including lack of confidentiality, lack of bed space, lack of transport to hospitals, shortages of qualified health workers, long queues, the criterion of treatment supporter, and serious side-effects now causing new forms of stigma (Chinsembu, 2009). ART of

*Corresponding author. E-mail: mhedimbi@unam.na, mhedimbi@yahoo.com.

HIV-infected patients has also been associated with the development of lipodystrophy (LD). LD is characterized by peripheral fat loss (lipoatrophy) and central fat accumulation which may result in thin facial pads, thin arms and legs, pot-bellies, or 'buffalo humps', leaving patients stigmatized (Lindegaard et al., 2004). Thus, while acknowledging that current antiretroviral drugs are vitally important in improving the quality and prolonging the life of HIV/AIDS patients, the drugs still have many disadvantages including resistance, toxicity, limited availability, high cost and lack of any curative effect (Vermani and Garg, 2002). These shortcomings of conventional ART continue to open new vistas in the use of ethnomedicinal plants and other natural products for the management of HIV/AIDS.

In many countries, the inclusion of anti-HIV ethnomedicines and other natural products in official HIV/AIDS policy is an extremely sensitive and contentious issue (Chinsembu, 2009). It is sensitive because anti-HIV ethnomedicines and other natural products can easily become a scapegoat for denial and inertia to roll-out ART (Chinsembu, 2009). It is also contentious because in various resource-poor settings, government-sponsored ART programmes discourage the use of traditional medicines, fearing that the efficacy of antiretroviral drugs may be inhibited by such natural products, or that their pharmacological interactions could lead to toxicity (Hardon et al., 2008; Chinsembu, 2009). Reliance on anti-HIV plants and other natural products can also lead to poor adherence to ART (Langlois-Klassen et al., 2007). Thus, many governments still have contradictory attitudes towards the use of anti-HIV plants and other natural products in the management of HIV/AIDS, discouraging them within ART programmes, and supporting them within other initiatives of public health and primary health care (Chinsembu, 2009).

In essence, many HIV-infected persons have access to antiretroviral drugs, but some still use ethnomedicinal plants and other natural products to treat opportunistic infections and offset side-effects from antiretroviral medication (Hardon et al., 2008). Medicinal plants and other natural products including mushrooms are used as primary treatment for HIV-related problems such as skin disorders, nausea, depression, insomnia, and body weakness (Babb et al., 2004). In the case of rural communities, formal biomedical services are also hardly accessible. Thus, whilst the majority of HIV/AIDS patients rely on ART, some still have faith in the use of traditional medicines. Understandably, HIV/AIDS patients are vulnerable in their choice of treatments (Hardon et al., 2008), such that some of them do vacillate from conventional ART programmes to traditional medicines and vice versa; they want to have the best of both worlds (Hardon et al., 2008; Chinsembu, 2009).

As early as 1989, the World Health Organization (WHO) had already voiced the need to evaluate ethnomedicines and other natural products for the manage-

ment of HIV/AIDS: "In this context, there is need to evaluate those elements of traditional medicine, particularly medicinal plants and other natural products that might yield effective and affordable therapeutic agents. This will require a systematic approach", stated a memorandum of the WHO (1989a). Thus, African governments expressed the need for a concerted, systematic and sustained effort at both local and regional levels to support and biochemically validate African traditional medicines (UNAIDS, 1998). To popularize this commitment, the Organization of African Unity (African Union) Heads of State and Government declared the period 2000-2010 as the Decade of African Traditional Medicine. In addition, the Director General of WHO, declared 31st August of every year as the African Traditional Medicine Day (Homsy et al., 2004). All these initiatives demonstrate the need to mainstream and institutionalize traditional medicine into the formal health care system.

The importance of investing in the high growth sectors of biotechnology and phytomedicine was also articulated in the founding document of the New Partnership for Africa's Development (NEPAD), and adopted by the African Biosciences Initiative (NEPAD, 2001; African Biosciences Initiative, 2005). Herbal medicines provide rational means for the treatment of many diseases that are obstinate and incurable in western systems of medicine (Vermani and Garg, 2002). Phytomedicines are regaining patient acceptance because they have fewer side effects, are relatively less expensive, are easy to use (Short, 2006), and have a long history of use (Vermani and Garg, 2002). Medicinal effects of plants tend to normalize physiological function and correct the underlying cause of the disorder (Murray and Pizzorno, 1999). Furthermore, medicinal plants are renewable in nature unlike the synthetic drugs that are obtained from non-renewable sources of basic raw materials such as fossil sources and petrochemicals (Samanta et al., 2000). Cultivation, gathering, and selling of medicinal plants can also be a source of income for poor families (Reihling, 2008).

Due to the renewed public interest in phytomedicines, NEPAD's Southern Africa Network for Biosciences (SANBio) has launched a flagship project to validate ethnomedicines for effective and affordable treatment of HIV/AIDS. Scientific validation of putative herbal plants used in the management of HIV/AIDS is done at the Council for Scientific and Industrial Research (CSIR) in Pretoria, South Africa. Prior to this NEPAD/SANBio initiative, there exisited only a handful of workers in Africa that had attempted to screen medicinal plants for anti-HIV active compounds: Moore and Pizza (1992); Iwu (1993); Boyd et al. (1994); Muanza et al. (1995); Matsuse et al. (1999); Bessong et al. (2005); Mills et al. (2005); Abere and Agoreyo (2006) ; Kisangau et al. (2007); Moore et al. (2007); Quattara et al. (2007); Igbinosa et al. (2009) and Klos et al. (2009).

Sub-Saharan Africa has rich plant biodiversity and a

long tradition of medicinal use of plants with over 3,000 species of plants used as medicines (Van Wyk and Gericke, 2000; Scott et al., 2004). Several of these plants may contain novel anti-HIV compounds. In the past decade, there has been a sustained bioprospective effort to isolate the active leads from plants and other natural products for preventing transmission of HIV and management of AIDS (Asres et al., 2001; Vermani and Garg, 2002). Screening of plants based on ethnopharmacological data increases the potential of finding novel anti-HIV compounds (Farnsworth, 1994; Fabricant and Farnsworth, 2001). Indigenous knowledge of medicinal plant use also provides leads towards therapeutic concept thereby accelerating drug discovery; this is now being called reverse pharmacology (Chinsembu, 2009; Kaya, 2009). Thus, it is important to search for novel antiretroviral agents which can be added to or replace the current arsenal of drugs against HIV (Klos et al., 2009).

Despite the rich African repertoire from which to select medicinal plants, traditional herbal medicines are still not well-researched (Mills et al., 2005), and African knowledge of herbal remedies used to manage HIV/AIDS is scanty, impressionistic and not well documented (Kayombo et al., 2007). Africa is also awash with fake AIDS cures (Amon, 2008). HIV/AIDS is a relatively new human disease, with few ethnobotanical treatments, but logical associations of treatments for other likely viral infections (such as hepatitis B) and closely linked disease states or symptoms (wasting, diarrhoea, lymphadenopathy, skin lesions, cough, and genital ulcers) can increase the prospect of finding new plant leads as potential anti-HIV agents (WHO, 1989ab; Cardellina and Boyd, 1995; Lewis and Elvin-Lewis, 1995). Natural products can be selected for biological screening based on ethnomedical use, random collection and a chemotaxonomic approach i.e. screening of species of the same botanical family for similar compounds. However, the follow-up and selection of plants based on literature leads would seem to be the most cost-effective way of identifying plants with anti-HIV activity (WHO, 1989a).

The general objective of this study was to carry out a desktop survey of the literature for plants and other natural products with anti-HIV activity. Such a desktop survey is an important prerequisite and starting point in the search for novel HIV/AIDS treatments in and outside Africa. The specific objectives of the current study were to search for: (a) plants and other natural products with known active compounds and mechanisms of action, and (b) plants and other natural products with known or unknown active compounds and/or known or unknown mechanisms of action against HIV and AIDS-related opportunistic infections.

METHODOLOGY

The key words "plants with anti-HIV activity" were searched in PubMed Central, the United States of America National Library of Medicine's digital archive of biomedical and life sciences journal literature. During the literature search which lasted 3 months, between 224 - 250 journal publications were reviewed. Wherever possible, families and species of plants and other natural products, their active compounds and modes of action, were listed in tables according to their taxonomic families, in alphabetical order. Evidence of use of plants and other natural products against HIV/AIDS was also documented.

RESULTS

Plants and other natural products with known active compounds and mechanisms of action

The search documented 40 plant families containing 65 plant species, two fungal families containing four mushroom species (*Ganoderma lucidum*, *G. frondosa*, and *G. pfeifferi*; and *Inonotus obliquus*), and one blue-green algae (*Nostoc ellipsosporum*), respectively, with known anti-HIV active compounds. Interestingly, some of the ethnopharmacological data that was downloaded (dating as far back as 1989) had not been included in previous reviews. The families (in alphabetical order) and species of plants and other natural products with known active compounds and their modes of action are listed in Table 1.

Plants and other natural products with known or unknown active compounds and/or known or unknown mechanisms of action against HIV and AIDS-related opportunistic infections

A total of 30 plant species from 15 families were reported to have activities against HIV or HIV related sicknesses, and also against other diseases such as cancer (*Mangifera indica*) (Muaza et al., 1995), fungal infections (*Fragaria virginiana* and *Potentilla simplex*) (Webster et al., 2008), inflammation, and general microbial opportunistic infections (Iwu, 1993) (Table 2). However, their active compounds or modes of actions were not well understood. Aqueous and ethanol extracts of *Baissea axillaries* (Abere and Agoreyo, 2006), *Melissa officinalis* (Geuenich et al., 2008), and *Leonotis leonurus* (Klos et al., 2009) were reported to posses activities against HIV-1.

Several unknown active components of plants belonging to the family Euphorbiaceae (*Jatropha curcas*, *J. multifida*, *Spirostachys africana* and *Trigonostema xyphophylloides*) were found to posses anti-HIV activities by inhibiting HIV-1 cell entry (Park et al., 2009). *Jatropha* species were also found to inhibit HIV-induced cytopathic effects (Igbinosa et al., 2009; Matsuse et al., 1999) but caused genotoxicity during ART (Muaza et al., 1995; Van den Beukel et al., 2008) (Table 2).

DISCUSSION

Several chemical compounds were found to interfere with

Table 1. Plants and other natural products with known active compounds and modes of action against HIV

Family species	Active constituents	Mechanism of action	References
Acanthaceae			
Andrographis paniculata	Aqueous extracts of leaves	Inhibition of HIV protease and reverse transcriptase.	Otake et al., 1995
	Diterpene lactones (andrographolide)	Inhibit HIV-infected cells from arresting in G2 phase in which viral replication is optimal. Inhibit cell-to-cell transmission, viral replication and syncytia formation in HIV-infected cells	Calabrese, 2000
Aceraceae			
Acer okamotoanum	Flavonoid gallate ester	Anti-HIV-1 integrase activity	Kim et al., 1998
Agaricaceae			
Lentinus edodes (Berk.) Singer	Sulfated lentinan	Prevent HIV-induced cytopathic effect	Suzuki et al., 1989
Amaryllidaceae			
Galanthus nivalis L.	Plant lectins: *G. nivalis* agglutinin (GNA),	Potent inhibitors that stop the spread of HIV among lymphocytes by targeting gp120 envelope glycoprotein; most prominent anti-HIV activity is found among MBLs;	Saidi et al., 2007
Hippeastrum hybrids	*Hippeastrum* hybrid agglutinin (HHA), and monocot mannose-binding lectins (MBLs)	GNA has specificity for terminal α(1-3)-linked mannose residues; HHA recognizes both terminal and internal α(1-3)- and α(1-6)-linked mannose residues	Cited in Balzarini et al., 2004
Anacardiaceae			
Rhus succedanea L.	Biflavonoids, robustaflavone and hinokiflavone	Strong inhibition of the polymerase of HIV-1 reverse transcriptase	Lin et al., 1997
Ancistrocladaceae			
Ancistrocladus korupensis	Michellamines A and B	Anti-HIV -1 and anti-HIV-2 activities. Act at early stage of the HIV life cycle by inhibiting reverse transcriptase and at later stages by inhibiting cellular fusion and syncytium formation	Boyd et al., 1994 Manfredi et al., 1991
Annonaceae			
Polyalthia suberosa	Lanostane-type triterpene, suberosol	Anti-HIV replication activity in H9 lymphocytes cells *in vitro*	Li et al., 1993
Apiaceae			
Lomatium suksdorfii	Suksdorfin	Suppress HIV-1 viral replication in H9 lymphocyte cells	Yu et al., 2007

Table 2. Contd.

Araliaceae			
Panax ginseng C.A. Meyer	-	Increases CD4/8 cells; has serious side effects	Sung et al 2005
Areschougiaceae			
Agardhiella tenera (J. Agardh) F. Schmitz	Sulfonated polysaccharides	inhibit the cytopathic effect of human immunodeficiency virus type 1 (HIV-1) and type 2 (HIV-2) in MT-4 cells	Witvrouw et al., 1994
Asphodelaceae			
Bulbine alooides	Aqueous and ethanol extracts	Extracts of *B. alooides*, *H. sobolifera*, Extracts of *B. alooides* retained HIV-1 protease inhibition after dereplication to remove non-specific tannins/ polysaccharides	Klos et al., 2009
Asteraceae			
Achyrocline satureioides (Lam.) DC (Marcela);	Two dicaffeoylquinic acids: 3,5-dicaffeoylquinic acid, and 1-methoxyoxalyl-3,5-dicaffeoylquinic acid	Potent and irreversible inhibition of HIV-1 integrase	Zhu et al., 1999; Robinson et al., 1996
Arctium lappa (Burdock)	Wedelolactone (a coumarin derivative); orobol (an isoflavone derivative)	Inhibit HIV-1 replication, block cell-to-cell transmission of HIV-1	Yao et al., 1992
Boraginaceae			
Arnebia euchroma (Royle) Jonst	Monosodium and mono-potassium salts of isomeric caffeic acid tetramer	Inhibitory activity against HIV replication in acutely infected H9 cells	Kashiwada, 1995
Cannabaceae			
Humulus lupulus	Xanthohumol	HIV-1 inhibitory activity as well as HIV-1-induced cytopathic effects, production of viral p24 antigen and reverse transcriptase in C8166 lymphocytes.	Wang et al., 2004
Celastraceae			
Tripterygium hypoglaucum	Triptonine A and Triptonine B	Exhibit potent *in vitro* anti-HIV activity	Duan et al., 2000
Celastrus hindsii	Celasdin B	Anti-HIV replication activity in H9 lymphocytes cells *in vitro* Inhibit HIV replication in H9 lymphocytes	Kuo and Kuo, 1997
Tripterygium wilfordii Hook F	Diterpene lactones (nortripterifordin)		Duan et al., 1999

Table 1 Contd.

Clusiaceae			
Callophyllum cordato-oblongum	Cordatolide A and B	Inhibitory activity against HIV-1 replication	Dharmaratne et al., 2002 Buckheit 1999
	(+)-calanolide A	Inhibit cytopathic effects of HIV-1 in T-cell lines, including both CEM-SS cells and MT-2 cells	Xu et al., 2000
Marila laxiflora	Laxofloranone	Novel non-nucleoside reverse transcriptase inhibitor with potent anti-HIV-1 activity	Bokesch et al., 1999
Symphonia globulifera	Guttiferone A	Inhibition of the cytopathic effects of in vitro HIV infection	Gustafson, 1992
Hypericum perforatum L.	Hypericin, 3-hydroxy lauric acid	Cytoprotection of CEM-SS cells from HIV-1 infection; inhibition of HIV-1 replication; anti-HIV activity with little or no cytotoxicity	Birt et al., 2009 Maury et al 2009
Combreataceae			
Combretum molle R.Br. ex G. Don	Gallotannin	Inhibits RNA-dependent-DNA polymerase activity of HIV-1 reverse transcriptase.	Bessong et al., 2005
Terminalia chebula	Gallic acid and galloyl glucose	Inhibits ribonuclease H activity of reverse tran.scriptase; also has HIV integrase inhibitory activity.	Ahn, 2002
Dipterocarpaceae			
Monotes africanus	Prenylated flavonoids,	HIV-inhibitory activity in XTT-based, whole cell screen	Meragelman et al., 2001
Vatica astrotricha	6,8-diprenylaromadendrin and 6,8-diprenylkaempferol Prostratin, a 12-deoxyphorbol	Inhibition of HIV-1 entry; blocks HIV-1 replication at the entry step	Park et al., 2009 Birt et al., 2009
Euphorbiaceae			
Homalanthus nutan (G. Forst.)	Prostratin, a 12-deoxyphorbol	Putative mechanisms are: down regulation of CD4 expression in CEM and MT-2 cells, interference in protein kinase C enzyme pathway.	Cited in review by Vermani and Garg, 2002
		Prostratin is a potent activator of HIV replication and expression in latently infected T-cells; hence it is used to flush out latent HIV from lymph nodes during antiretroviral therapy	Johnson et al., 2008; Cited in review by Gupta et al, 2005

Table 1. Contd.

Fabaceae			
Acacia auriculiformis A. Cunn. ex. Benth.	Saponins, alkaloids	Anti-HIV activity	Mandal et al 2005; Singh et al., 2005; Parekh and Chanda 2006 ;
Peltophorum africanum Sond.	Gallotannin	Inhibits RNA-dependent-DNA polymerase activity of HIV-1 reverse transcriptase; inhibits ribonuclease H activity of reverse transcriptase	Bessong et al., 2005
Ganodermataceae			
Ganoderma lucidum	Ganoderiol and Ganodermanontriol, Ganoderic acid.	Inhibition of HIV-1 induced cytopathic effect in MT-4 cells.	Lindequist et al., 2005
Ganoderma species including. *G. lucidum*, *G. frondosa*, and *G. pfeifferi*	Several triterpenes such as ganoderiol F (6a), ganodermanontriol (7a), and ganoderic acid B (8a); immunomodulators such as (1-6)-β-D-glucan, heteropolysaccharides, polysaccharide-protein complex, glycoproteins, (1-3)-β-D-glucan with (1-6)-β-D-glucosyl branches, complex mixture of polysaccharides and lignin, heteroglucans, glucoronoxylomannans, and lectins	Inhibits HIV-1 protease. Antiviral agents against HIV-1: ganoderiol F (6a) and ganodermanontriol (7a) inhibit HIV-1 induced cytopathic effect; ganoderic acid B (8a) inhibits HIV-1 protease. Lignins inhibit HIV protease; Sulfated (1-3)-β-D-glucan with (1-6)-β-D-glucosyl branches prevent HIV-induced cytopathic effect; polysaccharide-protein complexes inhibit HIV-1 gp120 binding to CD4 receptors and reverse transcriptase	Cited in review by Lindequist et al., 2005
Gentianaceae			
Swertia franchetiana	Flavonone-xanthone glucoside	Inhibits HIV-1 reverse transcriptase	Wang et al., 1994
Guttiferae			
Calophyllum teysmannii Miq.	(-)-calanolide B	Less activity than the A form	Cited in review by Gupta et al., 2005
Hymenochaetaceae			
Inonotus obliquus	Water-soluble lignins	Inhibit HIV-1 protease	Ichimura et al., 1998
Hypericaceae			
Garcinia speciosa	Protostanes, garcisaterpenes A and C	Inhibitory activity against HIV-1 reverse transcriptase	Rukachaisirikul, 2003

Table 1 Contd.

Species	Compound	Activity	Reference
Lamiaceae			
Melissa officinalis	Rosmarinic acid	Inhibit HIV-1 virions carrying different X4 and R5 HIV-1 Envs as well as the heterologous VSV-G, interfers with MoMLV infection; inhibit fusion of HIV-1 particles with cells.	Geuenich et al., 2008
Mentha piperita L. *Prunella vulgaris* L.	Sulfonated polysaccharides	Inhibit HIV-1 particles carrying R5 Envs; inhibit HIV-1 replication; target HIV-1 virion (virucidal).	Hauber et al., 2009; Yao et al., 1992;
Sideritis akmanii	Linearol	Anti-HIV replication in H9 lymphocyte cells	Bruno et al., 2002
Magnoliaceae			
Magnolia spp.	Neolignans e.g. magnolol 1 and honokiol 2	Antioxidant, antidepressant, induces apoptosis in tumor cells, weak anti-HIV-1 activity	Amblard et al., 2007
Menispermaceae			
Epinetrum villosum (Exell) Troupin	Cycleanine, a bisbenzylisoquinoline alkaloid	Acts against HIV-2 but is 10-times less active against HIV-1	Otshudi et al., 2005
Stephania cepharantha	Cepharanthine	Potently inhibit HIV replication	Ma et al., 2002
Musaceae			
Musa acuminata	BanLec, a jacalin-related lectin	Binds to glycosylated viral envelopes and blocks viral entry, hence is a good microbicide; potent inhibitor of HIV-1 replication	Swanson et al., 2010
Myrothamnaceae			
Myrothamnus flabellifolius (Welw.)	Polyphenols, gallotannins, 3,4,5-tri-O-galloylquinic acids	Polyphenols protect cell membranes against free radical-induced damage; gallotannins have anti-burn properties; 3,4,5-tri-O-galloylquinic acids have anti-HIV reverse transcriptase activity	Moore et al., 2005; Moore et al., 2007
Nostocaceae			
Nostoc ellipsosporum	Cyanovirin-N, an 11 KDa anti-HIV-1 protein	Inhibits HIV-1 replication through its vivid binding to HIV-1 gp120 and as a result, inactivates the viruses and blocks the fusion of viruses to the cell membrane	Gustafson et al., 1997; cited in Balzarini et al., 2004
Phyllanthaceae			
Phyllanthus niruri L.	Niruriside	Specific inhibitor of REV protein/RRE RNA	Qian-Cutrone, 1996
Physalacriaceae			
Flammulina velutipes (Curt.: Fries) Singer	Velutin	Inhibition of HIV-1 reverse transcriptase	Wang and Ng, 2001

Table 1 Contd.

Family / Species	Compound	Activity	Reference
Phytolaccaceae *Phytolacca Americana* L	Pokeweed antiviral protein (PAP)	PAP has broad spectrum antiviral activity against HIV; it is used as an antiviral microbicide; its anti-HIV-1 activity is superior *in vitro* compared to zidovudine (AZT)	Tumer et al., 1997; Uckun et al., 1998; D'cruz et al., 2004
Pinaceae *Pinus parviflora* Siebold & Zucc	PC6, an extract from cones, has potent immune modulatory activities	Inhibits replication of HIV-1 via reverse transcriptase and modification of microenvironment	Tamura et al., 1991
Pinaceae *Pinus parviflora* Siebold & Zucc	PC6, an extract from cones, has potent immune modulatory activities	Inhibits replication of HIV-1 via reverse transcriptase and modification of microenvironment	Tamura et al., 1991
Punicaceae *Punica granatum* L	PJ-S21	PJ-S21 inhibits binding of gp120 III-CD4 complexes to cells expressing CXCR4; inhibitor of X4 and R5 virus binding to the cellular receptor CD4 and co-receptors CXCR4/CCR5	Neurath et al., 2004
Rosaceae *Crataegus pinatifida*	Uvaol and ursolic acid	Inhibitory activity against HIV-1 protease	Min et al., 1999
Geum japonicum	Maslinic acid	Inhibitory activity against HIV-1 protease	Xu et al., 1996
Rubiaceae *Oldenlandia affinis*	Circulins	Anti-HIV activity	Cited in Jennings et al., 2001
Palicourea condensate	Palicourein	Inhibits the *in vitro* cytopathic effects of HIV-1 infection of CEM-SS cells	Bokesch et al., 2001
Rutaceae *Citrus paradisi*	6',7'-dihydroxy-bergamottin	Enhances bioavailability of HIV protease inhibitor (e.g. saquinavar) by inhibiting cytochrome P450 iso-enzyme 3A4 in liver and gut	Kupferschmidt et al., 1998
Citrus spp.	Limonin and nomilin	Inhibit HIV-1 protease. Inhibit the production of HIV-1 p-24 antigen in infected monocytes and macrophages	Battinelli et al., 2003
Clausena excavate	Limonoid (clausenolide-1-ethyl ether)	HIV inhibitory activity in 1A2 cell line in syncytium assay	Sunthitikawinsakul et al., 2003
Euodia roxburghiana (Cham.) Benth.	Buchapine	Protect CEM-SS cells from cytopathic effects of HIV-1 *in vitro*	McMormick et al., 1996
Toddalia asiatica (L.) Lam.	Nitidine	Inhibit HIV-reverse transcriptase	Tan et al., 1991

Table 1 Contd.

Sapindaceae			
Xanthoceras sorbifolia	Oleanolic acid	Inhibit HIV-1 replication in acutely infected H9 cells.	Sakurai, 2004
Schisandraceae			
Schisandra sphaerandra f.	Nigranoic acid	Inhibit HIV-1/2 reverse transcriptase	Sun, 1998
Schizymenia pacifica (Kylin) Kylin	Sulfated polysaccharide	Inhibit reverse transcriptase	Nakashima, 1987
Kadsura lancilimba	Triterpene lactone, lancilactone	Inhibitory activity against HIV replication in H9 lymphocytes	Chen et al., 1999
Symplocaceae			
Symplocos setchuensis	Harmine	Inhibit HIV replication in H9 lymphocyte cells.	Ishida et al., 2001
Theaceae			
Camellia japonica	Camellia-tannin H	HIV-1 protease inhibitory activity	Park, 2002
Camellia sinensis)	Polyphenol epigallocatechin-3-gallate	Inhibit semen-derived enhancer of virus infection (SEVI) activity and abrogates semen-mediated enhancement of HIV-1 infection	Hauber et al., 2009
Zingiberaceae			
Curcuma species including *C. longa* L.	Curcumin	Inhibits HIV-1 integrase, HIV-1 and HIV-2 protease, and HIV-1 Long Terminal Repeat-directed gene expression	Itokawa et al., 2008

HIV entry into cells while others were active against HIV reverse transcriptase, integrase, protease, and general replication. Some phytochemicals were also potent activators of HIV replication and expression in latently-infected T-cells, and others were known to inhibit syncytia formation (Table 1). Most of the entry inhibitors were lectins such as: agglutinins from *Galanthus nivalis* and *Hippeastrum* hybrid stopped the spread of HIV among cells (Saidi et al., 2007); BanLec, a jacalin-related lectin that binds to glycosylated viral envelopes blocked HIV-1 entry into cells (Swanson et al., 2010); cyanovirin, an 11

KDa protein isolated from *Nostoc ellipsosporum*, targeted gp120 proteins and blocked fusion of HIV-1 to lymphocyte membranes (Gustafson et al., 1997; Balzarini et al., 2004); glycoprotein complexes from *Ganoderma* mushrooms inhibited HIV-1 gp120 binding to CD4 cells (Lindequist et al., 2005); a code-named compound, PJ-S21, from *Punica granatum* inhibited the binding of gp120 to cells expressing CXCR4 receptors (Neurath et al., 2004); and *Phytolacca americana* pokeweed antiviral protein (PAP), a 29 KDa ribosome-inactivating protein that removes adenine from rRNA of prokaryotic and eukaryotic

ribosomes was found to be a potent microbicide (Turner et al., 1997; Uckun et al., 1998; D'cruz et al., 2004). Other active constituents included: diterpene lactones (Calabrese, 2000) and a coumarin named wedelolactone (Yao et al., 1992) inhibited cell-to-cell transmission of HIV-1; prostratin, a 12-deoxyphorbol, inhibited HIV-1 entry into lymphocytes (Park et al., 2009); and rosmarinic acid isolated from *Melissa officinalis* inhibited fusion of HIV-1 to cells (Geuenich et al., 2008). Twenty-eight different chemical compounds were known to be active against HIV reverse transcriptase (Table 1). Some HIV reverse

Table 2. Plants and other natural products with known or unknown active compounds and/or known or unknown mechanisms of action against HIV and AIDS-related opportunistic infections

Family *species*	Active constituents	Mechanism of action	References
Anacardiaceae			
Mangifera indica L.	-	Anti-cancer	Muaza et al., 1995
Apocynaceae			
Baissea axillaries Hua	Unknown aqueous and ethanol extracts contain alkaloids, tannins and cyanogenetic glycosides	Antimicrobial activity against clinical strains associated with HIV/AIDS-diarrhoea; shows toxicity at dose of 500 mg/kg	Abere and Agoreyo, 2006
Asteraceae			
Artemisia absinthium L.	Cardamonin, a known 2',4'-dihydroxy-6'-methoxychalcone	Anti-inflammatory, anti-cancer, antioxidant, antiviral, antifungal, antibiotic	Hatziieremia et al., 2006
Baccharis dracunculifolia	Brazilian propolis called Alecrim propolis and red coloured propolis from Cuba and Venezuela: contain phenolics, triterepenoids, isoflavonoids, prenylated benzophenones and a naphthoquinone epoxide	Antimicrobial activity and radical scavenging activity; an inseparable mixture of double bond isomers 3-methyl-2-butenyl and 3-methyl-3-butenyl is the active anti-HIV principle	Trusheva et al., 2006
Eclipta prostrate L. Hassk.		Anti-HIV-1 activity against integrase	Tewtrakul et al., 2007
Balanophoraceae			
Thonningia sanguinea Vahl	-	Stops HIV/AIDS-related diarrhoea, skin diseases and mycoses	Quattara et al., 2007
Betulaceae			
Alnus viridis DC *Betula alleghaniensis* Britt.	-	Anti-yeast activity	Webster et al., 2008
Cucurbitaceae			
Momordica charantia	1-monopalmitin, a simple monoglyceride with an abundant fatty acid chain	Major inhibitor of P-glycoprotein	Konishi et al., 2004

Table 2. Contd.

Euphorbiaceae			
Jatropha curcas L.	-	Inhibits HIV induced cytopathic effect; moderate cytoprotective effect against HIV.	Igbinosa et al., 2009; Matsuse et al., 1999
Jatropha multifida L.	-	Genotoxicity during HAART	Muaza et al., 1995
Spirostachys africana Sonder	-	Genotoxicity during HAART	Van den Beukel et al., 2008
Trigonostema xyphophylloides	-	Inhibition of HIV-1 entry	Park et al., 2009
Geraniaceae			
Pelargonium sidoides (DC), and *P. reniforme* (Curt)	-	Anecdotal ethnomedicinal uses reported for management of HIV/AIDS-related opportunistic infections in Southern Africa	Cited in review by Reihling, 2008
Fabaceae			
Castanospermum australe	Castanospermine	Blocks glycoprotein processing via inhibition of glucosidase l located in the endoplasmic reticulum.	Ruprecht, 1989
Peltophorum africanum	(+)-Catechin (flavonoid), bergenin (a C-galloyl-glycoside), betulinic acid	Used to treat diarrhoea, dysentery, sore throat wounds, HIV/AIDS, STIs,; betulinic acid had highest anti-HIV activity	Theo et al., 2009; Kashiwada et al., 2000
Sutherlandia frutescens	L-canavanine, GABA, and D-pinitol	L-canavanine has anti-viral activity against HIV and interacts with efflux of nevirapine; D-pinitol has been suggested as a treatment for wasting in cancer and AIDS patients though evidence is scanty	Brown et al., 2008
Guttiferae			
Garcinia kola Heckel	Biflavonoids, xanthones, benzophenones.	Anti-inflammatory, anti-microbial, antihepatotoxic, and antiviral activities	Iwu, 1993
Hypericum chinense L. var. salicifolium	Biyouyanagin A, which contains sesquiterpene, cyclobutane and spirolactone moieties	Significant activity against HIV	Tanaka et al., 2005

Table 2. Contd.

Hypoxidaceae			
Hypoxis hemerocallidea and *H. colchicifolia*	Hypoxoside, glycosides sterols, and	Hypoxoside is converted to its aglycone, *rooperol*, a potent antioxidant; also affects cytochrome P-450 system, P-glycoprotein, and the pregnane X receptor (PXR)	Cited in review by Mills et al., 2005
Hypoxis rooperi		Used by persons with moderate or advanced AIDS	Babb et al., 2004
Lamiaceae			
Melissa officinalis	Aqueous extracts	Exhibited a high and concentration-dependent activity against HIV-1 infection in immune cells; active against virions carrying diverse envelopes X4 and R5. Ethanol extract of *L. leonurus* inhibit HIV-1 by 33%; and	Geuenich et al., 2008
Leonotis leonurus	aqueous and ethanol extracts	*L. leonurus* inhibit HIV-1 reverse transcriptase in a non-specific way due to tannins/polysaccharides; Extracts of *B. alooides* and *L. leonurus* retained HIV-1 protease inhibition after dereplication to remove non-specific tannins/polysaccharides	Klos et al., 2009
Rosaceae			
Fragaria virginiana Duchesne	-	Antifungal potential	Webster et al., 2008
Potentilla simplex Michx.	-	Antifungal potential	Webster et al., 2008
Sterculiaceae			
Sterculia Africa (Lour) Fiori	-	Genotoxicity during HAART	Moshi et al., 2007

transcriptase inhibitors included: michellamines (Boyd et al., 1994); triterpene (Li et al., 1991); coumarins and isoflavone derivatives (Yao et al., 1992); caffeic acid tetramers (Kashiwada, 1995); (+)-calanolide A (Xu et al., 2000); hypericin and 3-hydroxy lauric acid (Birt et al., 2009); gallotannin (Bessong et al., 2005); flavonone-xanthone glucoside (Wang et al., 1994); linearol (Bruno et al., 2002); catechins 1-5 (Moore and Pizza, 1992); cepharanthine (Ma et al., 2002); galloylquinic acids (Moore et al., 2005; 2007); velutin (Wang and Ng, 2001); nitidine (Tan et al., 1991); oleanolic acid (Sakurai, 2004; nigranoic acid (Sun, 1998); sulfated polysaccharides (Nakashima, 1987); triterpene lactone (Chen et al., 1999); and harmine (Ishida et al., 2001).

Four of the identified active compounds were known to be HIV integrase inhibitors: flavonoid gallate ester from *Acer okamotaanum* of the Aceraceae family (Kim et al., 1998); dicaffeoyl-quinic acids from *Achyrocline satureioides* of the Asteraceae family (Robinson et al., 1996; Zhu et al., 1999); gallic acid and galloyl glucose from *Termainalia chebula* of the Combreataceae family (Bessong et al., 2005); and curcumin from *Curcuma* species in the Zingiberaceae family (Itokawa et al., 2008).

Six active compounds were observed to be HIV protease inhibitors: ganoderiol, ganodermanontriol, and ganoderic acid B (a triterpene) from *Gonoderma* mushrooms (Lindequist et al., 2005); lignins from *Gonoderma* and *Inonotus obliquus*, commonly known as chaga mushroom belonging to the Hymenochaetaceae family (Ichimura et al., 1998); uvaol and ursolic acid from *Crataegus pinatifida* (Min et al., 1999), and maslinic acid (a triterpene acid) from the plant *Geum japonicum* (Xu et al., 1996), both from the Rosaceae family; limonoids (Manners, 2007), including limonin and nomilin, secondary metabolites from citrus fruit species belonging to the Rutaceae family (Battinelli et al., 2003); and curcumin from *Curcuma* species (Itokawa et al., 2008). Curcumin was also shown to be active against HIV-1 integrase (Itokawa et al., 2008).

Many active compounds were known to inhibit general HIV replication (Table 1). Diterpene lactones from *Andrographis paniculata* (Otake et al., 1995) and *Tripterrygium wilfordii* (Duan et al., 1999); triterpene lactone and lancilactone from *Kadsura lancilimba* (Chen et al., 1999); biflavonoids from *Rhus succedanea* (Lin et al., 1997); lanostane-type triterpenes from *Polyalthia suberosa* (Li et al., 1993); suksdorfin from *Lomatium suksdorfii* (Yu et al., 2007); wedelolactone (a coumarin) and orobol (an isoflavone derivative) from *Arctium lappa* (Yao et al., 1992); caffeic acid tetramers from *Arnebia euchroma* (Kashiwada, 1995); celasdin from *Celestrus hindsii* (Kuo and Kuo, 1997); cordatolides from *Callophyllum cordato-oblongum* (Dharmaratne et al., 2002); hypericin and 3-hydroxy lauric acid from *Hypericum perforatum* (Birt et al., 2009); sulfonated polysaccharides from *Mentha piperita* and *Prunella vulgaris* (Hauber et al., 2009); linearol from *Sideritis akmanii* (Bruno et al., 2002); cepharanthine from *Stephania cepharantha* (Ma et al., 2002); cyanovirin from *Nostoc ellipsosporum* (Gustafson et al., 1997); oleanolic acid from *Xanthoceras sorbifolia* (Sakurai, 2004); and harmine from *Symplocos setchuensis* (Ishida et al., 2001). Prostratin, a 12-deoxyphorbol from *Homalanthus nutan*, was a potent activator of HIV replication and expression in latently-infected T-cells (Gupta et al., 2005; Johnson et al., 2008). Prostratin was therefore used to bring out latent HIV from lymph nodes so that the virus was exposed to lethal concentrations of the drugs (Gupta et al., 2005; Johnson et al., 2008).

Some active compounds were found to inhibit syncytia formation, a property of HIV that makes infected and healthy CD4 cells to fuse and form one giant cell with as many as 500 nuclei. Syncytia-inhibiting compounds included: diterpene lactones (Calabrese, 2000); michellamines A and B (Boyd et al, 1994; Manfredi et al., 1991); and limonoids (Sunthitikawinsakul et al., 2003). Eight natural compounds prevented HIV-induced cytopathic effect: sulfated lentinan (Suzuki et al., 1989); sulfonated polysaccharides (Witvrouw et al., 1994);

xanthohumol (Wang et al., 2004); cordatolides (Dharmaratne et al., 2002); laxofloranone (Bokesch et al., 1999); ganoderiol, ganodermamontriol, and ganoderic acid (Lindequist et al., 2005); sulfated (1-3)-β-D-glucan with (1-6)- β-D-glucosyl branches (Lindequist et al., 2005); and palicourein (Bokesch et al., 2001).

Some plant-derived compounds were reported to possess activities against HIV-related symptoms. Castanospermine from *Castanospermum australe* blocked glycoprotein processing via inhibition of glucosidase I located in the endoplasmic reticulum (Ruprecht, 1989). L-canavanine from *Sutherlandia frutescens* had anti-viral activity against HIV but interacted with the efflux of nevirapine (Brown et al., 2008). D-pinitol, also from *Sutherlandia frutescens,* had been suggested as a treatment for wasting in cancer and AIDS patients though evidence was scanty (Brown et al., 2008). (+)-Catechin (flavonoid), bergenin (a C-galloyl-glycoside) and betulinic acid from *Peltophorum africanum* were used to treat diarrhoea, dysentery, sore throat wounds, HIV/AIDS, and other sexually transmitted infections (Theo et al., 2009; Kashiwada et al., 2000).

Coumarins and naturally occurring benzopyrene derivatives from several plant species were reported to possess antioxidant, anti-inflammatory, antithrombotic, antiviral, anticarcinogenic, antiallergic, hepatoprotective, and anti-HIV properties (Kostova et al., 2006). Furthermore, 50 different compounds (belonging to tannins, terpenoids, flavonoids, flavones, alkaloids, coumarins, lignans, lignin-polysaccharide complexes, and lectins), found in 40 different plant species, were reported to inhibit adsorption, viral fusion, HIV reverse transcriptase, integrase, protease, snyctium formation, interference with cellular factors, and some unknown targets (Cowan, 1999).

Casternospermine, hypericin, pseudohypericin, interferons, glycyrrhizin, avarol and avarone inhibited replication of HIV-1 and other retroviruses (Lin et al., 1989). Gossypol, a polyphenolic bissesquiterpene isolated as a racemic mixture from cottonseed has selective activity against HIV-1 but the exact mechanism was unknown (Lin et al., 1989). Flavonoid derivatives inhibited HIV-1 reverse transcriptase (Moore and Pizza, 1992). Suramin inhibits HIV-1 reverse transcriptase in a non-specific way (Moore and Pizza, 1992). Sulphated polysaccharides were potent *in vitro* inhibitors of HIV-1 and HIV-2 adsorption, fusion or penetration, induced cytopathogenicity, and antigen expression (Talyshinsky et al., 2002). They also inhibited reverse transcriptase and RNAase H, essential enzymes for retrovirus replication (Talyshinsky et al., 2002).

A wide variety of organisms including single-celled microbes, insects and other invertebrates, plants, amphibians, birds, fish, and mammals including humans had been shown to possess antiviral peptides that suppress HIV gene expression. Magainin from frogs and cecropin from insects suppressed HIV gene expression (Jenssen et al., 2006). Dermaseptin from frogs disrupted HIV membranes while indolicidin from bovine inhibited

HIV integrase (Jenssen et al., 2006; Kim et al., 2002). Gordon and Ramanowski (2005) found that P113, isolated from human saliva, was a good mouth-rinse against *Candida albicans* in HIV/AIDS patients (Table 2). Plants such as *Viola yedoensis, Epimedium grandiflorum, Glycyrrhiza uralensis* and *Castanospermum australe* were used to manage HIV/AIDS opportunistic infections (WHO, 1989b), but their pharmacological interactions with ART were not well understood. Kisangau et al. (2007) documented 75 plant species (in 66 genera and 41 families) used in the management of HIV/AIDS opportunistic infections in one district of Tanzania, however their study did not report on the compounds that are active against HIV and did not report on their modes of action against HIV.

Conclusion

The literature survey revealed several anti-HIV active compounds such as terpenoids, coumarins, polyphenols, tannins, proteins, alkaloids, and biflavonoids that inhibit various steps of the HIV life cycle. These active compounds were isolated from 55 families of plants and other natural sources such as mushrooms (Ganodermaceae), cyanobacteria, and marine organisms. Phylogenetic analysis and other bioinformatics tools may shed light on unidentified but related plants and other organisms that may contain similar active compounds. Primary data of known active compounds and mechanisms of action were available from studies mostly done outside Africa. Most of the studies done in Africa were inconclusive about either the active compounds or mechanisms of action, or both. Pharmacological interactions of unknown active ingredients from herbal medications remain a source of great medical concern. Throughout the survey, it was clear that although Africa had a wealth of medicinal plants, most of the research on screening of plants and isolation of active compounds was carried out elsewhere in Asia, Europe and the Americas. Lack of long-term funding and infrastructure have conspired to exacerbate Africa's dependency on overseas laboratories for screening of plants for active compounds, *in vitro* tests against HIV, and mechanisms of action. Such a situation is a perfect recipe for biopiracy. There is also an urgent need to fast-track HIV/AIDS clinical trials of candidate drugs developed from novel compounds isolated from plants and other natural sources. This will ensure that the millions of people that require HIV/AIDS treatment will have access to newer, more effective, and less toxic drugs.

ACKNOWLEDGEMENTS

We are thankful to NEPAD/SANBio and UNAM's Research and Publications Committee for financial assistance towards research on traditional medicines for HIV/AIDS treatment in Namibia. We are also grateful to UNAM's Multi-Research Center (MRC) for financial support toward dissemination of research findings and for funding publication fees for this article. KC is chair of the UNAM/NEPAD Steering committee on Validation of traditional medicines for HIV/AIDS treatment in Namibia.

REFERENCES

Abere TA, Agoreyo FO (2006). Antimicrobial and toxicological evaluation of the leaves of Baissea axillaries Hua used in the management of HIV/AIDS patients. BMC Compl. Alten. Med. 6:22, doi:10.1 186/1472-6882-6-22.

African Biosciences Initiative (2005). Business Plan 2005-2010. NEPAD Office of Science and Technology, Pretoria.

Ahn MJ (2002). Inhibition of HIV-1 integrase by galloyl glucoses from *Terminalia chebula* and flavonol glucoside gallates from *Euphorbia pekinensis*. Planta Med. 68: 454–457.

Amblard F, Govindarajan B, Lefkove B, Rapp KL, Detoria M, Arbiser JL, Schinazi RF (2007). Synthesis, cytotoxicity and antiviral activities of new neolignans related to honokiol and magnolol. Bioorg. Med. Chem. Lett. 17(16): 4428-4431.

Amon JJ (2008). Dangerous medicines: unproven AIDS cures and counterfeit antiretroviral drugs. Globalization and Health, 4:5 doi:10.1186/1744-8603-4-5. http://www.globalizationandhealth.com/content/4/1/5.

Asres K, Bucar F, Kartnig T, Witvrouw M, Pannecouque C, De Clercq E (2001). Antiviral activity against human immunodeficiency virus type 1 (HIV-1) and type 2 (HIV-2) of ethnobotanically selected Ethiopian medicinal plants. Phytother. Res. 15: 62–69.

Babb DA, Pemba L, Seatlanyane P, Charalambous S, Churchyard GJ, Grant AD (2004). Use of traditional medicine in the era of antiretroviral therapy: experience from South Africa. Int. Conf. AIDS, Bangkok, Thailand, 11-16 July 2004.

Balzarini J, Van Laethem KV, Hatse S, Vermeire K, De Clercq ED, Peumans W, Van Damme E, Vandamme A-M, Bolsmstedt A, Schols D (2004). Profile of resistance of human immunodeficiency virus to mannose-specific plant lectins. J. Virology, 78(19): 10617-10627.

Battinelli L, Mengoni F, Lichtner M, Mazzanti G, Saija A, Mastroianni CM, Vullo V (2003). Effect of limonin and nomilin on HIV-1 replication on infected human mononuclear cells. Planta Med. 69: 910–913.

Bessong PO, Obi CL, Andreola ML, Rojas LB, Pouysegu L, Igumbor E, Meyer JJ, Quideau S, Litvak S (2005). Evaluation of selected South African medicinal plants for inhibitory properties against human immunodeficiency virus type 1 reverse transcriptase and integrase. J. Ethnopharmaco. 99(1): 83-91.

Birt DF, Widrlechner MP, Hammer KDP, Hillwig ML, Wei J, Kraus GA, Murphy PA, McCoy J, Wurtele, ES, Neighbors JD, Wiemer DF, Maury WJ, Price JP (2009). *Hypericum* in infection: identification of anti-viral and anti-inflammatory constituents. Pharm. Biol. 47(8): 774–782.

Bokesch HR, Pannell LK, Cochran PK (2001). A novel anti-HIV macrocyclic peptide from *Palicourea condensata*. J. Nat. Prod. 64: 249–250.

Boyd MR, Hallock YF, Cardellina II JH, Manfredi KP, Blunt JW, McMahon JB, Buckheit Jr RW, Bringmann G, Schaeffer M (1994). Anti-HIV michellamines from *Ancistrocladus korupensis*. J. Med. Chem. 37(12): 1740-1745.

Brown L, Heyneke O, Van Wyk JPH, Hamman JH (2008). Impact of traditional medicinal plant extracts on antiretroviral drug absorption. J. Ethnopharmaco. 119(3): 588-592.

Bruno M, Rosselli S, Pibiri I, Kilgore N, Lee KH. (2002). Anti-HIV agents from the *ent*-kaurane diterpenoid linearol. J. Nat. Prod. 65: 1594–1597.

Buckheit RW (1999). Unique anti-human immunodeficiency virus activities of the non-nucleoside reverse transcriptase inhibitors calanolide A, costatolide, and dihydrocostatolide. Antimicrob. Agents Chemother. 43: 1827–1834.

Calabrese C (2002). A phase I trial of andrographolide in HIV positive patients and normal volunteers. Phytother. Res. 14: 333–338.

Cardellina II JH, Boyd MR (1995). Phytochemistry of Plants Used in Traditional Medicine. In K. Hostettmann, A. Martson, M. Mallard, M Hamburger (eds) Proceedings of the Phytochemical Society of Europe. Oxford University Press: New York: pp. 81–94.

Chen DF, Zang SX, Wang HK (1999). Novel anti-HIV lancilactone C and related triterpenes from Kadsura lancilimba. J. Nat. Prod. 62: 94–97.

Chinsembu KC (2009). Model and experiences of initiating collaboration with traditional healers in validation of ethnomedicines for HIV/AIDS in Namibia. J. Ethnobio. Ethnomed. 5:30 doi:10.1186/1746-4269-5-30.

Cos P, Maes L, Berghe DV, Hermans N, Pieters L, Vlietinck A (2004). Plant substances as anti-HIV agents selected according to their putative mechanism of action. J. Nat. Prod. 67(2): 284-293.

Cos P, Vanden Berghe D, Bruyne TD, Vlietinck A (2004). Plant substances as antiviral agents: an update (1997–2001). Cur. Organ. Chem. 7: 1163–1180.

Cowan MM (1999). Plant products as antimicrobial agents. Clin. Microbio. Rev. 12(4): 564-582.

D'cruz OJ, Waurzyniak B, Uckun FM (2004). Mucosa toxicity studies of a gel formulation of native pokeweed antiviral protein. Toxicologic. Pathol. 32(2): 212-221.

Dharmaratne HRW, Tan GT, Marasinghe GPK, Pezzuto JM (2002). Inhibition of HIV-1 reverse transcriptase and HIV-1 replication by Callophyllum coumarins and xanthones. Planta Med. 68: 86–87.

Duan H, Takaishi Y, Bando M, Kido M, Imakura Y, Lee KH (1999). Novel sesquiterpene esters with alkaloid and monoterpene and related compounds from Tripterygium hypoglaucum: A new class of potent anti-HIV agents. Tetrahedron Lett. 40: 2969.

Duan H, Takaishi Y, Imakura Y, Jia, Y, Li D, Consentino LM, Lee KH (2000). Sesquiterpene alkaloids from Tripterigium hypoglaucum and Tripterygium wilfordii: A new class of potent anti-HIV agents. J. Nat. Prod. 63: 357–361.

Fabricant DS, Farnsworth NR (2001). The value of plants used in traditional medicine for drug discovery. Environ. Health Persp. 109: 69–75.

Farnsworth NR (1994). Ethnopharmacology and drug development. Ethnobotany, drug development and biodiversity conservation-exploring the linkages. In Ciba Foundation Symposium (vol. 185) Ethnobotany and the Search for New Drugs. John Wiley and Sons: Chichester. pp. 42–59.

Fletcher CV, Kakuda TN, Collier AC (2002). Bone and joint infections. In: JT Dipiro, RL Talbert, GC Yee, GR Matzke, BG Wells, LM Posey (eds) Pharmacotherapy- a pathophysiologic approach. Mcgraw-Hill Medical Publishing Division, United States of America, 5th edition, pp. 2151–2174.

Geuenich S, Goffine C, Venzke S, Nolkemper S, Baumann I, Plinkert P, Reichling J, Kepper OT (2008). Aqueous extracts from peppermint, sage and lemon balm leaves display potent anti-HIV activity by increasing the virion density. Retrovirology 5:27, doi:10.1186/1742-4690-5-27. http://www.retrovirology.com/contents/5/1/27

Gordon YJ, Romanowski EG (2005). A review of the antimicrobial peptides and their therapeutic potential as anti-infective drugs. Curr. Eye Res. 30(7): 505-515.

Gupta R, Gabrielsen B, Ferguson SM (2005). Nature's medicines: traditional knowledge and intellectual property management. Case studies from the national Institutes of Health (NIH), USA. Curr. Drug Discov. Technol. 2(4): 203-219.

Gustafson KR (1992). The guttiferones, HIV-inhibitory benzophenones from Symphonia globulifera, Garcinia livingstonei, Garcinia ovalifolia and Clusia rosea. Tetrahedron Lett. 48: 10093–10102.

Gustafson KR, Sowder RC, Henderson LE, Cardellina JH, McMahon JB, Rajamani U, Pannell LK, Boyd MR (1997). Isolation, primary sequence determination, and disulfide bond structure of cyanovirin-N, an anti-HIV (human immunodeficiency virus) protein from the cyanobacterium Nostoc ellipsosporum. Biochem. Biophys. Res. Commun. 238: 223-228.

Hardon A, Desclaux A, Egrot M, Simon E, Micollier E, Kyakuwa M (2008). Alternative medicines for AIDS in resource-poor settings: insights from exploratory anthropological studies in Asia and Africa. J. Ethnobiol. Ethnomed. 4:16 doi.10.1 186/1746-4269-4-16.

http://www.ethnobiomed.com/content/4/1/16.

Hatziieremia S, Gray A, Ferro VA, Paul A, Plevin R (2006). The effects of cardamonin on liposaccharide-induced inflammatory protein production and MAP kinase and NFκB signalling pathways in monocytes/ macrophages. Brit. J. Pharmaco. 149: 188-198.

Hauber I, Hohenberga H, Holstermanna B, Hunsteinb W, Hauber J (2009). The main green tea polyphenol epigallocatechin-3-gallate counteracts semen-mediated enhancement of HIV infection. PNAS USA. 106(22): 9033–9038.

Homsy J, King R, Tenywa J, Kyeyune P, Opio A, Balaba D (2004). Defining minimum standards of practice for incorporating African traditional medicine into HIV/AIDS prevention, care and support: a regional initiative in Eastern and Southern Africa. J. Altern. Complement. Med. 10(5): 905-910.

Ichimura T, Watanabe O, Maruyama S. (1998). Inhibition of HIV-1 protease by water-soluble lignin-like substance from an edible mushroom. Biosci. Biotechnol. Biochem. 62: 575-577.

Igbinosa OO, Igbinosa EO, Aiyegoro OA (2009). Antimicrobial activity and phytochemical screening of stem bark extracts from Jatropha curcas L. African Journal of Pharmacy and Pharmacology 3(2): 058-062. http://www.academicjournals.org/ajpp.

Ishida J, Wang HK, Oyama M, Cosentino ML, Hu CQ, Lee KH (2001). Anti-AIDS agents, anti-HIV activity of harman, an anti-HIV principle from Symplocos setchuensis, and its derivatives. J. Nat. Prod. 64: 958–960.

Itokawa H, Shi Q, Akiyama T, Morris-Natschke SL, Lee K-H (2008). Recent advances in the investigation of curcuminoids. Chinese Medicine 3:11. doi.10.1186/1749-8546-3-11. http://www.cmjournal.org/contents/3/1/11.

Iwu MM (1993). Handbook of African medicinal plants. CRC Press: Boca Raton, Florida.

Jennings C, West J, Waine C, Craik D, Anderson M (2001). Biosynthesis and insecticidal properties of plant cylotides: the cyclic knotted proteins from Oldenlandia affinis. PNAS USA. 98(19): 10614-10619.

Jenssen H, Hamill P, Hancock REW (2006). Peptide antimicrobial agents. Clinic. Microbio. Rev. 19(3): 491-511.

Johnson HE, Banack S, Cox AP (2008). Variability in content of the anti-AIDS drug candidate prostratin in Samoan populations of Homalanthus nutans. J. Nat. Prod. 71: 2041-2044.

Kashiwada Y (1995). Anti-AIDS agents. Sodium and potassium salts of caffeic acid triterpenes from Arnebia euchroma as anti-HIV agents. J. Nat. Prod. 58: 392–400.

Kashiwada Y, Nagao T, Hashimoto A, Ikeshiro Y, Okabe H, Cosentino ML, Lee K-H (2000). Anti-AIDS agents 38. Anti-HIV activity of 3-O-acyl ursolic acid derivatives. J. Nat. Prod. 63(12): 1619-1622.

Kaya HO (2009). Indigenous knowledge (IK) and innovation systems for public health in Africa. In FA Kalua, A Awotedu, LA Kamwanja, JDK Saka (eds) Science, technology and innovation for public health in Africa. NEPAD Office of Science and Technology: Pretoria. pp: 95-109.

Kayombo EJ, Uiso FC, Mbwambo ZH, Mahunnah RL, Moshi MJ, Mgonda YH (2007). Experience of initiating collaboration of traditional healers in managing HIV and AIDS in Tanzania. J. Ethnobio. Ethnomed. 3:6.

Kim HJ, Woo ER, Shin CG (1998). A new flavonol glycoside gallate ester from Acer okamotoanum and its inhibitory activity against human immunodeficiency virus-1 (HIV-1) integrase. J. Nat. Prod. 61: 145–148.

Kim HJ, Yu YG, Park H, Lee YS (2002). HIV gp41-binding phenolic components from Fraxinus sieboldiana var. angustata. Planta Med. 68: 1034–1036.

Kisangau DP, Lyaruu HVM, Hosea KM, Joseph CC (2007). Use of traditional medicines in the management of HIV/AIDS opportunistic infections in Tanzania: a case in the Bukoba rural district. J. Ethnobio. Ethnomed. 3: 29, doi:10.1 186/1746-4269-3-29.

Klos M, van de Venter M, Milne PJ, Traore HN, Meyer D, Oosthuizen V(2009). In vitro anti-HIV activity of five selected South African medicinal plant extracts. J. Ethnopharmaco. 124: 182-188.

Konishi T, Satsu H, Hatsugai Y, Aizawa K, Inakima T, Nagata S, Sakuda S, Nagasawa H, Shimizu M (2004). Inhibitory effect of a bitter melon extract on the P-glycoprotein activity in intestinal Caco-2 cells. Br. J. Pharmacol. 143: 379-387.

Kostova I, Raleva S, Genova P, Argirova R (2006). Structure-activity relationships of synthetic coumarins as HIV-1 inhibitors. Bioionorg. Chem. Applic. 1(9), doi 10.1155/BCA/2006/68274.

Kuo YH, Kuo LMY (1997). Antitumour and anti-AIDS triterpenoids from Celastrus hindsii. Phytochem. 44: 1275–1281.

Kupferschmidt HHT, Fattinger KE, Ha HR, Follath F, Krahenbuhl S (1998). Grapefruit juice enhances the bioavailability of the HIV protease inhibitor saquinavir in man. Brit. J. Clin. Pharmacol. 45: 355-359.

Langlois-Klassen D, Kipp W, Jhangri GS, Rubaale T (2007). Use of traditional herbal medicine by AIDS patients in Kabarole District, western Uganda. Am. J. Trop. Med. Hyg. 77: 757-763.

Lewis WH, Elvin-Lewis MPF (1995). Medicinal plants as sources of new therapeutics. Ann. Mo. Bot. Gard. 82:16–24.

Li HY, Sun NJ, Kashiwada Y, Sun L (1993). Anti-AIDS agents. Suberosol, a new C31 lanostane type triterpene and anti-HIV principle from Polyalthia suberosa. J. Nat. Prod. 56: 1130–1133.

Lin T-S, Schinazi R, Griffith BP, August ME, Eriksson BFH, Zheng D-K, Huang L, Prusoff WH (1989). Selective inhibition of human immunodeficiency virus type 1 replication by (-) but not the (+) enantiomer of gossypol. Antimicr. Agents Chemother. 33(12): 2149-2151.

Lin YM, Anderson H, Flavin MT, Pai YSH (1997). In vitro anti-HIV activity of biflavonoids isolated from Rhus succedanea and Garcinia multiflora. J. Nat. Prod. 60: 884–888.

Lindegaard B, Keller P, Bruunsgaard G, Pedersen BK (2004). Low plasma level of adiponectin is associated with stavudine treatment and lipodystrophy in HIV-infected patients. Clin. Exp. Immunol. 135(2):273-279.

Lindequist U, Niedermeyer THJ, Julich W-D (2005). The pharmacological potential of mushrooms: Evidence-based. Complim. Alternat. Med. 2(3): 285-299.

Liu JP, Manheimer E, Yang M (2005). Herbal medicines for treating HIV infection and AIDS. Cochrane Database Syst. Rev. 20(3): CD003937.

Ma CM, Nakamura N, Miyashiro H, Hattori M, Komatsu K, Kawahata T, Otake T (2002). Screening of Chinese and Mongolian herbal drugs for anti human immunodeficiency virus type-1 (HIV-1) activity. Phytother. Res. 16: 186–189.

Mandal P, Babu SPS, Mandal NC (2005). Antimicrobial activity of saponins from Acacia auriculiformis. Fitoterapia 76(5): 462-465.

Manfredi KP, Blunt JW, Cardellina JHI, McMahon JB, Pannell LK, Cragg GM, Boyd MR (1991). Novel alkaloids from the tropical plant Ancistrocladus abbreviatus inhibit cell killing by HIV-1 and HIV-2. J. Med. Chem. 34: 3402–3405.

Manners GD (2007). Citrus limonoids: analysis, bioactivity, and biomedical prospects. J. Agric. Food Chem. 55(21): 8285-8294.

Matsuse TI, Lim YA, Hattori M, Correa M, Gupta MP (1999). A search for anti-viral properties in Panamanian medicinal plants- the effect on HIV and essential enzymes. J. Ethnopharmacol. 64: 15-22.

Maury W, Price JP, Brindley M, Choonseok O, Neighbors JD, Wiemer DF, Wills N, Carpenter S, Hauck C (2009). Identification of light-independent inhibition of human immunodeficiency virus-1 infection through bioguided fractionation of Hypericum perforatum. Virol. J. 6:101. Doi:10.1186/1743-4222X-6-101.

McMormick JL, McKee TC, Cardellino JH, Boyd MR (1996). HIV inhibitory natural products. Quinoline alkaloids from Euodia roxburghiana. J. Nat. Prod. 59: 469–471.

Meragelman KM, McKee TC, Boyd MR (2001). Anti-HIV prenylated flavonoids from Monotes africanus. J. Nat. Prod. 64: 546–548.

Mills E, Cooper C, Seely D, Kanfer I (2005). African herbal medicines in the treatment of HIV: Hypoxis and Sutherlandia. An overview of evidence and pharmacology. Nutrit. J. 4:19 doi: 10.1 186/1475-2891-4-19

Min BS, Jung HJ, Lee JS (1999). Inhibitory effect of triterpenes from Crataegus pinatifida on HIV-1 protease. Planta Med. 65: 374–375.

Moore JP, Lindsey GG, Farrant JM, Brandt WF (2007). An overview of the biology of the desiccation-tolerant resurrection plant Myrothamnus flabellifolia. Ann. Bot. 99: 211-217.

Moore JP, Westall KL, Ravenscroft N, Farrant JM, Lindsey GG, Brandt WF (2005). The predominant polyphenol in the leaves of the resurrection plant Myrothamnus flabellifolius, 3,4,5 tri-O-galloylquinic

acid, protects membranes against desiccation and free radical-induced oxidation. Biochem. J. 385: 301-308.

Moore PS, Pizza C (1992). Observations on the inhibition of HIV-1 reverse transcriptase by catechins. Biochem. J. 288: 717-719.

Moshi M, Van den Beuke CPJ, Hamza OJM, Mbwambo ZH, Nondo ROS, Masimba PJ, Matee MIN, Kapingu MC, et al (2007). Brine shrimp toxicity evaluation of some Tanzanian plants used traditionally for the treatment of fungal infections. Afr. J. Trad. CAM 4(2): 219-225.

Muanza DN, Euler KL, Williams L, Newman DJ (1995). Screening for antitumor and anti-HIV activities of nine medicinal plants from Zaire. Pharmaceut. Bio. 33(2): 98-106.

Murray MT, Pizzorno JE (1999). Textbook of natural medicine. Churchill Living: China. pp.173-180.

Nakashima H (1987). Sulfation of polysaccharides generates potent and selective inhibitors of human immunodeficiency virus infection and replication in vitro. Japanese J. Cancer Res. 78: 1164-1168.

NEPAD [The New Partnership for Africa's Development] (2001). Founding document. Abuja, Nigeria.

Neurath AR, Strick N, Li Y-Y, Debnath AK (2004). Punica granatum (Pomegranate) juice provides an HIV-1 entry inhibitor and candidate topical microbicide. BMC Infect. Diseases 4(41) doi:10 1 186/1471-2334-4-41.

Otake T, Mori M, Ueba N, Sutardjo S, Kusumoto IT, Hattori M, Namba T (1995). Screening of Indonesian plant extracts for anti-human immunodeficiency virus-type1 (HIV-1) activity. Phytother. Res. 9: 6-10.

Otshudi AL, Apers S, Pieters L, Claeys M, Pannecouque C, De Clercq E, Van Zeebroeck A, Lauwers S, Frederich M, Foriers A (2005). Biologically active bisbenzylisoquinoline alkaloids from the root bark of Epinetrum villosum. J. Ethnopharmacol. 102(1): 89-94.

Parekh J, Chanda SV (2006). In vitro antimicrobial activity and phytochemical analysis of some Indian medicinal plants. Turk. J. Biol. 31: 53-58.

Park I-W, Han C, Song X, Green LA, Wang T, Liu Y, Cen C, Song X, Yang B, Chen G, He JJ (2009). Inhibition of HIV-1 entry by extracts derived from traditional Chinese medicinal herbal plants. BMC Compliment. Alternat. Med. 9:29. doi:10.1186/1472-6882-9-29.

Park JC (2002). Inhibitory effects of Korean medicinal plants and camellia tannin H from Camellia japonica on human immunodeficiency virus type-1 protease. Phytother. Res. 16: 422–426.

Pekala AZD (2007). Traditional medicines and HIV/AIDS. Medical J. Therapeutics Afr. 1(2): 94-95.

Qian-Cutrone J (1996). Niruriside, a new HIV REV/RRE binding inhibitor from Phyllanthus niruri. J. Nat. Prod. 59: 196–199.

Quattara B, Kra AM, Coulibaly A, Guede-Guina F (2007). Efficiency of ethanol extract of Thonningia sanguine against Cryptococcus neoformans. Sante. 17(4): 219-222.

Reihling HCW (2008). Bioprospecting the African renaissance: the new value of muthi in South Africa. J. Ethnobio. Ethnomed. 4:9 doi:10.1186/1746-4269-4-9,

Robinson Jr EW, Reinecke MG, Abdel-Malek S, Jia Q, Chow SA (1996). Inhibitors of HIV-1 replication that inhibit HIV integrase. PNAS USA. 93: 6326-6331.

Rukachaisirikul V (2003). Anti-HIV-1 protostane triterpenes and digeranylbenzophenone from trunk bark and stems of Garcinia speciosa. Planta Med. 69: 1141–1146.

Ruprecht RM (1989). In vitro analysis of castanospermine: a candidate antiretroviral agent. J. Acquir. Immune Defic. Syndr. 2: 149-157.

Saidi H, Nasreddine N, Jenabian M-A, Lecerf M, Schols D, Krief C, Balzarini J, Belec L (2007). Differential in vitro inhibitory activity against HIV-1 of alpha-(1-3)- and alpha-(1-6)-D-mannose specific plant lectins: implication for microbicide development. J. Translational Med. 5:28 doi: 10.1 186/1479-5876-5-28.

Sakurai N (2004). Anti-HIV agents. Actein, an anti-HIV principle from rhizome of Cimicifuga racemosa (black cohosh), and the anti-HIV activity of related saponins. Bioorg. Med. Chem. Lett. 14: 1329–1332.

Samanta MK, Mukherjee PK, Prasad MK, Suresh B (2000). Development of natural products. Easter.Pharmaci. pp: 23-27.

Scott G, Springfield EP, Coldrey N (2004). A pharmacognostical study of 26 South African plant species used as traditional medicines. Pharmaceutical Bio. 42: 186–213.

Sharma PC, Sharma OP, Vasudeva N, Mishra DN, Singh SK (2006). Anti-HIV substances of natural origin: an updated account. Nat. Prod. Radiance. 5(1):7 0-78.

Short RV (2006). New ways of preventing HIV infection: thinking simply, simply thinking. Philosoph. Transac.Royal Soc. Bio. Sci. 361: 811-820.

Singh IP, Bharate SB, Bhutani KK (2005). Anti-HIV natural products. Curr. Sci. 89(2): 269-290.

Sun H.-D (1996). Nigranoic acid, a triterpenoid from Schisandra sphaerandra that inhibits HIV-1 reverse transcriptase. J. Nat. Prod. 59: 525–527.

Sung H, Kang S-M, Lee M-S, Kim TG, Cho Y-K (2005). Korean red ginseng slows depletion of CD4 T-cells in human immunodeficiency virus type 1-infected patients. Clin. Diag. Lab. Immuno. 12(4): 497-501.

Sunthitikawinsakul A (2003). Anti-HIV-1 limonoid: First isolation from Clausena excavata. Phytother. Res. 17: 1101–1103.

Suzuki H, Okubo L, Yamazaki S, Suzuki K, Mitsuya H, Toda S (1989). Inhibition of the infectivity and cytopathic effect of the human immunodeficiency virus by water-soluble lignin in an extract of the culture medium of Lentinus edoded mycelia (LEM). Biochem Biophys. Res Commun. 160: 367-373.

Swanson MD, Winter HC, Goldstein IJ, Markovitz DM (2010). A lectin isolated from bananas is a potent inhibitor of HIV replication. J.Bio. Chem. 285: 8646-8655.

Talyshinsky MM, Souprun YY, Huleihel MM (2002). Anti-viral activity of red microalgal polysaccharides against retroviruses. Cancer Cell Int. 2(8) http://www.cancerci.com/conrent/2/1/8

Tamura Y, Lai PK, Bradley WG, Konno K, Tanaka A, Nonoyama M (1991). A soluble factor induced by an extract from Pinus parviflora Sieb et Zucc can inhibit the replication of human immunodeficiency virus in vitro. PNAS USA. 88: 2249-2253.

Tan GT, Pezzuto JM, Kinghorn AD, Hughes SH (1991). Evaluation of natural products as inhibitors of human immunodeficiency virus type 1 (HIV-1) reverse transcriptase. J. Nat. Prod. 54: 143–154.

Tanaka N, Okasaka M, Ishimaru Y, Takaishi Y, Sato M, Okakamoto M, Oshikawa T, Ahmed SU, Consentino LM, Lee K-H (2005). Biyouyanagin A, and anti-HIV agent from Hypericum chinense L. var. salicifolium. Org. Lett. 7(14): 2997-2999.

Tewtrakul S, Subhadhirasakul S, Cheenpracha S, Karalai C (2007). HIV-1 protease and HIV-1 integrase inhibitory substances from Eclipta prostrate. Phytother. Res. 21(11): 1092-1095.

Theo A, Masebe T, Suzuki Y, Kikuchi H, Wada S, Obi CL, Bessong PO, Usuzawa M, Oshima Y, Hattori T (2009). Peltophorum africanum, a traditional South African medicinal plant, contains an anti-HIV constituent, betulinic acid. The Tahoku J. Exp. Med. 217(2): 93-99.

Trusheva B, Popova M, Bonkova V, Somiva S, Marcucci CM, Miorin PL, Pasin FDR, Tsvetkova I (2006). Bioactive constituents of Brazilian red propolis. Evidence-based Comp. Alt. Med. 3(2): 249-254.

Tumer NE, Hwang D-J, Bonness M (1997). C-terminal deletion mutant of pokeweed protein inhibits viral infection but does not depurinate host ribosomes. PNAS USA 94: 3866-3871.

Tziveleka LA, Vagias C, Roussis V (2003). Natural products with anti-HIV activity from marine organisms. Curr. Topics Med. Chem. 3(13): 1512-1535.

Uckun FM, Chelstrom LM, Tuel-Ahlgren L, Dibirdik I, Irvin JD, Langlie M-C, Myers DE (1998). TXU (anti-CD7) pokeweed antiviral protein as a potent inhibitor of human immunodeficiency virus. Antimicro. Agents Chemother. 42(2): 383-388.

UNAIDS (2009). AIDS epidemic update 2009. In 2009 report on the global AIDS epidemic. Geneva: UNAIDS.

UNAIDS (2009). The regional workshop on adopting minimum standards of practice for THETA evaluation team. Participatory evaluation report. Innovation or reawakening? Roles of traditional healers in the management and prevention of HIV/AIDS in Uganda. Geneva: UNAIDS.

Van den Beukel CJ, Hamza OJ, Moshi MJ, Matee MI, Milkx F, Burger DM, Koopmans PP, Verweij PE, Schoonen WG, van der Ven AJ (2008). Evaluation of cytotoxic, genotoxic and CYP450 enzymatic competition effects of Tanzanian plant extracts traditionally used for treatment of fungal infections. Basic Clin. Pharmacol. Toxicol. 102(6): 515-526.

Van Wyk B, Gericke N (2000). People's plants: a guide to useful plants of South Africa. Briza Publications, South Africa, 1st edition. pp: 245-257.

Vermani K, Garg S (2002). Herbal medicines for sexually transmitted diseases and AIDS. J. Ethnopharmacol. 80: 49-66.

Wang HX, Ng TB (2001). Isolation and characterization of velutin, a novel low-molecular-weight ribosome inactivating protein from winter mushroom (Flammulina velutipes) fruiting bodies. Life Sci. 68: 2151-2158.

Wang JN, Hou CY, Liu YL, Lin LZ, Gil RR, Cordell GA (1994). Swertifrancheside, an HIV-reverse transcriptase inhibitor and the first flavone-xanthone dimer from Swertia franchetiana. J. Nat. Prod. 57: 211–217.

Wang Q, Ding ZH, Liu JK, Zheng YT (2004). Xanthohumol, a novel anti-HIV-1 agent purified from hops Humulus lupulus. Antiviral Res. 64: 189–194.

Webster D, Taschereau P, Belland RJ, Sand C, Rennie RP (2008). Antifungal activity of medicinal plant extracts; preliminary screening studies. J. Ethnopharmacol. 115(1): 140-146.

Witvrouw M, Este JA, Mateu MQ, Reymen D, Andrei, G, Snoeck R, Ikeda S, Pauwels R, Bianchini NV, Desmyter J, De Clercq E (1994). Activity of a sulfated polysaccharide extracted from the red seaweed Aghardhiella tenera against human immunodeficiency virus and other enveloped viruses. Antivir. Chem. Chemoth. 5: 297-303.

World Health Organisation (WHO) (1989a). In vitro screening of traditional medicines for anti-HIV activity:memorandum from a WHO meeting. Bull. World Health Organization 87: 613–618.

World Health Organisation (WHO) (1989b). Report of a WHO Informal Consultation on Traditional Medicine and AIDS: In Vitro Screening for Anti-HIV Activity. Global Prog. AIDS and Trad. Med. Programme, pp. 1–17.

Xu HX, Zeng FQ, Wan M, Sim KY (1996). Anti-HIV triterpene acids from Geum japonicum. J. Nat. Prod. 59: 643–645.

Xu ZQ, Flavin MT, Jenta TR (2000). Calanolides, the naturally occurring anti-HIV agents. Cur. Opinion in Drug Disc. Devel. 3(2): 155-166.

Yao XJ, Wainberg MA, Parniak MA (1992). Mechanism of inhibition of HIV-1 infection in vitro by purified extract of Prunella vulgaris. J. Virol. 187(1): 56-62.

Yu D, Morris-Natschke SL, Lee KH (2007). New developments in natural products-based anti-AIDS research. Med. Res. Rev. 27(1): 108-132.

Zhu K, Cordeiro ML, Atienza J, Robinson Jr EW, Chow S (1999). Irreversible inhibition of human immunodeficiency virus type integrase by dicaffeoylquinic acids. J. Virol. 73(4): 3309-3316.

15

Drug release and antimicrobial studies on polylactic acid suture

O. L. Shanmugasundaram[1]*, R. V. Mahendra Gowda[2] and D. Saravanan[3]

[1]Department of Textile Technology, KSR College of Technology, Tiruchengode-637 215, Tamil Nadu, India.
[2]VSB Engineering College, Karur-639 111, Tamil Nadu, India.
[3]Department of Biotechnology, KSR College of Technology, Tiruchengode-637 215, Tamil Nadu, India.

The aim of the present study was to develop drug releasing polylactic acid (PLA) suture. PLA suture was taken for the research work. The material was analyzed, its properties, such as tensile strength, elongation, knot strength and diameter. Biopolymers such as chitosan, sodium alginate, calcium alginate, and their blends were coated on the suture material. The polymer coated samples were subjected to Fourier transform infrared (FT-IR) analysis. Bacteria present in wound samples were found out using different biochemical methods. The results show that *Staphylococcus aureus and proteus* was found in the wound samples. Tetracycline hydrochloride, rifampin and chloramphenicol drugs were selected based on the antibiotic sensitivity test. These drugs were then incorporated into the polymer coated samples using immersion method. All the drug loaded samples were subjected to drug release study for about four days in static condition. Further, the drug loaded samples were subjected to antimicrobial test. Hence these sutures are quite suitable for wound healing in addition to wound closing.

Key words: Alginate, antimicrobial, bacteria, biopolymer, chitosan, suture.

INTRODUCTION

Suturing is performed for varied purposes in surgical field. Primary closure of tissues, which were separated by surgical procedure or accidental trauma, promotes a healing process and controls inactive bleeding (Shaw and Negus, 1996). Sutures that undergo rapid degradation in tissue, losing their tensile strength within 60 days, are considered absorbable sutures. Sutures that generally maintain their tensile strength for longer than 60 days are non-absorbable sutures (Rajendran and Anand, 2002). Multifilament suture materials consist of several filaments twisted or braided together, which give good handling and tying qualities (Jin-cheol et al., 2007).

Sutures must have good tensile strength, pliability and flexibility. Multifilament sutures may also be coated to help them pass relatively smoothly through tissue and enhance handling characteristics (Ben et al., 1999). Monofilament sutures are made up of a single strand of material. Because of their simplified structure, they encounter less resistance as they pass through tissue than multifilament suture materials (Karaca and Hockenberger, 2005). The monofilament suture resists the harboring microorganisms and ties smoothly, which can ease the judgment of the tightening of a knot. However, because it has relatively higher bending stiffness and the tendency to untie, it is hard to deal with it and to form stable knot (Gallup et al., 1990; Tomihata et al., 2005). Ideal suture materials should satisfy several requirements (Karaca and Hockenberger, 2005). They should have high tensile strength but lose strength at the same rate as the tissue gains strength, and should be easy to handle and form secure knots. They should be biologically inert; therefore, they should induce minimal tissue inflammation and should not promote infection. They should be able to stretch, accommodate wound edema and recoil to its original length with wound contraction. Balassa and Prudden (1984) reported that chitosan and alginate polymers are biodegradable and obtained from the natural origin and having bacteriostatic and fungistatic properties are particularly useful for wound treatment. Chitosan and alginate can act as antibacterial and antifungal activity (Choi et al., 2001).

*Corresponding author. E-mail: mailols@yahoo.com.

Prajapati and Sawant (2009) designed a formulation of polyelectrolyte complex film for transdermal delivery of antifungal agents for wound healing. Polyelectrolyte complex film is the film prepared out of the complexation of the two opposite charge polymers, chitosan as a cation and alginate as anion (Wang et al., 2002). A coating of Polyglactin 370 and calcium stearate, this lubricant coating gives Vicryl excellent handling and smooth tying properties (Moy et al., 1991).

The aim of the present research work was evaluating the drug loaded polymer coated sutures for their drug release characteristics and antibacterial activity against *Staphylococcus aureus* and *proteus*. PLA suture was coated with biopolymers followed by immobilization of antibiotic drugs used aiming at two layers of defense at the infected site.

MATERIALS AND METHODS

Materials

The polylactic acid multifilament yarn (140 denier) was procured from Haining Xin Gao Fibres Ltd, Zhejiang, China.

Methods

Scouring of material

Scouring is a process of removal of natural and added impurities present in textile materials, in order to improve the absorbency. Polylactic acid (PLA) material was scoured with 1% by weight of sodium hydroxide to eliminate impurities and washed with de-ionized water (Ching-Wen et al., 2008).

Measurement of tensile properties

Tensile properties of sutures are important for the practitioner making a knot. If the material is too weak and the knotting force is stronger than tensile strength of suture material, suture can easily break while tightening the knot. Therefore, it is essential to know the tensile strength of sutures. Fifty specimens of 7.5 cm in length were prepared from sample. Tensile properties were measured as per ASTM D2256 test method.

Measurement of diameter

The diameters of the sutures were measured at five different positions in the material using an Electronic Inspection Board (EIB). An average of hundred observations was taken for the measurement of diameter. Suture diameter is expressed in millimeters.

Measurement of knot strength

In the knot strength test, a single and simple surgeon knot was tied in the middle of the suture, that is, two ends of the suture was wrapped together at an angle of 360°. In order to have same knot tension in all the samples, the knots was tied mechanically on the

Instron by applying 1 N tension. Suture was clamped in between fixed the jaw and moving jaw of the Instron and pulled until they broke. The distance between the jaws was 7.5 cm, and the gauge speed was 5 mm/min. All tests were repeated five times.

Polymers coating on suture materials

Chitosan solution was prepared by stirring a dispersion of chitosan (8.0 g) in 2.0% (v/v) aqueous acetic acid solution at 60°C for 1 h. Then 2.0 g of sodium alginate polymer was added to the chitosan solution and stirred for 10 min. The polylactic acid materials were immersed with the solution for about 2 h. The material was then dried at 80°C for 5 min.

Calcium alginate solution was prepared by stirring a dispersion of calcium alginate (8.0 g) in 100 ml of sodium carbonate solution at 40°C for 30 min. Then 2.0 g of sodium alginate polymer was added to the solution and stirred for 10 min. The polylactic acid materials were immersed with the solution for about 2.5 h. The material was then dried at 70°C for 3 min.

The blends of chitosan, sodium alginate and calcium alginate polymer coated sample was prepared by processing the fabric in the similar way of chitosan-sodium alginate and calcium alginate coating as mentioned previously.

FT-IR

The polymer coated fabric samples were subjected to Infrared analysis. Infrared spectra were recorded using a Horiba FT-210 spectrophotometer with a potassium bromide pellet.

Incorporation of antibiotic drugs

About 250 mg of tetracycline hydrochloride, chloramphenicol and rifampicin drugs were dissolved separately in 10 ml of distilled water. The polymer coated samples were immersed in the drug solution and allowed to remain still for 24 h. Finally, the samples were washed with distilled water and dried at room temperature for two days. The drug add-on (%) was calculated using the following relationship:

Drug add-on (%) = $[(W_1 - W_2)/W_2] \times 100$

where W_1 is the weight of drug immobilized sample, and W_2 is the weight of un-immobilized sample.

Drug release study under static condition

The drug loaded samples with respective drug add-on were taken as such and suspended in 10 ml normal water for 24 h at room temperature. After 24 h, the supernatant from the sample tubes was tested for drug concentration using Hitachi UV visible spectrophotometer (U3210 Japan). The sample tubes were replenished with fresh water, which was tested for drug concentration the next day. The entire procedure was continued for 4 days.

Antibacterial test

The polymer coated and drug loaded samples were tested for antibacterial activity against *S. aureus* and *Proteus* bacteria according to agar diffusion standard test method. The nutrient agar medium was prepared by dissolving 3.7 g in 100 ml distilled water

Table 1. Measurement of suture properties.

Property	PLA suture
Tenacity (g/tex)	650.98
Elongation (%)	8.37
Knot strength (g/tex)	349.1
Elongation with knot (%)	18.0
Diameter (mm)	0.187

Table 2. Polymer goating on suture materials.

Name of the sample	Name of the polymer code	Weight of the sample after coating (g)	Polymer add-on (g)	Polymer add-on (%)
Polylactic acid	P_1	5.9	2.9	96
	P_2	5.7	2.7	90
	P_3	5.8	2.8	93

Sample weight 3 g; P_1: chitosan + sodium alginate polymer; P_2: calcium alginate + sodium alginate polymer; P_3: chitosan + sodium alginate + calcium alginate;

and the conical flask was tightly closed by using non absorbent cotton (Shanmugasundaram and Mahendra, 2011).

RESULTS AND DISCUSSION

Measurement of tensile properties

The tensile properties of PLA suture is given in Table 1. It can be observed that PLA suture has high tensile strength and elongation. Its strength and elongation is quite suitable for making a drug releasing suture material. PLA suture has 0.187 mm in diameter.

Evaluation of knot strength

Suture materials have to maintain tissue unification during the healing process without slippage untying and breaking. Changes in tensile properties of suture after knotting, is given in Table 1. The results show that knot strength and elongation was decreased to a larger extent due to the internal fibre fracture. These materials are less stiff but have a higher coefficient of friction. Multifilament suture generally has greater tensile strength and better pliability and flexibility than monofilament suture. This type of suture handles and ties well. These all knots slip to some degree, regardless of the type of suture material. Knotting induces decrease in mechanical properties. Suture failure occurs most frequently at the knot since local stresses weaken the fibre. Therefore, the US Pharmacopeia (USP) specifies minimum knot-pull tensile strength requirements for sutures. This suture material

meets the USP standards.

Polymers coating

The polymer add-on on suture is given in Table 2. From the table, it can be inferred that the polymer add-on was good in almost all the samples. It ranges from 2.8 to 2.9 g, respectively. The percentage add-on was calculated based on the material weight and it was mentioned in Table 2. It shows that add-on percentage vary from 92 to 96. In general, as the polymer add-on increases, its antibacterial and wound healing property increases.

FT-IR (Fourier transform infrared spectroscopy) analysis

Figure 1 illustrates the FT-IR spectra of chitosan-sodium alginate polymer coated sample. It is observed that a broad band occurs at 3300 cm^{-1}, a relatively sharp band at 2950 cm^{-1} and a fingerprint region from 1600 to 2950 cm^{-1}. Resolution of two peaks was observed in the FT-IR spectrum of the chitosan-sodium alginate coated fabric. It is also observed that the band at 3450 cm^{-1} in chitosan polymer is shifted to 3300 cm^{-1} in chitosan-sodium alginate polymer coated sample.

Figure 2 illustrates the FT-IR spectra of calcium-sodium alginate polymer coated sample. It can be inferred that a broad band occurs at 3290 cm^{-1}, a relatively sharp band at 2950 cm^{-1} and a fingerprint region from 1600 to 2950 cm^{-1}. Resolution of two peaks is observed in the FT-IR spectrum of the calcium-sodium alginate coated fabric.

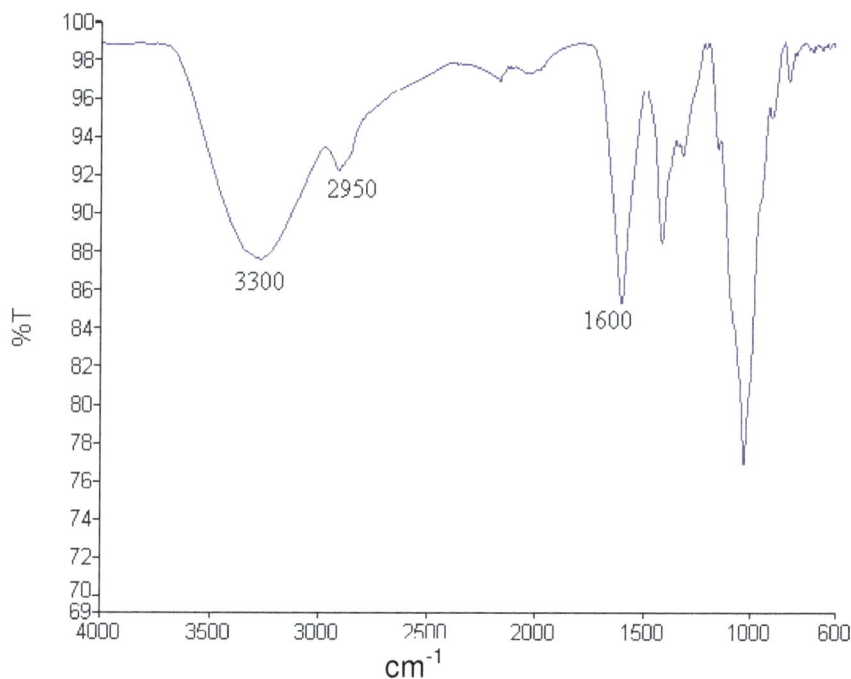

Figure 1. FTIR spectra of chitosan + sodium alginate polymer coated sample.

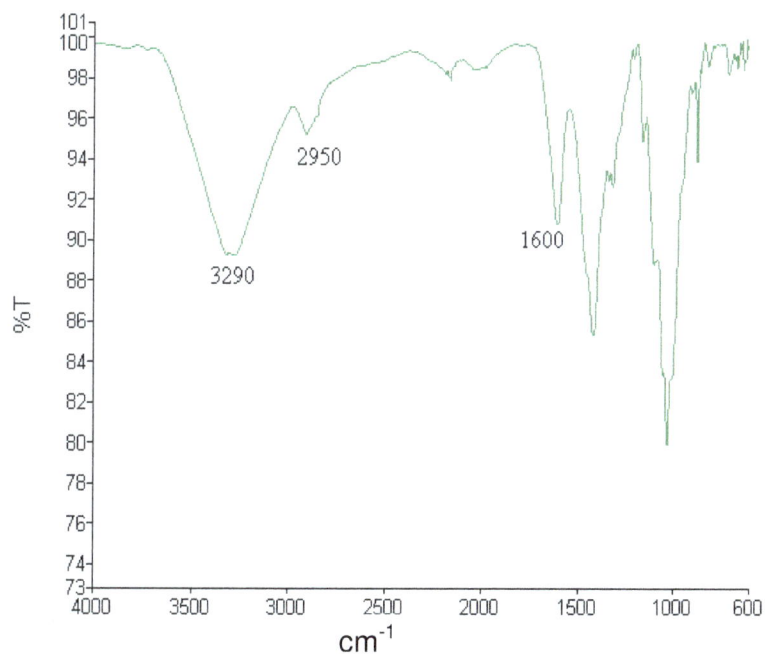

Figure 2. FTIR spectra of calcium alginate + sodium alginate polymer coated sample.

Figure 3 illustrates the FT-IR spectra of chitosan-sodium alginate-calcium alginate polymer coated sample. It can be seen that a broad band occurs at 3290 cm^{-1}, a relatively sharp band at 2950 cm^{-1} and a fingerprint region from 1550 to 2950 cm^{-1}.

Incorporation of antibiotic drugs

The drug add-on is given in Table 3. It was found that the drug add-on percentage various from 88 to 92. The main purpose of adding medicine is to inhibit the bacterial

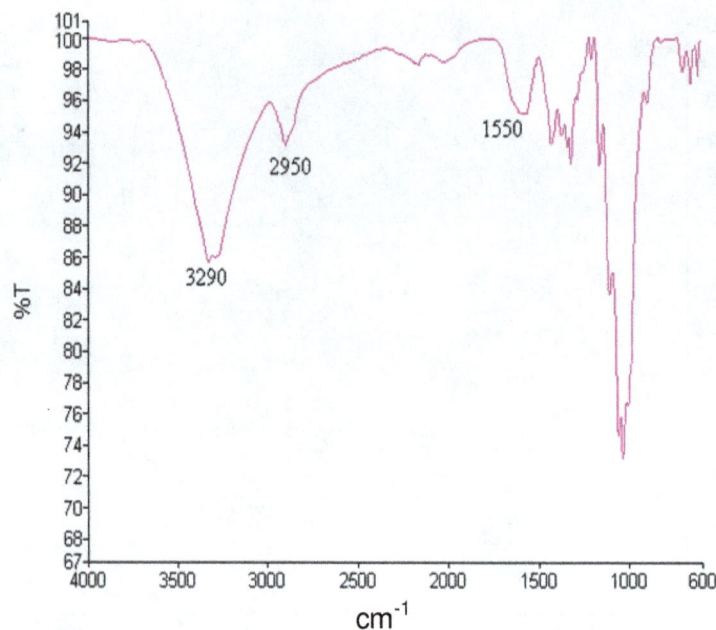

Figure 3. FTIR spectra of chitosan + calcium alginate + sodium alginate polymer coated sample.

Table 3. Incorporation of antibiotic drugs on polymer coated sutures.

Sample code	Name of the drug	Weight of the sample after coating (g)	Drug add-on (mg)	Drug add-on (%)
	Rifampin (D$_1$)	1.230	230	92
P$_1$-Polylactic acid	Tetracycline hydrochloride (D$_2$)	1.220	220	88
	Chloramphenicol (D$_3$)	1.215	215	86
	Rifampin (D$_1$)	1.220	220	88
P$_2$- Polylactic acid	Tetracycline hydrochloride (D$_2$)	1.220	220	88
	Chloramphenicol (D$_3$)	1.225	225	90
	Rifampin (D$_1$)	1.230	230	92
P$_3$- Polylactic acid	Tetracycline hydrochloride (D$_2$)	1.215	215	86
	Chloramphenicol (D$_3$)	1.220	220	88

P$_1$: Chitosan + sodium alginate polymer; P$_2$: calcium alginate + sodium alginate polymer; P$_3$: chitosan + sodium alginate + calcium alginate.

growth, thereby facilitating the rate of wound healing processes. Antibiotic is a substance or compound that kills or inhibits the growth of bacteria. It belongs to the broader group of antimicrobial compounds used to treat infections caused by microorganisms, including fungi and protozoa. These antibiotics diffuse out from antibiotic-containing disks and inhibit the growth of bacteria resulting in a zone of inhibition. Rifampin is used as antibacterial drug to treat or prevent infections that are proven or strongly suspected to be caused by bacteria. It inhibits DNA-dependent RNA polymerase activity in susceptible cells. Specifically, it interacts with bacterial RNA polymerase but does not inhibit the mammalian enzyme. Chloramphenicol is an antimicrobial agent, used to combat serious infections where other antibiotics are either ineffective or contraindicated. It can be used against gram-positive cocci and bacilli and gram-negative aerobic and anaerobic bacteria.

Drug release study

Current efforts in the area of drug delivery include the development of targeted delivery in which the drug is only

Table 4. Drug release characteristics of sutures.

Name of the sample	Weight of the sample (g)	Drug release (mg)		
		1st day	2nd day	3rd day
P_1-D_1-PLA	1.230	95	65	35
P_1-D_2- PLA	1.220	90	70	30
P_1-D_3- PLA	1.215	95	60	25
P_2-D_1- PLA	1.220	90	70	30
P_2-D_2- PLA	1.220	85	75	35
P_2-D_3- PLA	1.225	95	60	40
P_3-D_1- PLA	1.230	90	70	30
P_3-D_2- PLA	1.215	90	70	25
P_3-D_3- PLA	1.220	95	60	35

P_1: Chitosan + sodium alginate polymer; P_2: calcium alginate + sodium alginate polymerr; P_3: chitosan + sodium alginate + calcium alginate; D_1: rifampin drug; D_2: tetracycline hydrochloride drug; D_3: chloramphenicol drug.

Table 5. Evaluation of antibacterial activity of drug loaded suture samples.

Name of the sample	Zone of inhibition (cm)					
	S. aureus			Proteus		
	1st day	2nd day	4th day	1th day	2nd day	4th day
P_1-D_1-PLA	3.2	3.4	3.5	3.3	3.5	3.6
P_1-D_2- PLA	3.2	3.3	3.4	3.2	3.2	3.3
P_1-D_3- PLA	3.2	3.3	3.4	3.2	3.4	3.6
P_2-D_1- PLA	3.3	3.4	3.5	3.3	3.5	3.6
P_2-D_2- PLA	3.2	3.4	3.6	3.3	3.6	3.8
P_2-D_3- PLA	3.0	2.9	3.0	3.0	3.2	3.5
P_3-D_1- PLA	3.4	3.6	3.8	3.5	3.7	4.0
P_3-D_2- PLA	3.4	3.7	3.9	3.6	3.8	4.1
P_3-D_3- PLA	3.2	3.3	3.4	3.3	3.5	3.7

P_1: Chitosan + sodium alginate polymer; P_2: calcium alginate + sodium alginate polymer; P_3: chitosan + sodium alginate + calcium alginate; D_1: rifampin drug; D_2: tetracycline hydrochloride drug; D_3: chloramphenicol drug.

active in the target area of the body and sustained release formulation. The drug release study was carried out for about 4 days. The release characteristic of the drug is given in Table 4 and Figure 8. The release characteristics show that a high amount of drug was released on the first day in all the samples. It ranges from 85 to 95 mg. A minimum amount of drug (85 mg) was released from the calcium alginate and sodium alginate polymer coated with chloramphenicol drug loaded PLA suture and a maximum amount of drug (95 mg) was released from the chitosan, sodium alginate and calcium alginate coated with tetracycline hydrochloride drug loaded PLA sample. On the second day, the release of drug ranges from 60 to 75 mg. This study was continued on the fourth day; it shows that 25 to 40 mg of drug was released from each sample. It was also observed from the table that the drug imbibed on the surface of the material is released quicker initially and at the end of

fourth day of the study, the suture is left with residual drug. This indicates that there will be antibacterial activity shown by the suture at the infected site even after four days.

Antibacterial test

The drugs loaded PLA sutures were taken for antimicrobial effect against S. aureus and Proteus bacteria by agar diffusion test. This study was carried out for about four days. It can be seen from Table 5 and Figures 4 to 7. The zone of inhibition is greater than 3.0 cm in all samples. The antibacterial activities of the PLA samples show that the zone of inhibition is 3.0 to 3.4 and 3.0 to 3.6 cm on the first day against S. aureus and Proteus bacteria. The zone of inhibition was increased on the second day in all the samples. The study was

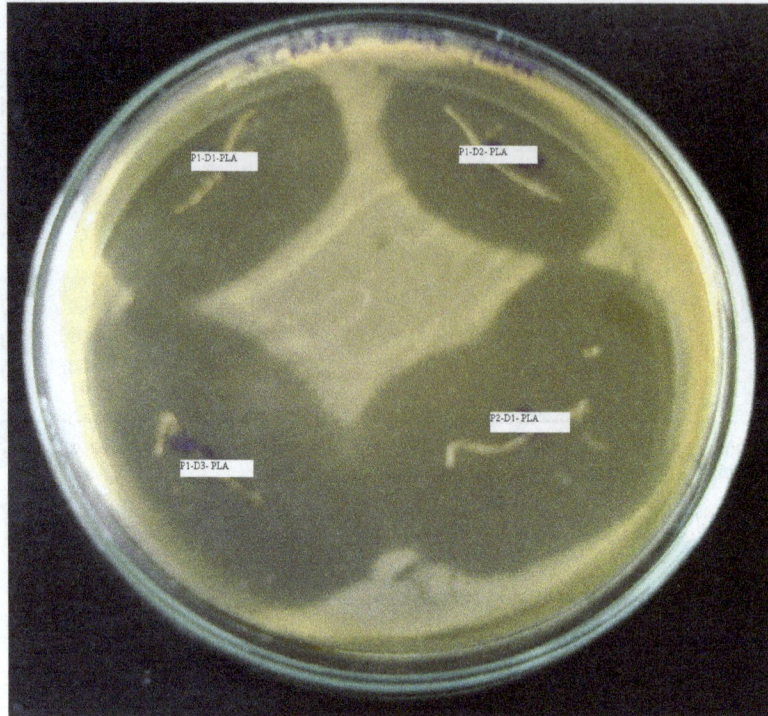

Figure 4. Zone of inhibition of polymer coated with drug immobilized samples against *Proteus* bacteria. P_1: chitosan + sodium alginate polymer; P_2: calcium alginate + sodium alginate polymer; P_3: chitosan + sodium alginate + Calcium alginate; D_1: rifampin drug; D_2: tetracycline hydrochloride drug; D_3: chloramphenicol drug.

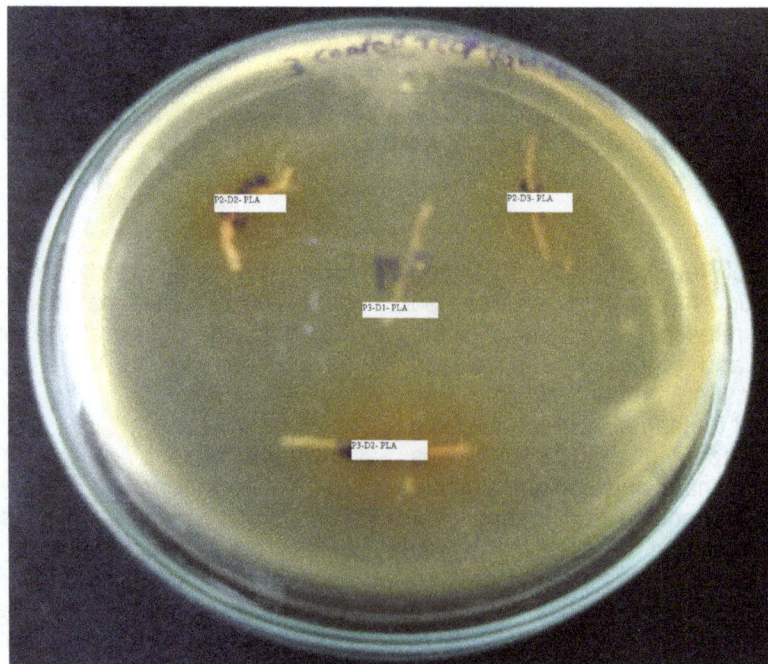

Figure 5. Zone of inhibition of polymer coated with drug immobilized samples against *Proteus* bacteria. P_1: chitosan + sodium alginate polymer; P_2: calcium alginate + sodium alginate polymer; P_3: chitosan + sodium alginate + Calcium alginate; D_1: rifampin drug; D_2: tetracycline hydrochloride drug; D_3: chloramphenicol drug.

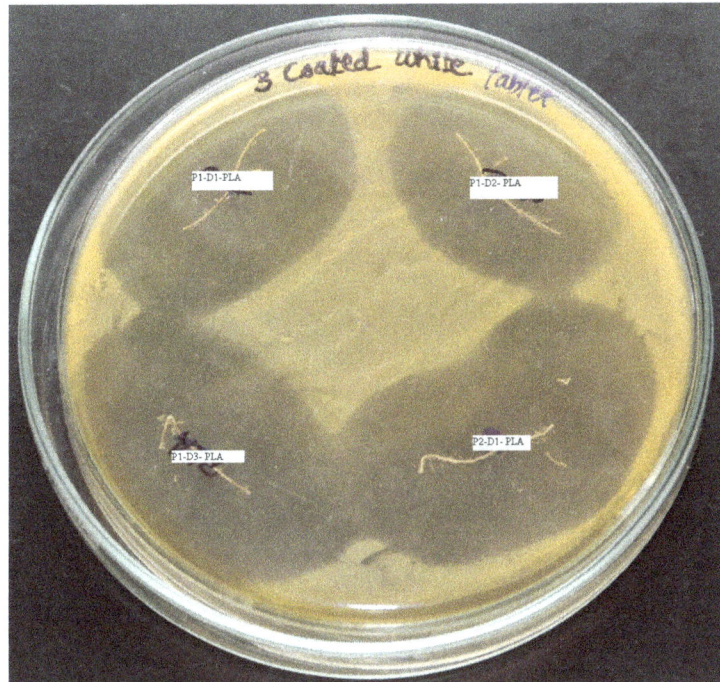

Figure 6. Zone of inhibition of polymer coated with drug immobilized samples against *S. aureus* bacteria. P_1: chitosan + sodium alginate polymer; P_2: calcium alginate + sodium alginate polymer; P_3: Chitosan + sodium alginate + calcium alginate; D_1: rifampin drug; D_2: tetracycline hydrochloride drug; D_3: chloramphenicol drug.

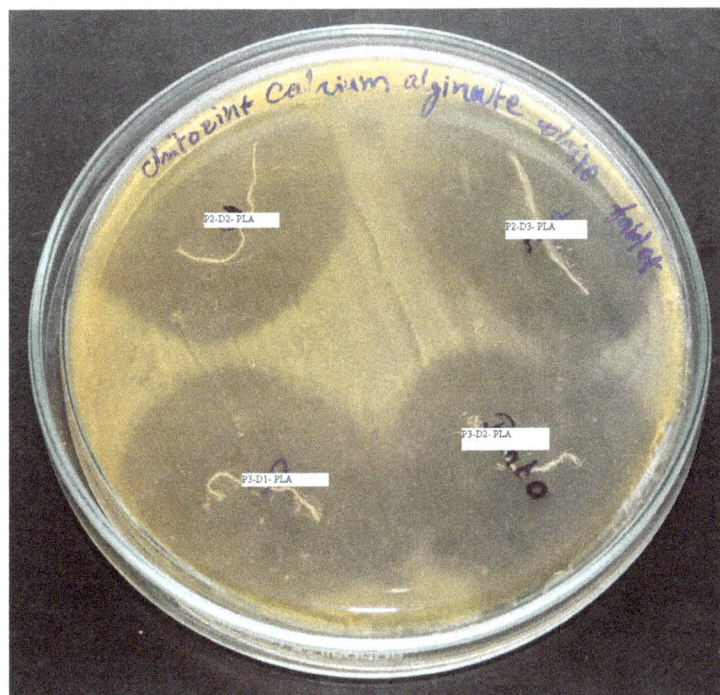

Figure 7. Zone of inhibition of polymer coated with drug immobilized samples against *S. aureus* bacteria. P_1: chitosan + Sodium alginate polymer; P_2: calcium alginate + sodium alginate polymer; P_3: chitosan + sodium alginate + Calcium alginate; D_1: rifampin drug; D_2: tetracycline hydrochloride drug; D_3: chloramphenicol drug.

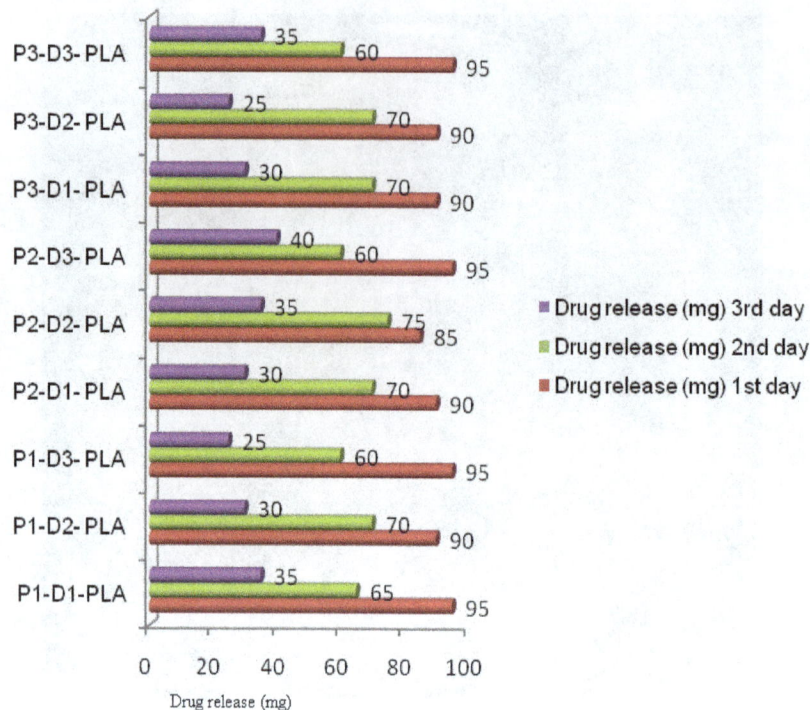

Figure 8. Drug release characteristic of polylactic acid sutures. P_1: Chitosan + sodium alginate polymer; P_2: calcium alginate + sodium alginate polymer; P_3: chitosan + sodium alginate + calcium alginate; D_1: rifampin drug; D_2: tetracycline hydrochloride drug; D_3: chloramphenicol drug.

continued on the fourth day; it shows the zone of inhibition is 3.0 to 3.9 and 3.3 to 4.1 cm against the two bacteria. The strongest antibacterial activity is found in the chitosan, sodium alginate and calcium alginate coated with tetracycline hydrochloride and chloramphenicol drug immobilized PLA samples.

Conclusions

In this research work, PLA material was used for the preparation of drug releasing sutures. The suture properties were studied as per ASTM standard methods. Antimicrobial polymers such as chitosan, sodium alginate, calcium alginate and their mixture were coated separately on the suture material in order to improve the wound healing and antibacterial property. Three types of antibiotic drugs were selected based on the antibiotic sensitivity test and incorporated into the polymer coated materials. From the study, it was found that the polymer add-on and drug add-on percentage was normally good in all the samples. The drug loaded samples were subjected to drug release study. It was found that all coatings showed a continuous drug release for about four days. The drug loaded PLA samples show an excellent antibacterial activity against *S. aureus* and *Proteus* bacteria. Hence it is concluded that the drug loaded PLA sutures are quite suitable for

wound healing in addition to wound closing.

REFERENCES

Balassa LL, Prudden JF (1984). Applications of chitin and chitosan in wound healing acceleration, in chitin, chitosan and related enzymes. Academic Press. San Diego, pp. 296-305.

Ben AS, Chakfe N, Le Magnen, Beaufigeau M, Adolphe D, Geny B, Akesbi S Riepe G (1999). Influence of crimping textile polyester vascular prostheses on the fluid flow kinetics. Eur. J. Vas. Endov. Surg., 18: 9-12.

Ching-Wen L, Chun-Hsu Y, Yueh-Sheng C, Tsung-Chih H, Jia-Horng L, Wen-Hao H (2008). Manufacturing and properties of PLA absorbable surgical suture. Text. Res. J., 78: 958-960.

Choi BK, Kim KY, Yoo YJ, Oh SJ (2001). *In Vitro* antimicrobial activity of a chitooligosaccharide mixture against actinobacillus. Int. J. Antimicr. Agent, 18: 553-557.

Gallup DG, Nolan TE, Smith RP (1990). Primary mass closure of midline incision with a continuous polyglyconate monofilament absorbable suture. Obstetr. Gynecol., 76: 872-875

Jin-Cheol Kim, Yong-Keun Lee et al., ??? Please provide names of other authors (2007). Comparison of tensile and knot security properties of surgical sutures. J. Mat. Sci.: Mat. Medi. 18: 2363-2366.

Karaca E, Hockenberger AS (2005). Investigating changes in mechanical properties and tissue reaction of silk, polyester, polyamide and polypropylene sutures *in vivo*. Text. Res. J., 75: 297-300.

Moy RL, et al.,??? Please provide names of other authors (1991). Commonly used suture materials in skin surgery. Am. Fam. Phys., 44: 2123-2125.

Prajapati G, Sawant K (2009). Polyelectrolyte complex of chitosanalginate for local drug delivery. Inter. J. Chem. Tech. Res., 1: 643-648.

Rajendran S, Anand S (2002). Developments in medical textile. J. Text Inst., 32: 7-10.

Shanmugasundaram OL, Mahendra GRV (2011). Development and characterisation of bamboo gauze fabric coated with polymer and drug for wound healing. Fibers Polym., 12: 15-20.

Shaw J, Negus TW (1996). A prospective clinical evaluation of the longevity of resorbable sutures in oral mucosa. Br. J. Oral Maxillofacial Surg., 34: 252-256.

Tomihata K, Suzuki M, Tomiya N (2005). Handling characteristics of poly(L-lactide-co-caprolactone) monofilament suture. Bio-med. Mater. Eng., 15: 381-385.

Wang L, Khor E, Wee A, Lim LY (2002). Chitosan-alginate PEC membrane as a wound dressing:Assessment of incisional wound healing. J. Biomed. Mater. Res., 63: 610-618.

16

Enhancing thermostability of the biocatalysts beyond their natural function *via* protein engineering

Shelly Goomber[1], Pushpender K. Sharma[1,2]*, Monika Sharma[1], Ranvir Singh[3] and Jagdeep Kaur[1]

[1]Department of Biotechnology, Sector 14, Panjab University, Chandigarh 160014, India.
[2]Indian Institute of Sciences Education and Research, S.A.S. Nagar, Sector-81, Mohali, Punjab, 140306, India.
[3]National Centre for Human Genome Studies and Research (NCHGSR), Sector 14, Panjab University, Chandigarh 160014, India.

Majority of the naturally occurring enzymes lacks essential features required during the harsh conditions of the industrial processes, because of their less stability. Protein engineering tool offers excellent opportunity to improve the biochemical properties of these biocatalysts. These techniques further help in understanding the structure and function of the proteins. Most common methods employed in protein engineering are directed evolution and rational mutagenesis. Several research groups have utilized these methods for engineering the stability/activity of diverse class of enzymes. *In silico* tools further plays an important role in designing better experimental strategy to engineer these proteins. The availability of vast majority of data on protein thermostability will enable one to envisage the possible factors that may contribute significantly in maintaining the protein structure and function during various physical conditions. This review discusses the common method employed in protein engineering along with various molecular/computational approaches that are being utilized for altering protein activity, along with important factors associated with these processes.

Key words: Computational database, rational, directed evolution, thermostability, hydrophobicity, three dimensional structure, configuration, biocatalysts.

INTRODUCTION

Almost all industrial processes are carried out at high temperature; therefore there is pressing need to discover new thermostable enzymes/or modify existing enzymes, using protein engineering approach. Thermostable enzymes are widely employed during manufacturing of detergents, food processing, production of high fructose corn syrup etc. (Crabb and Mitchinson, 1997). Previously, Turner et al. (2007) has also provided a valuable insight into the use of thermostable enzymes in biorefining. Industrial processes performed at high temperature possesses following major advantages that is, increased rate of reaction, less microbial contamination, increased solubility of the substrates etc. A recent progress made in protein engineering approach has resulted in modification of several biocatalysts for enhanced thermostability. Two most common methods that are employed *in vitro* evolutions of thermostable enzymes are: directed evolution, which do not require any structural or mechanistic information and rational designing, that require prior knowledge of the three dimensional structure, as depicted in the flow chart (Figure 1).

A number of factors are associated with protein thermostability and mainly include increased hydrogen bonding, increased percentage of hydrophobic residues and less percentage of thermo labile residues etc. (Sadeghi et al., 2006; Russell et al., 1997; Bogin et al., 1998; Kumar et al., 2000). Among these factors, the hydrophobicity in protein is known to direct configurationally complexities of folding and unfolding (Kauzmann,

*Corresponding author. E-mail: pushpender@iisermohali.ac.in.

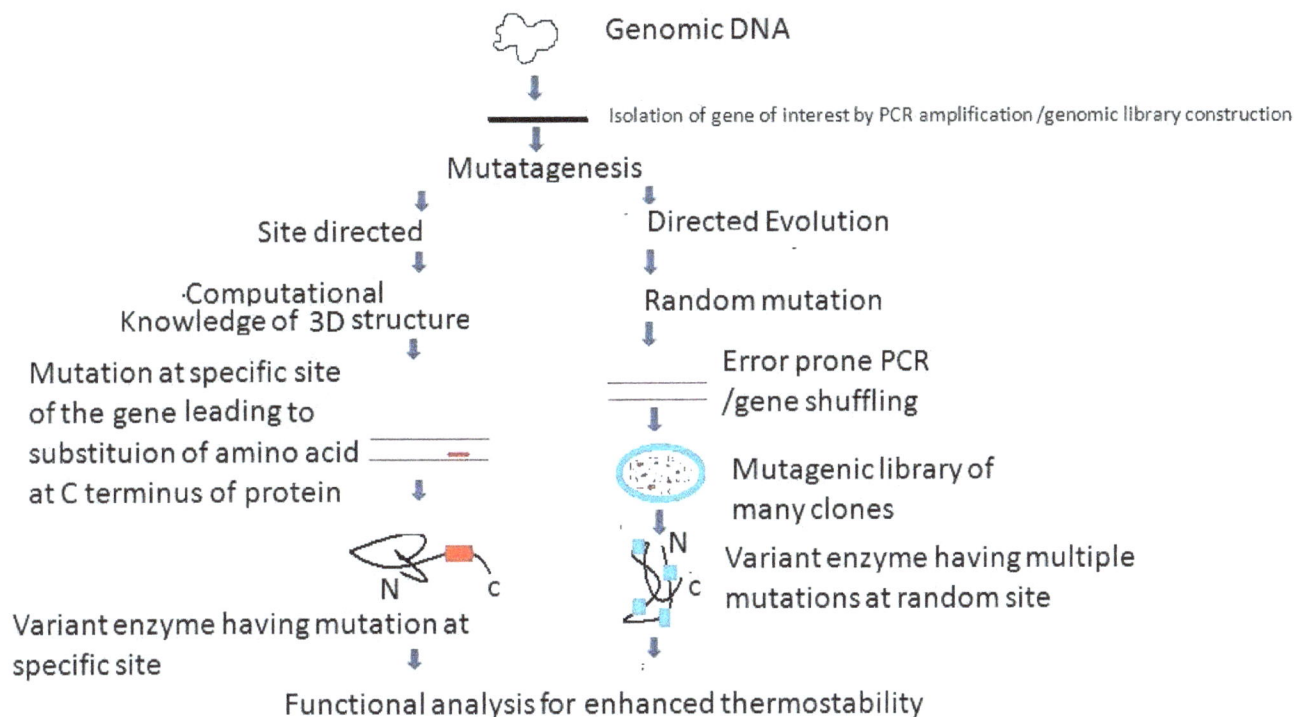

Figure 1. Flow chart representing methodology involved in mutagenesis.

1959; Franks, 2002). In addition to this, stability of a folded protein is modulated by weak interactions that is, vander Waals interactions, H-bonding, salt bridges, aromatic-aromatic interactions and disulfide bonds etc. (Feller, 2010). Interestingly, despite difference in protein sequences and structures of extremophiles and its counterpart mesophile, there is only marginal difference in their free energy of stabilization. Therefore, no general strategy of stabilization has yet been established for specifying a structure that is carrying a particular function (Jaenicke and Bohm, 1998; Leisola and Turunen, 2007). In this review, we are primarily discussing methods (directed evolution/rational/computational) in modifying industrially relevant enzymes for enhanced thermostability. This review further shed light into various factors responsible for enhanced thermostability.

DIRECTED EVOLUTION

Directed evolution is routinely used by several research groups in creating enzymes with improved properties. Directed evolutions methods that is, error prone polymerase chain reaction (PCR) and gene shuffling generate arbitrary mutations, and resulted in positive/negative or neutral mutations. Despite lack of prior knowledge of the structure, the mutations obtained using these methods has produced a number of thermostable enzyme variants. Here, we have provided

few examples such as, p-nitrobenzylesterase (Spiller et al., 1999), subtilisin E (Zhao and Arnold, 1999; Miyazaki et al., 2000), lipase (Ahmad et al., 2008), N-carbamyl-D-amino acid amidohydrolase (Yu et al., 2009), histone acetyltransferase (Leemhuis et al., 2008), asparaginase (Kotzia and Labrou, 2009) amylase (Kim et al., 2003). Interestingly, thermostabilizing mutation created by directed evolution sometime resulted in rare mutation, and cannot be rationalize structurally. Never the less, these mutations contribute a lot in understanding structure and function of the enzymes.

Furthermore, data gathered from several studies point out that mutation at protein surface are important in enhancing thermostability. Below, we are discussing few examples where mutation at surface has resulted in enhanced protein thermostability. An interesting example is generation of a variant subtilisin E from mesophilic bacteria *Bacillus subtilis* that showed remarkable increase in its optimum temperature (approximately 17°C) compared to WT. Moreover, the variant protein showed identical optimum enzyme activity, as observed with thermophilic homologue thermitase, produced by *Thermoactinomyces vulgari* (Zhao and Arnold, 1999). In another case study, seven arbitrary mutations were created in a maltogenic amylase by deoxyribonucleic acid (DNA) shuffling which enhanced optimum temperature of the enzyme by 15°C, compared to wild type enzyme. O ut of these, the mutations A398V and Q411L stabilized the enzyme by enhancing inter domain hydrophobic interactions, while other two mutations R26Q and P453L resulted in increased H bonding (Kim et al., 2003).

RATIONAL APPROACHES

Here, prior knowledge of the three dimensional structures is prerequisite for selecting useful mutations. There are number of rational for designing useful mutations, we are discussing few examples, first case where improved residual packing of protein structure (lipase) of a *Bacillus subtilis* resulted in enhanced protein thermostability. Here, glycine and alanine residues were targeted to improve packing of protein structures and consequently site directed mutagenesis was done to generate variants A38V, G80A and G172A. Interestingly, the packing at the interior of the protein enhanced the stability (Abraham et al., 2005). In second case, twelve mutations were created in lactate oxidase, based on following parameters; mutation that decrease enthalpy and increase free energy of difference between folded and unfolded state, changing amino acids at the interface, which can resist irreversible denaturation, replacement of hydrophobic pocket residues, for example, alanine to valine/leucine.

So far, out of these twelve deliberately designed rational mutations, only one had a positive effect on thermostability (Kaneko et al., 2005). Nevertheless, several other mutations were created by considering number of factors that had improved thermostability of mutant protein to greater extent than the wild type, for example, entropic stabilization by mutating Gly→Ala, X-Pro (Matthews et al., 1987), enhanced secondary structure propensity (Zhao and Arnold, 1999), increased H-bonding (Kim et al., 2003), hydrophobic interactions (Song and Rhee, 2000), electrostatic interactions (Blasco et al., 2000), helix capping by introducing residues that interact with α- helix dipole (Nicholson et al., 1988, 1991), disulfide bridges (Wang et al., 2006), salt bridges (Serrano et al., 1990), aromatic interactions (Burley and Petsko, 1985).

DIRECTED/RATIONAL

It is worth to discuss here that designing thermostable mutations rationally is difficult, because many of these are neutral and destabilizing. Additionally, despite several efforts, no universal strategy could be generalized for creating stable mutations. Furthermore, the molecular understanding of proteins is narrow and predicting mutation on rational basis is not so robust (Kaneko et al., 2005). On the other side, though the success of getting a desired mutation is less in directed evolution strategy, this technique has resulted in evolution of favorable mutations that further enriched understanding of structure and function of a protein to some extent and can also

direct in scheming rational mutation. The major limitation for directed evolution approach is lack of efficient screening for selection of desired variants.

COMPUTATIONAL METHODS IN PREDICTING STABILITY (SEQUENCES/STRUCTURE)

Computational methods have contributed significantly in designing effective and favorable mutations (Kraemer-Pecore et al., 2001; Gordon et al., 1999). With the rapid expansion of structural bioinformatics (RCSB) and protein data bank (PDB), there will be enormous benefit for structural engineer in creating vast libraries of desired variants (Johannes and Zhao, 2006). In addition, a number of software have been developed which can predict structure of the thermostabilizing mutation (Dantas et al., 2003). Furthermore, software like homology derived secondary structure of proteins (HSSP) is useful computational tool that distinguish conserved and variable regions of sequential database. On specifying PDB code, it scans multiple sequences and identifies evolutionary conserved amino acid positions which are projected at three dimensional levels (Patrick and Firth, 2005). The methods were employed previously in predicting three mutations with a modeled enzyme that improved melting temperature of protein by 10°C and half life of the protein to 30 fold at 50°C (Korkegian et al., 2005).These structural predictions can efficiently be used in designing high temperature adapting mutations. Computational analysis of protein structures is very promising in designing mutations. Basu and Sen (2009) also suggested the use of computational tools for designing thermostabilizing mutations. The structure based protein designing had successfully been used for producing a thermostable papain (a plant cysteine protease) by performing mutations in the inter domain region, to enhance H-bond, salt bridge interactions and to reduce the flexibility. These all factors contributed significantly in enhancing the thermostability of papain mutant whose half life was extended by 94 minutes at 60°C (Choudhury et al., 2010). In summar y, for designing thermo stabilizing mutations, physical interactions among residues at atomic level should be optimized, while steric constraints should be avoided during substitution. Recently, a fully automatic protocol (Rosetta $_{VIP}$) was developed which can select point mutation that improve the quality of core packing and may provide enhanced thermodynamic stability to the protein (Borgo and Havranek, 2012).

ROLE OF HYPERTHERMOPHILLIC GENES/ HOMOLOGUES IN PREDICTING STABLE MUTATIONS

Alternatively, thermostable proteins can be generated by

introducing conserve amino acid residues from their ancestral hyperthermophilic homologues. Let us discuss few such examples where this approach was utilized successfully for obtaining thermostable proteins for example, a thermostable isocitrate dehydrogenase from *Cladococcus noboribetus* (Iwabata et al., 2005) and isopropyl malate dehydrogenase from *Thermus thermophilus* (Watanabe et al., 2006). In another case, several ancestral residues were incorporated in β-amylase of a *Bacillus circulans* which enhanced the thermostability of mutant protein compared to wild type amylase (Yamashiro et al., 2010). In one interestingly example, a mesophilic triosephosphate isomerase was converted into a super stable enzyme, without losing its catalytic power on replacing its highly conserved residues (Williams et al., 1999). During *in vitro* evolution of *Bacillus* lipase (lip A) a remarkable 15℃ shift in melting temperature and approximately 20℃ shift in thermal denaturation was observed.

Interestingly, multiple sequence alignment of this lipase suggests that all these mutations were present in at least one of its natural homologue (Ahmad et al., 2008). In another case study, a lipase with six mutations was generated from *Bacillus* by replacing glycine to valine/alanine. Out of these six, three mutants (A38V, G80A, and G172A) showed enhanced stability and were found to be conserved (Abraham et al., 2005). Random evolution methods in combination with semi rational consensus approach generated a thermostable xylanase XT6 from *Geobacillus stearothermophilus* (Zhang et al., 2010). Recently, we also reported one thermostable mutant where the altered amino acid of metagenomic lipase replaces conserved amino acids (Sharma et al., 2012). Hence, it becomes evident from aforementioned examples that laboratory evolved thermostable enzymes in majority of the cases, substituted those residues which were already existed in their natural homologues and may be helpful in predicting thermostable mutations.

MAJOR FACTORS CONTRIBUTING TOWARDS STABILIZING MUTATIONS

For successful engineering of thermostable proteins, we need first to understand the factors associated with the thermostability. Interestingly, the structural characterization of several thermo stabilizing mutations revealed that they are generally affected by secondary structure, solvent accessibility of the important residues and hydrophobicity profile of a protein. Below, we attempted to shed light on controlling role of these factors in protein stability.

1. Thermostability of a protein is highly influenced by secondary structure that is, helix, sheet, coil or random turns. Previous studies have shown that most remarkable thermostabilizing mutations generally lie in the random

coils, turns and loops. On the other side, the helix and sheets are less cooperative for such substitutions, may be because of less plasticity in these regions. Some of the studies carried out previously reflect that most of these mutations are present in the loop and random coils, e.g. mutation G195E in galactose oxidase (Sun et al., 2001) lipase (Acharya et al., 2004; Ahmad et al., 2008) subtilisin E (Zhao, 1999) horse radish peroxidase (Ryan and O'Fágáin, 2008), α glucosidase, (Zhou et al., 2010), L-asparaginase (Kotzia and Labrou, 2009), *N*-carbamyl-D-amino acid amidohydrolase (Yu et al., 2009).

2. Next, the presence of hydrophobic residues at the interior of proteins helps in maintaining the structural integrity. They determine the folding process of protein (Dill, 1990), interestingly, sequence alignment from various proteins revealed that the hydrophobic core regions in proteins is conserved throughout evolution, therefore suggesting possible role of these residues in maintaining stability and function (Di et al., 2003). The conserve nature of the proteins at the interior may be attributed to low substitution rate due to selective pressure (Lim and Sauer, 1989, 1991; Smith and Raines, 2006). An example is ubiquitin, where random mutation has revealed that specific core packing arrangements are critical for maintaining stability (Finucane and Woolfson, 1999). Furthermore, most of the residues essential for function of a bovine pancreatic ribonuclease (RNase A) were found buried in hydrophobic core (Smith and Raines, 2006). In another study, out of 15 serine residues of an RNase, only one that is, Ser75 found to be buried, conserved and critical for stability (Johnson et al., 2007). Many studies have proven that mutation within core of proteins are destabilizing, because of formation of cavity that resulted in loss of vander waals interactions (Xu et al., 1998; Ratnaparkhi and Varadarajan, 2000; Kono et al., 2000; Chakravarty et al., 2002; Vlassi et al., 1999; Beadle and Shoichet, 2002).

3. It is further suggested that surface residues have high solvent accessibility relative to average residue of protein and contributes significantly in stabilization (Leemhuis et al., 2008). Replacing a charged residue at the protein surface must fit criteria of optimized local interactions, while minimizing repulsive interactions; otherwise such substitution may result in destabilization of the protein structure, as observed in ribonuclease H1 of the *E. coli* (You et al., 2007). Recently, a cellobiohydrolase from *Talaromyces emersonii* (*Te* Cel7A) showed enhanced protein thermostability, when five disulphide bonds were engineered at surface of the protein (Voutilainen et al. 2010). Furthermore, two human acylphosphatase enzymes (AcPh and Cdc42 GTPase) were made thermostable by bringing in residues that increase the surface charge interactions (Gribenko et al., 2009). Typically, polar residues are preferred at protein surface; however current studies pointed out that even hydrophobic interaction can contribute to stability for example, on replacing a charge residue with hydrophobic

one (arginine) at the surface of helical protein, enhances protein stability (Spector et al., 2000), another such example is enhancement of thermostability in family 11 of xylanases from *Bacillus subtilis* (Miyazaki et al., 2006). Briefly, such an observation may be attributed to flexibility of protein at the surface.

CASE STUDIES FROM OUR LABORATORY

Our laboratory is working in field of directed evolution/site directed mutagenesis for the past 10 years. Recently, we reported a highly thermostable lipase mutant (Sharma et al., 2012). The mutation was observed on the surface of protein having substitution N355K. Interestingly, the mutant lipase showed 144 fold enhancement in the thermostability compared to WT lipase. Moreover the catalytic efficiency was also enhanced 20 folds. During biophysical characterization, we found that mutant retained its secondary structure at high temperature compared to wild type enzyme, simultaneously, the molecular dynamics of the generated three dimensional structure revealed increase in the hydrogen bonding. In another case, Khurana et al. (2010) reported a mutant of a *Bacillus* lipase that showed enhanced half of the mutant 3 folds at 50°C. The kinetic parameters of the mutant enzyme were significantly altered. The mutation was observed on the part of helix which is exposed to the solvent and away from the catalytic triad. This study demonstrated that replacement of a solvent exposed hydrophobic residue (Ile) in WT to a hydrophilic residue (Thr) in mutant might impart thermostability to the protein.

CONCLUSION

Biochemical nature of proteins is largely determined by number of weak non-covalent interactions which make stability of protein highly unpredictable. Thermostable variants generated by directed evolution and rational approaches can cast light into molecular behavior of proteins. Consequently, with the availability of massive sequence, structure database and *in silico* tools along with robust molecular techniques, it is becoming easy for protein biochemists to alter the function of the biocatalysts beyond their natural functions. Furthermore, the protein engineering hold promises to unravel the mechanism underlying biochemical and molecular function of proteins.

ACKNOWLEDGEMENTS

Authors wish to thank University Grant Commission, New Delhi India, for granting Junior and Senior Research Fellowship to SG, and Council of Scientific and Industrial, New Delhi India, for granting Senior Research Fellowship to PKS.

REFERENCES

Abraham T, Pack SP, Yoo YJ (2005). Stabilization of *Bacillus subtilis* Lipase A by increasing residual packing. Biocatal. Biotran. 23:217-224.

Acharya P, Rajakumara E, Sankaranarayanan R, Rao NM (2004) Structural basis of selection and thermostability of laboratory evolved *Bacillus subtilis* lipase. J. Mol. Biol. 341:1271-1281.

Ahmad S, Kamal MZ, Sankaranarayanan R, Rao NM (2008). Thermostable *Bacillus subtilis* Lipases: *In vitro* evolution and structural insight. J. Mol. Biol. 381:324-340

Basu S, Sen S (2009) Turning a mesophilic Protein into a thermophilic one: A computational approach based on 3D structural features. J. Chem. Inf. Model. 49:1741-1750.

Beadle BM, Shoichet BK (2002). Structural bases of stability–function tradeoffs in enzymes. J. Mol. Biol. 321:285-296.

Blasco GG, Aparicio JS, Gonzalez B, Hermoso JA, Polaina J (2000) Directed evolution of beta glucosidase from *Paenobacillus polymyxa* to thermal resistance. J. Biol. Chem. 275:13708-13712

Bogin OM, Peretz Y, Hacham Y, Burstein Y, Korkhin Y, Kalb J, Frolo F (1998). Enhanced thermal stability of *Clostridium beijerinckii* alcohol dehydrogenase after strategic substitution of amino acid residues with prolines from the homologous thermophilic *Thermoanaerobacter brockii* alcohol dehydrogenase. Protein Sci. 7:1156-1163.

Borgo B, Havranek JJ (2012). Automated selection of stabilizing mutations in designed and natural proteins. Proc. Natl. Acad. Sci. 109(5):1494-1499.

Burley SK, Petsko GA (1985). Aromatic–aromatic interaction: a mechanism of protein structure stabilization. Science 229:23-28.

Chakravarty S, Bhinge A, Varadarajan R (2002). A procedure for detection and quantitation of cavity volumes in proteins. Application to measure the strength of the hydrophobic driving force in protein folding. J. Biol. Chem. 277:31345-31353.

Choudhury D, Biswas S, Roy S, Dattagupta JK (2010). Improving thermostability of papain through structure-based protein engineering. Prot. Eng. Des. Select. 23(6):457-467.

Crabb WD, Mitchinson C (1997). Enzymes involved in the processing of starch to sugars. *Trends* Biotechnol. 15:349-352.

Dantas G, Kuhlman B, Callender D, Wong M, Baker D (2003). A large scale test of computational protein design: folding and stability of nine completely redesigned globular protein. J. Mol. Biol., 332:449-460.

Di NAA, Larson SM, Davidson AR (2003). The relationship between conservation, thermodynamic stability and function in the SH3 domain hydrophobic core. J. Mol. Biol. 333:641-655.

Dill KA (1990). Dominant forces in protein folding. Biochemistry 29:7133-7155.

Feller G (2010). Protein stability and enzyme activity at extreme biological temperatures. J. Phys. Condens. Matter 22:323101.

Finucane MD, Woolfson DN (1999). Coredirected protein design. II. Rescue of a multiply mutated and destabilized variant of ubiquitin. Biochemistry 38:11613-11623.

Franks F (2002). Protein stability: The value of 'old literature'. Biophys. Chem. 96(2/3):117-127.

Gordon DB, Marshall SA, Mayo SL (1999). Energy functions for protein design. Curr. Opin. Str. Biol. 9:509.

Gribenko AV, Patel MM, Liu J, McCallum SA, Wang C, Makhatadze GI (2009). Rational stabilization of enzymes by computational redesign of surface charge– charge interactions. Proc. Natl. Acad. Sci. 106:2601-2606.

Iwabata H, Watanabe K, Ohkuri T, Yokobori S, Yamagishi A (2005). Thermostability of ancestral mutants of *Caldococcus noboribetus* isocitrate dehydrogenase. FEMS Microbiol. Lett. 243:393-398.

Jaenicke R, Bohm G (1998). The stability of proteins in extreme environments. Curr. Opin. Str. Biol. 8:738-748.

Johannes T, Zhao H (2006). Directed Evolution of Enzymes and Biosynthetic Pathways. Curr. Opin. Microb. 9:261-267.

Johnson RJ, Lin SR, Raines RT (2007). Genetic selection reveals the role of a buried, conserved polar residue. Prot. Sci. 16:1609-1616.

Kaneko H, Minagawa H, Shimada J (2005). Rational design of thermostable lactate oxidase by analyzing quaternary structure and prevention of deamidation. Biotechnol. Lett. 27:1777-1784.

Kauzmann W (1959). Some factors in the interpretation of protein denaturation. Adv. Prot. Chem., 14: 1-63

Khurana J, Singh R, Kaur J (2010). Engineering of Bacillus lipase by directed evolution for enhanced thermal stability: effect of isoleucine to threonine mutation at protein surface. Mol. Biol. Rep. DOI 10.1007/s11033-010-9954.

Kim YW, Choi JH, Kim JW, Park C, Kim JW, Cha H (2003). Directed evolution of Thermus maltogenic amylase toward enhanced thermal resistance. Appl. Environ. Microbiol. 69:4866-4874.

Kono H, Saito M, Sarai A (2000). Stability analysis for the cavity-filling mutations of the Myb DNA-binding domain utilizing free-energy calculations. Prot. Str. Funct. Genet. 38:197-209.

Korkegian A, Black ME, Baker D, Stoddard BL (2005). Computational thermostabilization of an enzyme, 308:857-860.

Kotzia GA, Labrou NE (2009). Engineering thermalstability of asparaginase by directed evolution. FEBS J. 276:1750-1761.

Kraemer-Pecore CM, Wollacott AM, Desjarlais JR (2001). Computational protein design. Curr. Opin. Chem. Biol. 5:690-695.

Kumar S, Ma B, Tsai CJ (2000). Electrostatic strengths of salt bridges in thermophilic and mesophilic glutamate dehydrogenase monomers. Proteins, 38: 368-383.

Leemhuis H, Nightingale KP, Hollfelder F (2008). Directed evolution of a histone acetyltransferase – enhancing thermostability, whilst maintaining catalytic activity and substrate specificity. FEBS J. 275: 5635–5647

Leisola M, Turunen O (2007). Protein engineering: opportunities and challenges. App. Microbiol. Biotechnol. 75: 1225-32.

Lim WA, Sauer RT (1991). The role of internal packing interactions in determining the structure and stability of a protein. J. Mol. Biol. 219: 359-376.

Lim WA, Sauer RT (1989). Alternative packing arrangements in the hydrophobic core of lambda repressor. Nature, 339:31–36.

Matthews BW, Nicholson H, Becktel WJ (1987). Enhanced protein thermostability from site-directed mutations that decrease the entropy of unfolding. Proc. Natl. Acad. Sci. USA 84:6663–6667.

Miyazaki K, Takenouchi M, Kondo H, Noro N, Suzuki M, Tsuda S (2006). Thermal stabilization of Bacillus subtilis family-11 xylanase by directed evolution. J. Biol. Chem. 281:10236-10242.

Miyazaki K, Wintrode PL, Grayling RA, Rubingh DN, Arnold FH (2000). Directed evolution study of temperature adaptation in a psychrophilic enzyme. J. Mol. Biol. 297:1015-1026.

Nicholson H, Anderson DE, Dao-pin S, Mathews BW (1991). Analysis of the interaction between charged side chains and the alpha-helix dipole using designed thermostable mutants of phage T4 lysozyme. Biochemistry 30:9816-9828.

Nicholson H, Becktel WJ, Matthews BW (1988). Enhanced protein thermostability from designed mutations that interact with alpha-helix dipoles. Nature 36:651-656.

Patrick WM, Firth AE (2005). Strategies and computational tools for improving randomized protein libraries. Biomol. Eng. 22:105-112

Ratnaparkhi GS, Varadarajan R (2000). Thermodynamic and structural studies of cavity formation in proteins suggest that loss of packing interactions rather than the hydrophobic effect dominates the observed energetics. Biochemistry 39:12365-12374.

Russell RJ, Ferguson JM, Hough DW, Danson MJ (1997). The crystal structure of citrate synthase from the hyperthermophilic archaeon Pyrococcus furiosus at 1.9A resolution. Biochemistry 36:9983-9994.

Ryan BJ, O'Fágáin C (2008). Effects of mutations in the helix G region of horseradish peroxidase. Biochemistry 9:1414-1421.

Sadeghi M, Naderi-Manesh H, Zarrabi M, Ranjbar B (2006). Effective factors in thermostability of thermophilic proteins. Biophys. Chem. 119:256-270.

Serrano L, Horovitz A, Avron B, Bycroft M, Fersht AR (1990). Estimating the contribution of engineered surface electrostatic interactions to protein stability by using double-mutant cycles. Biochemistry 2:9343-9352.

Sharma PK, Kumar R, Kumar R, Mohammad O, Singh R, Kaur J (2012a). "Engineering of a metagenome derived lipase towards thermal tolerance: effect of asparagine to lysine mutation on the protein surface" Gene, 10: 491(2):264.

Smith BD, Raines RT (2006). Genetic selection for critical residues in ribonucleases. J. Mol. Biol. 362:459-478.

Song JK, Rhee JS (2000). Simultaneous Enhancement of Thermostability and Catalytic Activity of Phospholipase A₁ by Evolutionary Molecular Engineering Appl. Environ. Microbiol. 66(3):890-894.

Spector S, Wang M, Carp SA, Robblee J, Hendsch OZS, Fairman R, Tidor OB, Raleigh DP (2000). Rational modification of protein stability by the mutation of charged surface residues. Biochemistry 39:872-879

Spiller B, Gershenson A, Arnold FH, Stevens RC (1999). A structural view of evolutionary divergence. Proc. Natl. Acad. Sci. USA, 96: 12305–12310

Sun L, Petrounia IP, Yagasaki M, Bandara G, Arnold FH (2001). Expression and stabilization of galactose oxidase in Escherichia coli by directed evolution. Protein Eng., 14: 699-704.

Turner P, Mamo G, Karlsson EN (2007). Potential and utilization of thermophiles and thermostable enzymes in biorefining Microbial Cell Factories, 6: 9

Vlassi M, Cesareni G, Kokkinidis M (1999). A correlation between the loss of hydrophobic core packing interactions and protein stability. J. Mol. Biol. 2:817-827.

Voutilainen SP, Murray PG, Tuohy MG, Koivula A (2010). Expression of Talaromyces emersonii cellobiohydrolase Cel7A in Saccharomyces cerevisiae and rational mutagenesis to improve its thermostability and activity. Prot. Eng. Des. Sel. 23:69-79.

Wang Y, Fuchs E, da Silvac R, McDaniel A, Seibele JC, Ford C (2006). Improvement of Aspergillus niger glucoamylase thermostability by directed evolution. Starch 58:501-508.

Watanabe K, Ohkuri T, Yokobori S, Yamagishi A (2006). Designing thermostable proteins: ancestral mutants of 3-isopropylmalate dehydrogenase designed by using a phylogenetic tree. J. Mol. Biol. 355:664-674.

Williams JC, Zeelen JP, Neubauer G, Vriend G, Backmann J, Michels PAM, Lambeir AM, Wierenga RK (1999). Structural and mutagenesis studies of leishmania triosephosphate isomerase: a point mutation can convert a mesophilic enzyme into a superstable enzyme without losing catalytic power. Protein Eng. 12:243-250.

Xu J, Baase WA, Baldwin E, Matthews BW (1998). The response of T4 lysozyme to large-tosmall substitutions within the core and its relation to the hydrophobic effect. Protein Sci. 7: 158-177.

Yamashiro K, Yokobori SI, Koikeda S, Yamagishi A (2010). Improvement of Bacillus circulans b-amylase activity attained using the ancestral mutation method. Prot. Eng. Des. Sel. 23:519-528.

You DJ, Fukuchi S, Nishikawa K, Koga Y, Takano K, Kanaya S (2007). Protein thermostabilization requires a fine-tuned placement of surface-charged residues. J. Biochem. 142:507-516.

Yu H, Li J, Zhang D, Yang Y, Jiang W, Yang S (2009). Improving the thermostability of N-carbamyl-D-amino acid amidohydrolase by error-prone PCR. Appl. Microbiol. Biotechnol. 82:279-285.

Zhang ZG, Yi ZL, Pei XQ, Wua ZL (2010). Improving the thermostability of Geobacillus stearothermophilus xylanase XT6 by directed evolution and site-directed mutagenesis. Bioresour. Technol. 101:9272-9278.

Zhao H, Arnold FH (1999). Directed evolution converts subtilisin E into a functional equivalent of thermitase. Prot. Eng. 12:47-53.

Zhou C, Xue Y, Ma Y (2010). Enhancing the thermostability of α-glucosidase from Thermoanaerobacter tengcongensis MB4 by single proline substitution J. Biosci. Bioeng. 110(1):12-17.

17

Antigen specific cellular response in patients with hepatitis C virus infection and its association with HLA alleles

Chinmaya Mahapatra[1]* and Anuradha S. Tripathy[2]

[1]Dr. D. Y. Patil Biotechnology and Bioinformatics Institute, Dr. D. Y Patil University, Pune, India.
[2]Immunology Research Group, National Institute of Virology, Indian Council of Medical Research, Pune, India.

Host genetic diversity is believed to contribute to the spectrum of clinical outcomes in hepatitis C virus (HCV) infection. The present study aimed at finding out the frequencies of HLA class I alleles of HCV infected individuals from Western India (Maharashtra State). Forty-three clinically characterized anti-HCV positive patients from Maharashtra were studied for HLA A, B and C alleles by PCR- sequence specific primer (SSP) typing method and compared with 67 and 113 ethnically matched anti-HCV negative healthy controls from Western India (Maharashtra State). The study's analysis reveals an association of HLA alleles A*03 (OR = 16.69, EF, 0.44, P = 7.9E-12), A*32 (OR = 1474, EF 0.21, P = 1.8E-9), HLA B*15 (OR = 14.11, EF 0.39, P = 2.18E-10), B*55 (OR = 12.09, EF 0.07, P = 0.005), Cw*16 (OR = 7.45, EF 0.12, P = 0.001) and Cw*18 (OR = 402, EF 0.05, P = 0.003), with HCV infection chronicity, while its results suggest that the establishment of viral persistence in patients is due to a failure of the immune response and is associated with HLA class 1 allele (mainly, A*03, A*32, B*15, B*55, Cw*16 and Cw*18 restricted individuals), as indicated by the absence of a significant T-cell response. Thus, this proves that associated haplotype influence HCV infection as a host genetic factor.

Key words: Hepatitis C virus (HCV), human leukocyte antigen - HLA A*03, A*32, B*15, B*55, Cw*18, Western India.

INTRODUCTION

Hepatitis C virus (HCV) is an important blood borne pathogen that is eliminated from the host in approximately 15% of acutely infected individuals, but persists in the remaining 85% (Anuradha et al., 2009). HCV is responsible for a wide spectrum of chronic liver lesions ranging from minimal to cirrhosis or hepatocellular carcinoma (HCC) and fatal outcome. Both virus-related factors such as viral heterogeneity and replicative activity (Silini et al., 1995) and the host determinants such as lack of efficient immune responses (Spengler et al., 1996) are involved in the pathogenesis of chronic hepatitis (Nathalie and Patrick, 2000). Further, the liver damage in HCV infected patients is probably associated with direct cytopathic effects and immune mediated mechanisms (Abdul, 2008). Though the exact basis for the differential clinical presentation of HCV infection is not fully understood, the viral load and genotype have been reported to influence the prognosis. The observation of different clinical presentations despite the same source of infection led to the recognition of the importance of host genetic factors in disease manifestations (Yee, 2004). In an Irish cohort, of the 704 women infected with HCV from contaminated anti-D immune globulin, 390 (55%) became persistently infected (Yee, 2004). MHC class I and class II antigens are central to the host immune response and thus are ideal candidate genes to investigate the associations with HCV. Classes I and II HLA are encoded by the most polymorphic genes that present antigens to CD8+ cytotoxic T cells and D4+ helper T cells, respectively. Polymorphisms in binding regions of these

---

*Corresponding author. E-mail: cmbbsr@gmail.com.

molecules determine antigenic specificities and the strength of the immune response to a given pathogen (Wang, 2003). Moreover, during cellular immune response, HLA class I molecules may present HCV epitopes to cytotoxic T cells, resulting in a protective immune response. The human leukocyte antigen is a crucial genetic factor that initiates and regulates immune responses by presenting foreign or self-antigens to T lymphocytes. Certain HLA alleles have been shown to influence the outcome of chronic HCV infections (Just, 1995; Thio et al., 1999)), while various HLA alleles have been linked with either persistence or clearance of the virus. Several studies have aimed to identify the involvement of HLA with different outcomes of HCV infection, but the results have not been consistent. Moreover, the literature review revealed that the prevalence of HCV infection was significantly low in the Indian population (Jain et al., 2003; Chowdhury et al., 2003; Irshad et al., 1995). So far, no data on HLA association with HCV infection have been reported from Western India. Hence, this study was undertaken, to assess specific cellular response against different HCV antigens in the peripheral blood of HCV infected patients to assess whether or not there is any genetic basis for the involvement of the HLA class I alleles with the disease course and outcomes in HCV infection

MATERIALS AND METHODS

Study population

The study was carried out at NIV, Pune. A total of 43 anti-HCV positive individuals selected from different parts of Maharashtra, over a period of three years (2009 and 2010), were included in this study. All these individuals were referred to NIV, Pune for diagnosis of hepatitis C infection by molecular and serological testing. All patients were tested for HLA A, B and C alleles. Of these patients, 33 were HCV RNA positive. Among these positives, 10 were on maintenance haemodialysis, 8 were voluntary blood donors and the remaining 15 were suffering from chronic liver diseases. Furthermore, patients suffering from chronic liver diseases had elevated levels of serum alanine aminotransferase (ALT), leaving aside 8 of 15 patients suffering from chronic liver disease, while none of the anti-HCV individuals were hospitalized. All of them were tested negative for both hepatitis B surface antigen (HBsAg) and anti-HIV antibodies. The control group consisted of 67 healthy unrelated normal Maharashtrian subjects for HLA A, B and C. Controls were negative for anti-HCV antibodies, HBsAg and anti-HIV antibodies.

Serological testing

All the samples were screened for the presence of anti-HCV antibodies using enzyme linked immunosorbent assay (ELISA, Ortho Clinical diagnostics, Inc. IIIrd generation, New Jersey, USA), while anti-HIV antibodies and HBsAg were screened by ELISA (Lab systems HIV EIA and Surase B-96, General Biological Corporation, Taiwan). Serum ALT was measured by a commercial kit (Span Diagnostics Ltd, India) and all tests were done according to the manufacturer's protocol. HCV RNA was detected by reverse

transcription nested polymerase chain reaction (RT-PCR) using primers located in the highly conserved 5' NCR region (Bukh et al., 1992).

HLA typing

Genomic DNA was extracted from frozen peripheral blood mononuclear cells by Qiagen Blood mini kit (Germany). Molecular typing was carried out by polymerase chain reaction-sequence specific primer (PCR-SSP) method (Olerup SSP AB, GenoVision, Inc, Sweden) utilizing allele specific primers along with the control primers to identify the respective alleles. The allele specific primers were provided in the kit (Olerup SSP HLA-A-B-C SSP Combi Tray). The primer set contained 5' and 3' primers for grouping the HLA-A *0101 to *8001 alleles, 5' and 3' primers for grouping the B*0702 to *8302 alleles and 5' and 3' primers for grouping the Cw*0102 to Cw*1802 alleles. The amplified products were visualized under UV following agarose gel electrophoresis, and the interpretations of the alleles were based on Helmberg SCORE programme (update V3.118 KIT software GenoVision, Inc, Sweden).

PBMC isolation

Blood samples were drawn in K3 EDTA tubes. Ficoll-hypaque solution was taken as 1:2 proportions in 15 ml tube. The blood sample was layered from vaccutainer on ficoll by adding it to the tube from the wall of the tube. The tube was centrifuged at 2000 RPM for 40 min. After centrifugation was over, 4 distinct bottom layer RBCs were observed, then Ficoll, PBMC (white opaque ring or buffy coat) and the uppermost plasma layers (7 to 8 ml RPMI) were taken with a centrifuge tube. The plasma layer was removed by pasture pipette collected into a tube and kept at +4°C, while PBMC was collected by pasture pipette suspended in RPMI for first washing at 800 to 1000 RPM in the centrifuge for 10 min. After centrifugation, PBMC was removed from the centrifuge and the medium was discarded. Fresh RPMI (7 ml) were added into the tube and the PBMC were re-suspended again and centrifuged for a second wash at 800 to 1000 RPM for 10 min. After the second washing had removed the tubes from the centrifuge and the medium had been discarded, and after 5 ml of fresh RPMI had been added and the PBMC had been suspended in it, 20 μl of this suspension is taken in a micro well plate, and 20 μl of Trypan Blue dye mix is added to this mixture. Subsequently, 10 μl is loaded on Heamocytometer and the PBMC is counted in the WBC chamber under microscope. After counting of gels was completed, in the centrifuge tube at 1000 RPM for 5 min, the supernatant was discarded. PBMC was suspended in complete medium (RPMI 1640 + 10% FBS) in such a manner that each ml contained 10^6 PBMC kept in incubator at 37°C.

Nested PCR of 5' non-coding region of hepatitis C virus

QIAamp Viral RNA Mini Kit combines the selective binding properties of a silica-gel based membrane with the speed of micro spin or vacuum technology and is ideally suited for simultaneous processing of multiple samples. The sample was first lysed under highly denaturing conditions to inactivate RNase and to ensure isolation of intact viral RNA. Buffering conditions were then adjusted to provide optimum binding of the RNA to the QIAamp membrane, and the sample was loaded onto the QIAamp mini spin column. The RNA binds to the membrane, and the contaminants are efficiently washed away in two steps using two different wash buffers. High

Table 1. Primers used for HCV diagnosis.

Primer	Sequence
External forward (JENS1)	5' ACT GTC TTC ACG CAG AAA GCG TCT AGC CAT 3'
External reverse (JENS 2)	5' CGA GAC CTC CCG GGG CAC TCG CAA GCA CCC 3'
Internal forward (JENS 3)	5' ACG CAG AAA GCG TCT AGC CAT GGC GTT AGT 3'
Internal reverse (JENS 4)	5' TCC CGG GGC ACT CGC AAG CAC CCT ATC AGG 3'

Table 2. Reverse transcription PCR mix.

Components	μl
D W	77
TAQ 10X with 25 mM MgCl$_2$	10
Primer 1 (Jens 1) 10 μM	5
Primer 2 (Jens 2) 10 μM	5
AMV RT 10 u/ μl (Promega)	1
Rnasin 40 u./ μl (Promega)	0. 5
TAQ polymerase (Perkin Elmer) 5u./ μl	0. 5
dNTP's (Promega) 25 mM	1
Total volume	100

Table 3. Reaction conditions for RT-PCR.

Steps	Temperature (°C)	Time (min)	Cycles
1	94	5.00	
2	94	1.00	
3	55	1.00	
4	72	1.00	For 35 cycles
5	72	3.00	
6	4	Store	

quality RNA is eluted in a special RNase free buffer ready for direct use or safe storage. The purified RNA is free of protein, nucleases and other contaminants and inhibitors. This RNA is used for cDNA preparation and subsequent amplification of the cDNA by PCR. This amplified product is then analyzed by gel electrophoresis. The region amplified (for qualitative or diagnostic PCR) is 5'NCR with 255 base pairs (5' NCR is the conserved region throughout the genome of HCV and is also among the major 6 HCV genotypes available) (Tables 1 to 6).

Reverse transcription PCR

100 μl of the RT-PCR was added to the dry pellet mixture. It was flicked to mix well and then spin down. Subsequently, the mixture was transfered to the PCR tubes.

The PCR mixture for nested PCR or 2nd round PCR

10 μl of the 1st PCR was added and mixed with the 2nd PCR mixture.

Gel electrophoresis

About 2.0% agarose (Sigma) gel was prepared in 1× TAE buffer (pH 8.0) and 8 μl of the PCR product was run by adding 2 μl of 6X loading dye to it. The gel was run by using Power pack (Biorad model 3000 xi) at 60 V for 45 min with 100 bp DNA ladder as marker. The gel was observed on transilluminator (Transilluminator UVP) and gel documentation system (Syngene). However, the use of the machine should be noted with details in the Logbook provided. Nested PCR products (PCR amplicons) were analyzed by electrophoresis on 2% TAE ultra pure agarose (invitrogen) gel and was visualized by ethidium bromide staining on a U. V. transilluminator. Subsequently, the image was captured in a gel documentation system. The gel was sliced from the region showing the band of the PCR product and was used for further processing of the cycle sequencing.

Elispot assay

PVDF membranes of plate were wet with 70% ethanol by adding 50 μl/ well. The plate was incubated for 5 min at room temperature in

Table 4. PCR mixture for nested PCR.

Components	µl
D W	36.25
Taq 10 X Buffer (promega)	5.00
PRIMER 3 (Jens 3) 10µM	1.50
PRIMER 4 (Jens 4) 10µM	1.50
Taq polymerase (promega)	0.25
DNTP's (promega)	0. 5
cDNA from 1st round PCR	5.00
Total volume	50.00

Table 5. Thermal conditions for nested PCR of 5' NCR of hepatitis C virus.

Reaction conditions		
Temperature (°C)	Time (min)	Cycles
42	60	
94	5	**35**
94	1	
55	1	
72	1	

decant ethanol solution. Plates were washed with 1X PBS thrice and 300 µl/ well were added to the coated HA or IP plate (Millipore Multiscreen) with 100 ul/ well antibody captured in PBS (for example, 5 ug/ml Mabtech anti-human IFN-gamma antibody). The solution was incubated overnight at 4°C in decant antibody solution. Plates were washed with 1X PBS thrice and 300 µl/ well were added to the block with 200 ul/ well of heat inactivated FBS (GIBCO) for at least 4 h at 37°C. When preparing the effector cells, the PBMC from blood sample were isolated by the Ficoll Hypaque density gradient method. They were washed in RPMI 1640 medium and counted by hemocytometer. Later, they were resuspended at a final concentration of 1×10^6 cells/ml in a complete medium containing RPMI + 10% FBS. The plate out assay was done in decant blocking medium and PBMC (recommendation: 1×10^5 cells/well) was gently plated out in 100 ul RPMI medium/well, where 50 ul HCV core peptide/ well was added to the solution. It was incubated for 24 h (for example, IFN-gamma assay) in the control sample or 48 h (for example, INF- gamma assay) in the HCV infected individual at 37°C and 5% $CO2$. Later, the cells were discarded and the plates were washed twice with 1X PBS and 4 times with 0.05% Tween 20+1X PBS (PBST).

A total of 100 ul/well biotinylated detection antibodies (for example, 1 ug/ml Mabtech biotinylated anti-human IFN-g antibody) were added to the PBS/0.5% BSA and were incubated for at least 3 h at room temperature. Afterwards, the biotinylated detection antibodies were discarded and the plates were washed twice with 1X PBS and 4 times with 0.05% Tween 20+1X PBS (PBST). DPBS (10 ml) was taken in Petri dish and 1 drop each of reagents A and B of VECTASTAIN and 10 µl of Tween 20 were added to it to prepare avidin. Subsequently, 10 ml of the mixture was filtered by 0.45 µ syringe filter, while avidin-enzyme-complex was discarded. It was washed thrice with PBS/0.05% of Tween 20, and was followed by a 3 times washing with plain 1X PBS. Afterwards, 100 ul AEC substrate per well was added and incubated for 4 min, while the spot development was stopped under tap water. The substrates

were removed under drain, while the excess liquid were removed from wells. The wells were dried thoroughly with a paper towel and the plate was allowed to dry overnight in darkness before the membrane was removed with ELI-Puncher Kit (ZellNet Consulting). Spots were counted (recommendation: KS Elispot Automated Reader System from Carl Zeiss) and the enzyme-complex (recommendation: from Vector Laboratories) was left at room temperature for about 30 min prior to use.

Furthermore, 100 ul of avidin-enzyme-complex/well was added to the solution and incubated for 1 h at room temperature. In preparing the substrate, 2.5 ml DMF was taken in a 15 ml tube cover with aluminium foil and 1 tablet of AEC was added to it. 47.5 ml of the D-Water (SIGMA) was taken in 50 ml tube and 280 µl sodium acetate, 180 µl acetic acid and 25 µl hydrogen peroxide were added to the tube. As a result, 2.5 ml DMF was mixed with AEC in this mixture.

Human leukocyte antigen MHC class 1 ABC -PCR ssp

Major histocompatibility complex (MHC) is the most polymorphic gene cluster known in man. Using conventional serology, one can determine HLA-A, HLA-B and HLA-C specificities. Serology has several limitations including cross reactivity, complement problems, nonavailability of good reagents and difficulties in phenotype assignment in cases where the expression of LA molecule is low or absent. Furthermore, the most available HLA anti-sera are developed in Caucasoid, and often, reagents from other ethnic groups cannot be used with the same degree of accuracy. The development of molecular typing procedures has considerably enhanced the study's capabilities of typing HLA alleles at the DNA level; thus, the HLA specific primers coated in plate amplified the DNA. This shift from serology to sequencing has been rapid. These technologies are more sensitive and allow the typing of several subtypes of commonly known alleles in HLA class 1 and class 2 loci. The molecular techniques are based on polymerase chain

Table 6. HLA distribution in anti-HCV positive individuals and controls from Maharashtra, India.

HLA	Anti-HCV positives (n = 43) AF (%)	Controls (n = 67) AF (%)	OR	Ki2	EF	PF	P value
A*03	47.91	5.22	16.69	12.12	0.44		7.9E-12**
A*11	2.08	11.94	0.15			0.10	0.040
A*24	6.25	23.88	0.21			0.18	0.007
A*26	2.08	0.74	2.82		0.01		
A*31	4.16	6.71	0.60				
A*32	22.91	0.00	1474.00		0.21		1.8E-8**
A*33	2.08	8.95	0.21			0.06	
A*66	6.25	1.49	4.40		0.04		
A*68	6.25	4.47	1.42				
B*07	10.41	9.70	1.08				
B*08	6.25	2.23	2.91		0.03		
B*15	43.75	5.22	14.11	10.42	0.39		2.18E-10**
B*27	4.16	2.98	1.41				
B*40	8.33	23.88	0.28				0.020
B*44	8.33	17.16	0.43			0.17	
B*51	8.33	4.47	1.93			0.09	
B*55	8.33	0.74	12.09	1.81	0.07		0.005
B*57	2.08	0.00	134.00		0.01		0.090
Cw*01	6.25	1.49	4.40		0.04		0.080
Cw*02	4.16	3.73	1.12				
Cw*03	4.16	10.44	0.37				
Cw*04	16.66	16.41	1.01				
Cw*06	6.25	19.40	0.27				0.030
Cw*07	12.50	22.38	0.49				
Cw*08	2.08	0.00	134.00		0.01		0.090
Cw*12	8.33	5.22	1.64				
Cw*14	2.08	2.98	0.69				
Cw*15	14.58	12.68	1.17				
Cw*16	14.58	2.23	7.45		0.12		0.001
Cw*18	6.25	0.00	1.64		0.05	0.02	0.003
Cw*19	2.27	12.68	0.49				

*EF, etiological fraction or attributable risk; PF, preventive fraction; Ki2, Chi-square with Yates correction; OR, odds ratio; **Very highly significant P value, obtained by multiplying with the no. of alleles; AF, allele frequency.

reaction (PCR) which is capable of amplifying a defined stretch of DNA (of several hundreds of base pairs) into several fold copies. The most commonly used PCR based HLA DNA typing techniques in tissue typing laboratories are: PCR based sequence specific oligonucleotide probe (PCR-SSOP) hybridization, PCR based sequence specific primer (PCR-SSP) typing and PCR based restriction fragment length polymorphism (PCR-RFLP).

The sequence based typing (SBT) method and the ARMS-PCR technique, were used for HLA-class 1 typing. Out of these, the PCR based sequence specific primer (PCR-SSP) typing was used.

In order to set up HLA-PCR, DNA was extracted from the blood sample. As a result, the entire blood cell or PBMCs were suspended in 200 µl phosphate buffer saline. Then, 200 µl of the lysis buffer was added after 40 µl of Quiagen protease had already been added. After that, the solution was mixed by vortexing for 15 s and then spinning was done. This was followed by incubation for 10 min at 60°C. Subsequently, 200 µl of absolute ethanol was added

to it and centrifugation was done briefly to remove drops from inside the lid. Then, the mixture was added to the QIAamp spin column. The cap was closed and centrifugation was done for 1 min at 8000 rpm RT. Consequently, the QIAamp spin column was placed in a 2 ml collection tube containing filtrate. Then, QIAamp spin column was carefully opened and 500 µl of AW1 buffer was added to it. The cap was closed and centrifuged for 1 min at 8000 rpm RT. Afterwards, the collection tube containing filtrate was discarded. Further, 500 µl of AW2 buffer was added without wetting the rim; however, the cap was closed and centrifugation was done for 3 min at 14000 rpm RT. Again, the collection tube containing filtrate was discarded, and the column was placed in a clean 1.5 ml micro centrifuge tube. The column was opened and 100 µl of elution buffer AE was added to it. Incubation was done for 5 min at RT and then centrifugation was done for 1 min at 8000 rpm RT. The eluted solution was collected in an eppendrof and this was followed by agarose gel electrophoresis in order to check the quality of DNA

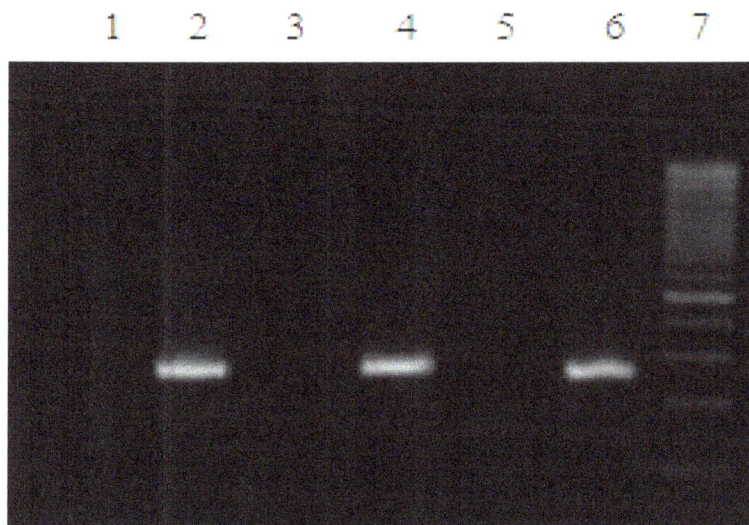

Figure 1. HCV RNA results of hepatitis C patients along with the positive control (Well 7 containing 255 bp band). Lane 1: Negative control 1; Lane 2: Specimen 1; Lane 3: Negative control 2; Lane 4: Specimen 2; Lane 5: Negative control 3; Lane 6: Positive control; Lane 7: 1 kb ladder (Fermentas).

that was thus obtained. For the quantitative analysis, the absorbance of the eluted sample was noted at 260 and 280 nm, respectively. Therefore, the obtained DNA was stored at 4 °C and the standard ratio of OD at 260 nm / OD at 280 nm was 1.8. If the ratio is more than 1.8, it can be interpreted that there is a contamination of RNA in DNA, but if it is less than 1.8, it is said that there is protein contamination. However, A260 of 1 denotes 50 µg DNA / ml of water. Thus, in order to measure the concentration of the eluted DNA sample at 260 nm, the following formula was used:

Concentration = 50 µg/ml * A260 * dilution factor

From this, the total amount of DNA in the study's sample was gotten by multiplying the concentration obtained with the total volume of eluted sample. Then the eluted sample was taken in setting up HLA-PCR. For a case where the PCR reaction mix is required, this contains nucleotides, buffer, glycerol and cresol red.

This PCR mix (Olerup SSP^{TM}) has a total amount of 312 µl without Taq. It was transferred to an ependroff and then 8.3 µl of the Taq was added to it. A-B-C-SSP Combi tray (coated with primers specific for MHC1 alleles) was taken, and well numbers 24, 72 and 96 were marked as a negative control which contained 7 µl of distilled water + 3µl of PCR mix. From the 320.8 µl PCR mix, 311.3 µl was taken and added, with 607 µl distilled water + amount of DNA, to 7.3 µg/100 µl of the sample, while 10 µl from this mix was added to the remaining wells. This tray was then transferred to thermal cycler, following the adjustment of cycle parameters:

1. Initial denaturation at 95 °C for 3 min.
2. First 10 cycles:

(i) Denaturation at 95 °C for 30 s.
(ii) Annealing and extension at 65 °C for 50 s.

3. Last 25 cycles.
(i) Denaturation at 95 °C for 30 s.
(ii) Annealing at 62 °C for 50 s.
(iii) Extension at 72 °C for 30 s.
Keep the PCR product at 4 °C.

To run the PCR product, a 2% agarose gel is prepared and a 96 well gel is cast. After that, a 2 ul of loading dye is added to each well of the combi tray and loaded into the sample. Subsequently, the gel is run at 120 v for 20 min; thereafter, the gel image is saved in the gel doc system.

Statistical analysis

The phenotype frequencies, odds ratios (OR), probability value, Chi-square with Yates correction, aetiological and preventive fraction were estimated using the available data-base and computer programme (Shankarkumar et al., 2002). Since each individual was tested for several HLA alleles and the same data were used for comparing frequency, it was possible that one of the alleles would by chance deviate significantly. To overcome this error, the P value was corrected by the use of Bonferroni inequality method (Dunn, 1961), that is, by multiplying with the number of alleles compared. The alleles were determined with the Helmberg SCORE V3.118T software supplied along with the kit.

RESULTS

It was observed that among the HLA A locus, the frequencies of HLA A*03, A*26, A*32 and A*66 alleles increased, although the increase of the first two alleles were highly significant (Figures 1 - 10). Similarly, A*11, A*24 and A*33 alleles decreased among patients when compared to controls. Among the HLA B locus, the frequencies of alleles HLA B*08, B*15, B*55 and B*57 increased, though increase in the allele frequency of B*15 was highly significant. In the same way, allele B*40 decreased. Among the HLA C locus, the frequencies of HLA Cw*01, Cw*08, CW*16 and Cw*18 alleles increased, while Cw*06 decreased in HCV infected individuals when compared to controls. Hence,

HLA   A B C Class I Alleles typing by PCR-SSP method

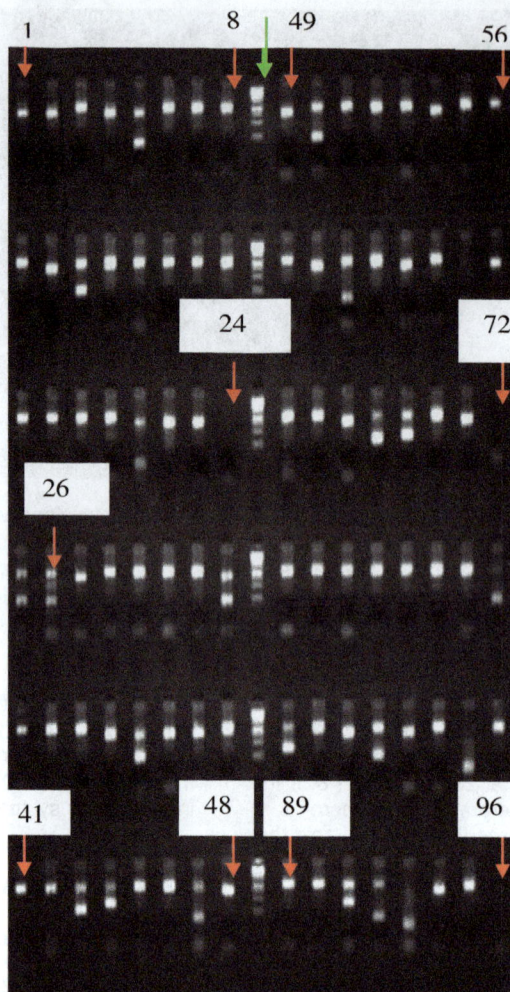

Figure 2. An agarose gel electrophorogram of a control individual. Some well numbers have been mentioned on the picture (numbers in yellow). The wells marked with red arrows are those corresponding to negative controls and have hence not lighted up. The column of wells marked by the green arrow has the 1 Kb DNA ladder. The lower bands in the ladder correspond to the 250 bp mark and the penultimate band corresponds to the 500 bp mark. Bands from 1 to 23 represent HLA alleles; bands from 25 to 71 represent HLA B alleles and bands from 73 to 95 represent HLA C alleles.

irrespective of the HCV RNA status (Figure 1), the entire patients' blood samples were immediately processed for IFN-gamma specific ELISPOT for which viable PBMCs were required. However, HLA Class I allele typing was carried out only for these HCV positive patients.

DISCUSSION

One of the striking features of HCV infection is the very high rate of development of chronicity (Hans and Michael,

1996). Approximately, 15% of infected patients successfully eliminate the virus, while others develop chronic infection with a wide spectrum of disease. Some will remain asymptomatic whereas others may have a more severe course leading to cirrhosis or hepatocellular carcinoma. There are evidences that immune mechanisms contribute to control the HCV infection. In the host immune reaction against viral infections, HLA alleles play a vital role in modulating the immune responses (Chaoyang et al., 2008). Hence, this study was designed to examine the frequencies of HLA

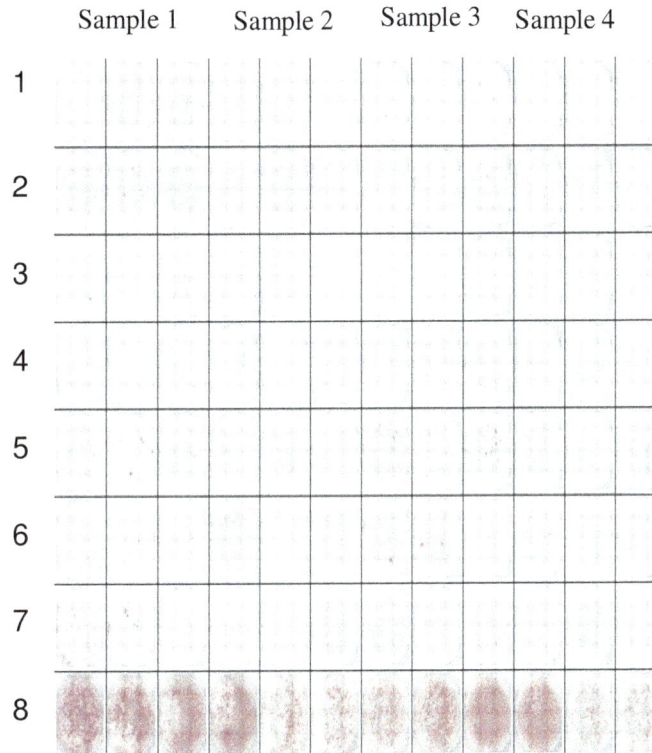

Figure 3. T cell response to various HCV antigen/peptides in healthy individuals as assessed by IFN-γ ELISPOT assay. (1) Negative control; (2) Medium control; (3) NS3 antigen; (4) NS5 antigen; (5) Core antigen ; (6) Pool 7 peptides ; (7) Pool 4 peptides and (8) Positive control.

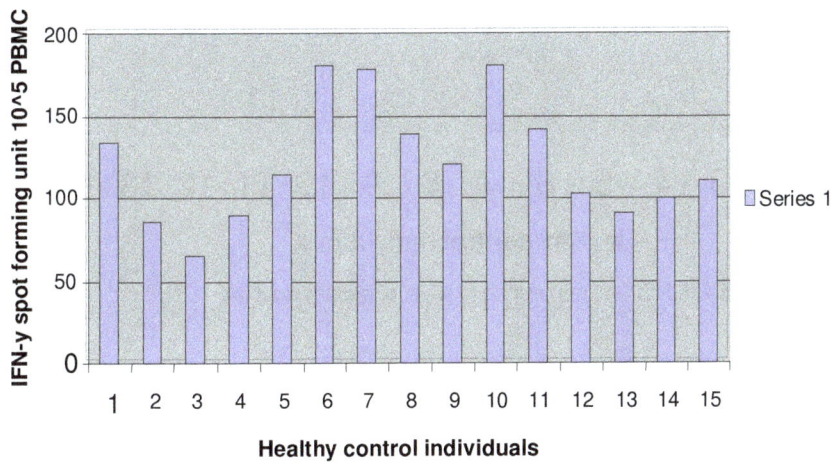

Figure 4. Average spot forming cells in positive control wells of healthy controls.

class I. The Major findings of the present study were a significant increase among the allele frequencies of HLA A*03, A*32, HLA B*15, B*55, Cw*16 and Cw*18. HLA-A3, HLA-B35 and HLA-B46 significantly increased in chronic HCV carriers compared with the controls in the Korean population (Yoon et al., 2005). In an Egyptian population Zekri et al. (2005) observed the HLA class I alleles of A28, A29 and B14 to be significantly encountered in HCV positive cases than negative cases.

An association of the HLA-B 27 with spontaneous HCV clearance has also been reported (Silini et al., 1995). Thio et al. (1999) have reported an association of

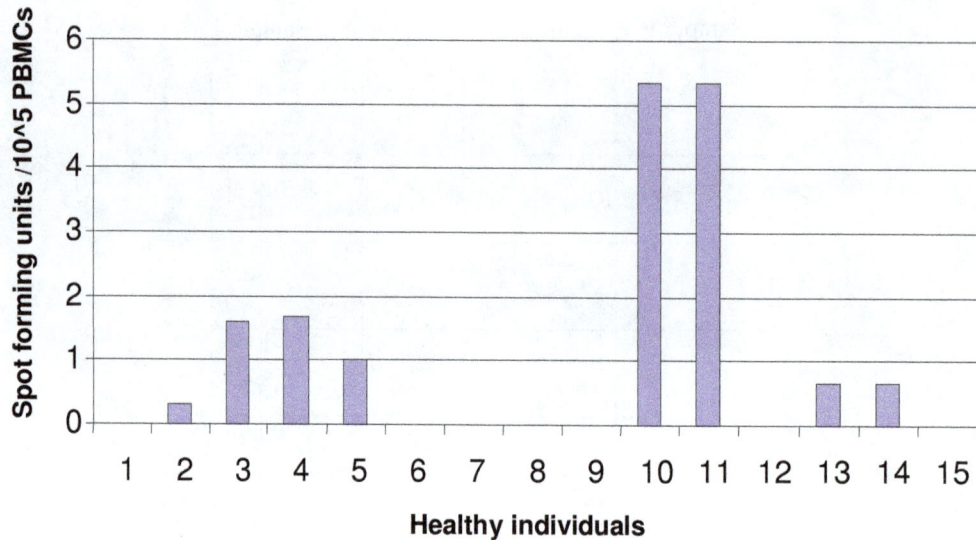

Figure 5. Average spot forming cells in cell control wells of healthy controls.

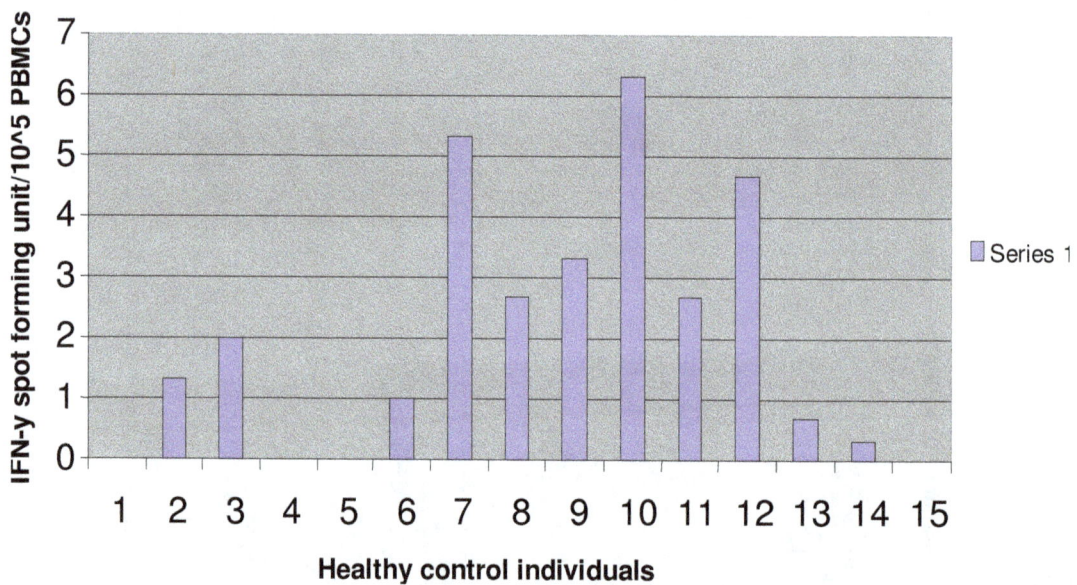

Figure 6. Average SFCs in NS3 antigen added to wells of healthy controls.

Cw*0102 with HCV clearance in Caucasians and of A*2301 and Cw*04 with HCV persistence in both African-Americans and Caucasians. However, allele A*32 was observed among HCV antibody positive individuals from western India. An increased frequency of haplotype HLA A*11 and Cw *04 in viraemic HCV patients was reported in a white population in Ireland (Andrew et al., 2000). Moreover, an association of B*15 allele with HCV infection was observed in this study. In a European population, Romero-Gomez et al. (2003) have reported the association of HLA - B*44 and have sustained HCV response to ribavirin/interferon combination therapy. Among Irish population, McKierman et al. (2004) have

reported an increased frequency of B*08 and B*54 in those with chronic HCV infection when compared to those without the infection. These results suggest that HLA association with HCV infection involves both class I antigens. When taken together, the study's results suggest that HLA associations with hepatitis C infection vary in relation to the ethnicity of the population studied. Nevertheless, differences in antigen frequency of selected HLA class I and II alleles between normal subjects and in hepatitis C infected individuals from Maharashtra suggest that a susceptibility factor may contribute towards acquiring hepatitis C virus infection. Most of the healthy individuals had HLA A *03 allele, but

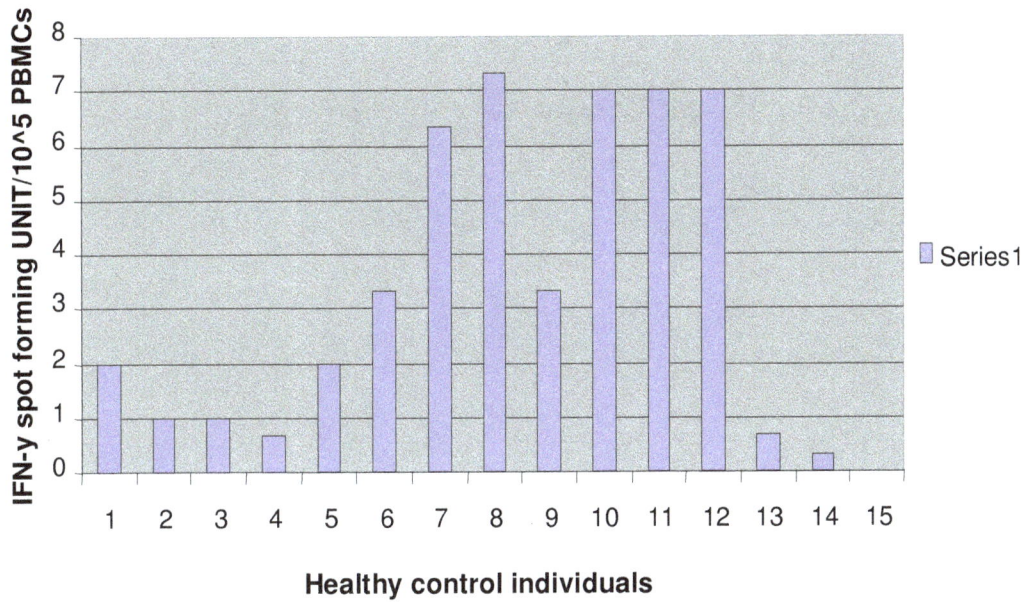

Figure 7. Average spot forming cells in NS5 antigen added to wells of healthy controls.

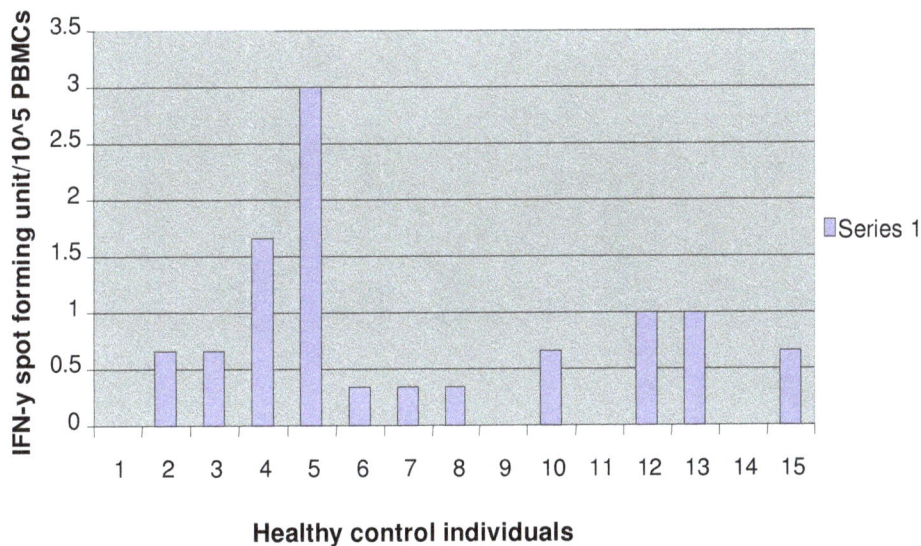

Figure 8. Average spot forming cells in pool 7 peptide added to wells of healthy controls.

none of the healthy individuals had any T cell response against any of the HCV core peptides. Out of the three HCV infected patients, two showed CTL response against the core antigen.

None of the hepatitis C patients responded to any of the core peptides. As such, it is interesting to note here that one of the HCV infected individuals had HLA A*03 allele, thus specifying the hypothesis that people with restricted HLA A*03 show substantial viral clearance and low level of pathogenesis (Silini et al., 1995).

Out of these 4 patients, 3 were positive for HCV RNA and two had a previous history of viral hepatitis. The negative sample could not be screened for anti HCV antibody by Elisa, because of the non-availability of the kit during the study period. One HCV infected patient could have been missed by this way, since there are reports that some hepatitis C cases are only antibody positive.

One of the three confirmed patients was a new case in the dialysis unit who had only recently started undergoing dialysis and who had shown CTL response against core antigen, thus showing that he might be chronically infected. This needs further verification with liver histology and other clinical parameters. The rest two

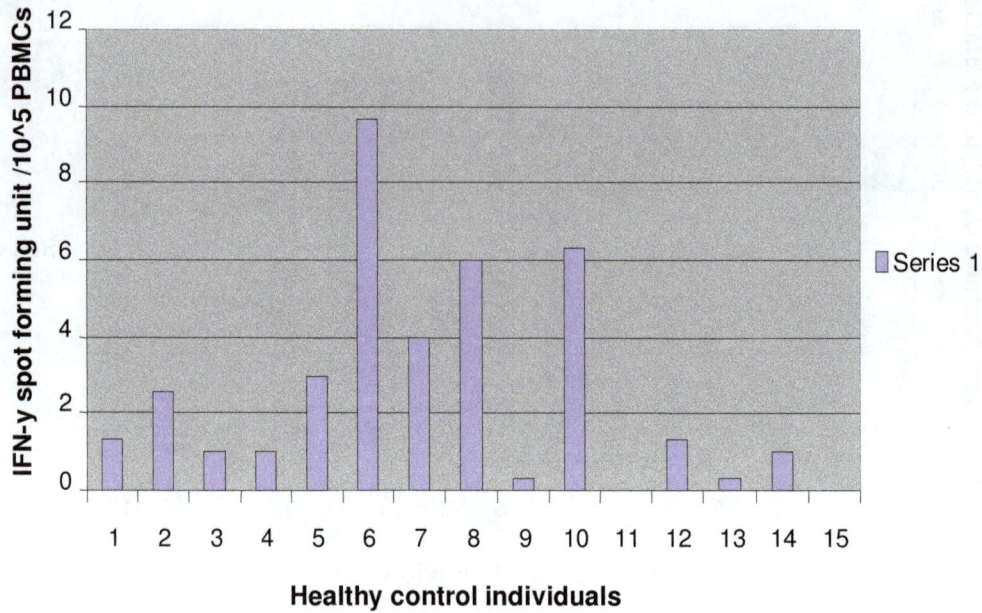

Figure 9. Average spot forming cells in core antigen added to wells of healthy controls.

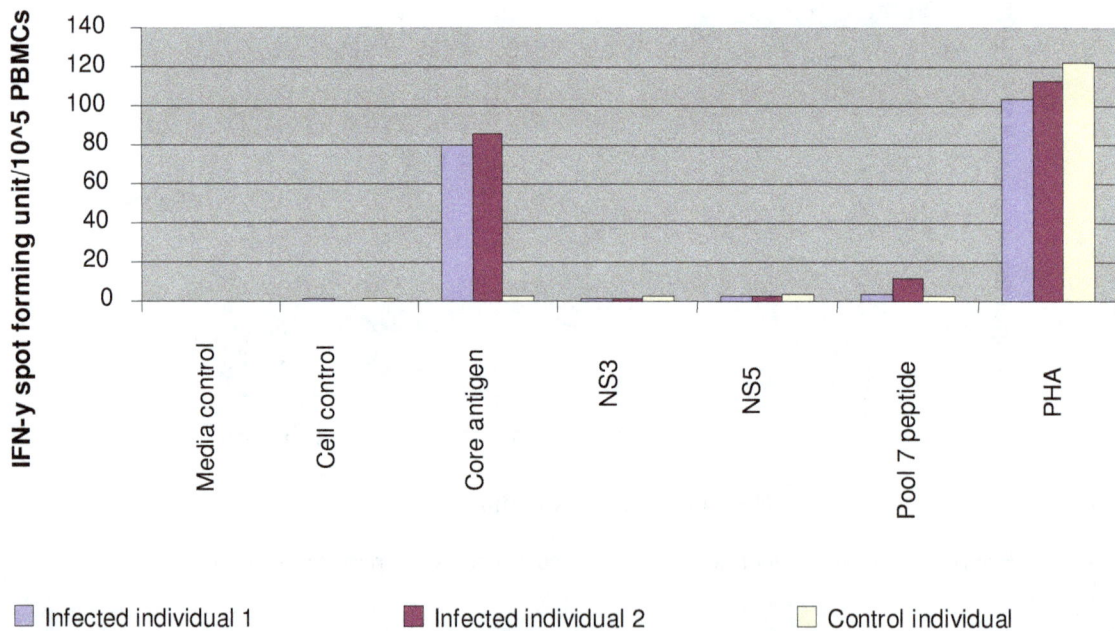

Figure 10. HCV core peptides, NS3, NS5 and core antigen not showing T cell response in healthy controls, but showing reactivity in the HCV infected individual.

patients who were chronically infected did not have CTL response against HCV core peptide, and this might be due to the high viral load that leads to defective T cell response.

There are reports that peripheral blood lymphocytes (PBLs) from HCV-infected patients without cirrhosis respond to NS3 and core proteins, producing predominantly IFN-γ. The study's data partially support the aforesaid report where none of the study's samples had cirrhosis. In contrast, PBLs from HCV-infected patients did not respond to NS3, but to the core protein, suggesting a selectively altered immune state during cirrhosis (Anthony et al., 2001). Also, the study's data provide support for the notion that HCV-specific IFN-γ-producing immunity is important in the pathogenesis of progressing HCV-related disease.

Thus, an essential process for resolution of viral infections is the efficient recognition and elimination of intracellular virus. Recognition of viral antigens in the form of short peptides associated with HLA class I molecule is a major task of CD8[+] cytotoxic T lymphocytes (Andrew et al., 2000). In this study, the frequency of the HLA class I alleles have been evaluated in patients with hepatitis C, although HLA-A3, A24, B51, -B52, -B55, -B56, -B61, B70, -Cw1, -Cw3 and -Cw4 are less frequent in patients with chronic hepatitis C(GracËa et al., 1998) The adaptive immune response is crucial for spontaneous resolution of acute hepatitis C virus (HCV) infection, in that it also constitutes the driving force for viral escape (Weseslindtner et al., 2009). For acutely HCV-infected dialysis patients, little is known about the host response and its impact on prognosis. A recent study in the year 2009, with four acutely infected dialysis patients have reported a robust CD4+ and CD8+ T-cell response and its association with transient control of infection, while in the other patients, weak responses correlated with persistently high viremia (Nasser and Paul Klenerman, 2007).

Despite the presence of CD8+ T-cell effectors, the establishment of viral persistence in the special patient group may be due to a failure of the adaptive immune system (Pawlotsky, 2003). This study supports this report as shown by the absence of any T-cell response by two chronic HCV infected individuals. In one of the authors study, it was not appropriate at all to propose/ predict any hypothesis with only three hepatitis C patients. The hepatitis C patient who responded to the core antigen needs to be followed up. The fact that the hepatitis C patient who was a CTL responder did not have HLA Class I allele suggests that this peptide might be an immunodominant CTL epitope that is recognized universally, irrespective of HLA restriction.

Conclusion

There are reports that peripheral blood lymphocytes (PBLs) from HCV-infected patients without cirrhosis respond to NS3 and core proteins, producing predominantly IFN-γ. These data provide support for the notion that HCV-specific IFN-γ-producing immunity is important in the pathogenesis of progressing HCV-related disease and partially supports the aforesaid report where none of the study's samples had cirrhosis. In contrast, peripheral blood lymphocytes from HCV-infected patients did not respond to NS3, but to the core protein, suggesting a selectively altered immune state during cirrhosis. The study's results suggest that the establishment of viral persistence in patients is due to a failure of the immune response and is associated with HLA class 1 allele that are mainly A*03 restricted individuals. As indicated by the absence of a significant T-cell response, the HCV core peptide, HCV core antigen, HCV NS3 antigen and the HCV NS5 antigen could be used for defining the prognosis of an HCV infected individual.

ACKNOWLEDGMENT

The authors acknowledge the financial support from the Indian Council of Medical Research (ICMR), New Delhi.

REFERENCES

Abdul MA (2008). Current Challenges in Hepatitis C. TAJ, 21(1): 93-96.

Andrew KS, David AP, Annette O, Anthony DK, Rodney EP (2000). Cytotoxic T Lymphocyte Responses to Human Immunodeficiency Virus: Control and Escape. Stem Cells, 18: 230-244.

Anthony DD, Post AB, Valdez H, Peterson DL, Murphy M, Heeger PS (2001). ELISPOT analysis of hepatitis C virus protein-specific IFN-gamma-producing peripheral blood lymphocytes in infected humans with and without cirrhosis. Clin. Immunol., 99(2): 232-40.

Anuradha ST, Shankarkumar U, Mandeep SC, Kanjashkya G, Vidya AA (2009). Association of HLA alleles with hepatitis C infection in Maharashtra, western India (November). Indian J. Med. Res., 130: 550-555

Bukh J, Purcell RH, Miller RH (1992). Importance of primer selection for the detection of hepatitis C virus RNA with the polymerase chain reaction assay. Proc. Natl. Acad. Soc., 89: 187-91.

Chaoyang W, Xiang H, Hongfang M, Ningbo H, Congwen W, Ting S, Yanhong Z, Liping S, Qingjun M, Hui Z (2008). Hepatitis C Virus Infection Downregulates the Ligands of theActivating Receptor NKG2D. Cell. Mol. Immunol., 5(6): 475-478.

Chowdhury A, Santra A, Chaudhuri S, Dhali GK, Chaudhuri S, Maity SG, Naik TN, Bhattacharya SK, Mazumder DN (2003). Hepatitis C virus infection in the general population: a community based study in West Bengal, India. Hepatol. 37: 802-809.

Dunn OJ (1961). Multiple comparisons among means. Am. J. Stat. Assoc., 56: 52-64.

GracËa PH, Alves PR, JoseÂ MC, Cristina P, AnunciacËaÄo R, Benvindo J, Roger W, Maria De S (1998). Major histocompatibility complex class I associations in iron overload:evidence for a new link between the HFE H63D mutation, HLA-A29,and non-classical forms of hemochromatosis. Immunogenet., 47: 404±410.

Hans LT, Michael PM (1996). Mode of hepatitis C virus infection, epidemiology, and chronicity rate in the general population and risk groups. Dig. Dis. Sci., 41(12): 27S-40S.

Irshad M, Acharya SK, Joshi YK (1995). Prevalence of hepatitis C virus antibodies in the general population and in selected groups of patients in Delhi . Indian J. Med. Res., 102: 162-164.

Jain A, Rana SS, Chakravarty P, Gupta RK, Murthy NS, Nath MC, Gururaja S, Chaturvedi N, Verma U, Kar P (2003). The prevalence of hepatitis C virus antibodies among the voluntary blood donors of New Delhi, India. Eur. J. Epidemiol., 18: 695-697.

Just JJ (1995). Genetic predisposition to HIV-1 infection and acquired immune deficiency virus syndrome: A review of the literature examining associations with HLA. Hum. Immunol., 44:156-69.

McKierman SM, Hagan R, Curry M, McDonald GS, Kelly A, Nolan N, Walsh A, Hegarty J, Lawlor E, Kelleher D(2004). Distinct MHC class I and II alleles are associated with hepatitis C viral clearance, originating from a single source. Hepatol (2004). Distinct MHC class I and II alleles are associated with hepatitis C viral clearance, originating from a single source. Hepatol. 40: 108-14.

Nasser S, Klenerman P (2007). CD4+ T cell responses in hepatitis C virus infection. World J. Gastroenterol., 13(36): 4831-4838.

Nathalie B, Patrick M (2000). Natural history of hepatits C and the impact of anti-viral therapy. Trends Exp. Clin. Med., 10(1): 4-18.

Pawlotsky JM (2003). The nature of interferon-[alpha] resistance in hepatitis C virus infection. Curr. Opin. Infect. Dis., 16(6): 587-592.

Romero-Gomez M, Gonzalez-Escribano MF, Torres B, Barroso N,Montes-Cano MA, Sanchez-Munoz D, Núñez-Roldan A, Aguilar-Reina J. (2003). HLA class I B44 is associated with sustained response to interferon+ ribavirin therapy in patients with chronic hepatitis C. Am. J. Gastroenterol., 98: 1621-1626.

Shankarkumar U, Devraj JP, Ghosh K, Mohanty D (2002). Seronegative spondarthritis and human leukocyte antigen association. Br. J. Biomed. Sci., 59: 38-41.

Silini E, Bono F, Cividini A, Cerino A, Bruno S, Rossi S,Belloni G, Brugnetti B, Civardi E, Salvaneschi L(1995). Differential distribution of hepatitis C virus genotypes in patients with and without liver function abnormalities .Hepatol. (1995). Differential distribution of hepatitis C virus genotypes in patients with and without liver function abnormalities. Hepatol., 21: 285-290.

Thio CL, Carrington M, Marti D, O'Brien SJ, Vlahov D, Nelson KE, Astemborski J, Thomas DL (1999). Class II HLA alleles and hepatitis B virus persistence in African Americans. J. Infect. Dis., 179: 1004-1006.

Wang RF (2003). Identification of MHC class II-restricted tumor antigens recognized by CD4+ T cells. Methods, 29(3): 227-235.

Weseslindtner L, Neumann-Haefelin C, Viazov S, Haberstroh A, Kletzmayr J, Aberle JH, Timm J, Ross SR, Klauser-Braun R, Baumert TF, Roggendorf M, Thimme R, Holzmann H (2009). Acute infection with a single hepatitis C virus strain in dialysis patients: Analysis of adaptive immune response and viral variability. J. Hepatol., 50(4): 693-704.

Yee LJ (2004). Host genetic determinants in hepatitis C virus infection. Genes Immun., 5: 237-245

Yoon SK, Han JY, Pyo C-W, Yang JM, Jang JW, Kim CW, Chang UI, Bae SH, Choi JY, Chung KW, Sun HS, Choi HB, Kim TG (2005). Association between human leukocytes antigen alleles and chronic hepatitis C virus infection in the Korean population. Liver Intl., 25: 1122-1127.

Zekri AR, El-Mahallawy HA, Hassan A, El-Din NH, Kamel AM. HLA alleles in Egyptian HCV genotype-4 carriers . Egypt J. Immunol., 12: 77-86.

18

Sourcing starter cultures for *Parkia biglobosa* fermentation I: Phylogenic grouping of *Bacillus* species from commercial 'iru' samples

Aderibigbe E. Y.[1]*, Visessanguan W.[2], Sumpavapol P.[3] and Kongtong K.[2]

[1]Department of Microbiology, University of Ado-Ekiti, P. M. B. 5363, Ado-Ekiti, Nigeria.
[2]Food Biotechnology Research Unit, National Center for Genetic Engineering and Biotechnology (BIOTEC), Pathumthani, Thailand.
[3]Department of Food Technology, Faculty of Agro-Industry, Prince of Songkla University, Hat Yai, Songkhla, 90112, Thailand.

Twenty five isolates of *Bacillus* species were obtained from ten commercial samples of 'iru' (fermented African locust bean), after heat-treatment at 80°C for 30 min. All the isolates were Gram-positive, sporeformer, catalase-positive, non-motile rods. The phylogenic relationship between the isolates was studied by repetitive PCR fingerprinting using the $(GTG)_5$ primer, referred to as $(GTG)_5$-PCR fingerprinting and 16S rRNA gene sequencing analyses. The $(GTG)_5$-PCR fingerprinting resulted in grouping of the isolates into 12 phylogenetic groups. Based on the 16S rRNA gene sequence analysis, most of the *Bacillus* isolates were found to be closely related to *Bacillus subtilis*, while strain 8B was closely related to *Bacillus licheniformis*.

Key words: 'Iru', *Bacillus*, starter culture, African locust bean, *Parkia biglobosa* fermentation, rep-PCR fingerprint, 16S rRNA sequencing.

INTRODUCTION

'Iru' is an indigenous protein-rich soup condiment produced by fermenting the cotyledons of African locust bean (*Parkia biglobosa*). It is consumed mostly by the local rural dwellers as protein supplement in the diets in many West African countries. 'Iru' is known as 'soumbala' in Burkina Faso (Ouoba et al., 2003) and 'afitin' in Benin Republic (Azokpota et al., 2006) and 'dawadawa' by the Hausa-speaking ethnic groups in West Africa (Odunfa and Adewuyi, 1985a). 'Iru' is not only consumed as a soup flavoring food additive, but also serves as a cheap meat substitute amongst poor families. A variant of 'dawadawa' is obtained by fermenting soybean (*Glycine max*) (Terlabie et al., 2006). In some Southwest Nigerian towns/ villages, 'iru' is used as a local remedy in the treatment of eye infections (Aderiye and Laleye, 2003).

The production of 'iru' being a traditional art, the fermentation is initiated by chance inoculation of natural microflora and thus the products vary considerably in quality and shelf-life. The ammonical flavor that develops in the post-fermentation product has been a major factor why the 'elites' in urban centers do not consume the product. Recently, Osho et al. (2010) conducted a comparative study on the microbial load in some local fermented foods in Nigeria, including 'iru'.

Microorganisms associated with the fermentation are mostly strains of *Bacillus subtilis* group, while *Lactobacillus* sp. and *Staphylococcus epidermidis* are present in lower numbers (Odunfa, 1981a). The optimum conditions (temperature and time) required for fermentation are 35°C and 36 h respectively (Odunfa and Adewuyi, 1985a). Starter culture experiments have proved that strains belonging to the *B. subtilis* group are responsible for the fermentation (Odunfa and Adewuyi,

*Corresponding author. E-mail: esther_aderibigbe@yahoo.co uk.

Table 1. Sources of commercial 'iru' samples in Southwest Nigeria.

S/No.	Place of sample collection	Sample code
1	Iloro-Ekiti (Ekiti State)	S1
2	Oye-Ekiti (Ekiti State	S2
3	Isan-Ekiti (Ekiti State)	S3
4	Ado-Ekiti (Ekiti State)	S4
5	Ikare-Akoko 1 (Ondo State)	S5
6	Ikare-Akoko 2 (Ondo State)	S6
7	Ikare-Akoko 3 (Ondo State)	S7
8	Rore (Kwara State)	S8
9	Iru-pete (Lagos State)	S9
10	Ikare-Akoko 3 (Ondo State)	S10

1985b). The strains of *Bacillus species* (*B. subtilis, Bacillus licheniformis* and *Bacillus pumilus*) differed in their growth and extracellular enzymes production in broth medium (Aderibigbe and Odunfa, 1990). Some indigenous fermented foods have enjoyed technological advancement and standardization by the use of proven strains as starter cultures. These include soy sauce, natto and tempeh. A key factor in industrialization of 'iru' production is the development of proven strains of bacteria as starter cultures for the fermentation (Latunde-Dada, 1995). This will ensure standardization of product, having longer shelf-life and high hygienic quality. Hence, the purpose of this study was to isolate strains of *B. subtilis* group from commercial samples of 'iru', confirm their phylogenic relationship, screen them through starter culture experiments and evolve a new strains which have the potentials of being developed into starter cultures for industrial scale production of 'iru'. This paper reports on the isolation and characterization of strains of *Bacillus* species in commercial 'iru' samples.

MATERIALS AND METHODS

Sources of 'iru' samples

'Iru' samples were bought from reputable local vendors in Southwest Nigeria, whose products are judged as of good quality (by visual observation and perception of ammonia odor threshold). The place from where the samples were collected and codes given to them are shown in Table 1. The samples were kept in sterile 50 ml polystyrene bottles and stored at -20°C until their analysis.

Culturing, Isolation and identification of bacterial isolates from Iru samples

One gram of Iru sample was weighed and transferred into 9 ml of sterile 0.1% peptone water, mixed by vortexing and heated in a water bath at 80°C for 30 min. Microbes were isolated by means of the serial dilution plating on nutrient agar medium (NA, Difco) containing 0.3% (w/v) beef extract, 0.5% (w/v) peptone and 1.5% (w/v) agar, pH 7.2. The plates were incubated at 35°C for 18 h. A pure culture was obtained by repeated transfers of individual

colonies on nutrient agar medium as mentioned previously. Number of colonies and their characteristics were recorded. The standard methods of Gram-staining, motility test (wet-mount) and catalase test were performed. The modified method described by Hamouda et al. (2002) was used to identify the spore formers. The reference strains used were *Bacillus amyloliquefaciens* KCTC 1660[T], *B. subtilis subsp. subtilis* KCTC 3135[T], *Bacillus vallismortis* KCTC 3707[T], *Bacillus licheniformis* KCTC 1918[T], and *Bacillus mojavensis* KCTC 3706[T]. Unless otherwise stated, strains were grown in the same medium and cultivated at 37°C for 24 h.

Taxonomic grouping of *Bacillus* isolates through molecular methods

Extraction and purification of DNA

A single colony was inoculated into 5 ml nutrient broth (NB) and incubated overnight at 35°C. The starter culture (5 ml) was used to inoculate 50 ml NB in 250 ml conical flask and incubated with shaking at 200 rpm at 35°C overnight. Cultures were centrifuged at 10,000 rpm (4°C) for 10 min and pellets were re-suspended in 5 ml of NB. A 1.3 ml of cells' suspension was added into a 1.5 ml microtube and centrifuged at 11,000 rpm (4°C) for 1min. Total genomic DNA was extracted by using a Wizard® Genomic DNA Purification kit (Promega, Madison, WI, USA), according to the manufacturer's instructions, with minor modification by Plengvidhya et al. (2004). In this modification, the cell lysis solution contains 5 µl of mutanolysin (2500 U/ml; Sigma, St. Louis, MO, USA) in addition to 10 µl of lysozyme (10 mg/ml, Sigma, USA). The DNA quality was determined by electrophoresis on 1% agarose gel. After electrophoresis, the gels were stained in 0.5X TAE containing 0.5 µg/ml of ethidium bromide (Fluka) for 15 min. Pictures of the gels were digitally captured using the Bioimaging System GeneGenius (SynGene, Cambridge, England). The DNA concentration was measured at 260 nm with spectrophotometer. For rep-PCR, the DNA was adjusted to 50 ng/µl with TE buffer.

Rep-PCR genomic fingerprinting

DNA extracted from selected *Bacillus* isolates were subjected to rep-PCR analysis using primer (GTG)$_5$ (5'-GTGGTGGTGGTGGTG-3') as described by Sumpavapol et al. (2010). Amplicons were separated on 1% LE Seakem® agarose (BME, Rockland, ME, U.S.A) in 0.5X TBE at 120 V for 2 h and 40 min (2 x 1 h: 20 min) prior to stain in 0.5X TBE containing 5 µg/ml ethidium bromide

Table 2. Microbial load and cultural characteristics of *Bacillus* species in commercial 'iru' samples.

Sample Code	Microbial load (x 10^7 cfu/g)	Cultural characteristics of bacterial isolates on NA					
		Isolate code	Color	Margin/edge	Elevation	Surface	Size
S1	21	1A	Cream	Lobate	Flat	Dry	++++
		1B	"	"	"	"	++++
S2	37	2A	"	"	"	"	++++
		2B	"	"	"	"	++++
S3	7.5	3A	"	Lobate/Rhizoid	"	"	+++
		3A-2	"	Lobate/Rhizoid	"	"	+++
		3B	"	Lobate	"	"	++++
S4	42	4A	"	Rhizoid	"	"	+++
		4B	Cream/White	"	"	"	++++
S5	8.5	5A	Cream	"	"	"	+++
		5B	"	"	"	"	+++
S6	24	6A	"	"	"	"	+++
		6B	"	"	"	"	+++
S7	29.5	7A	Cream	Lobate	Flat	Dry	+++
		7B	"	Rhizoid	"	Dry/Glossy	+++
S8	9.0	8A	"	"	"	"	++++
		8B	"	Entire	Raised/ Flattened	Glossy	++++
		8B-1	"	Rhizoid	"	"	++++
		8C	"	Lobate/Rhizoid	Flat	"	++++
S9	48	9A	"	"	"	Dry	++++
		9B	"	"	"	"	++++
		9C	"	Lobate	"	"	++++
S10	11	10A	"	Lobate/Rhizoid	"	"	++++
		10B	"	Rhizoid	"	"	+++
		10C	"	"	"	"	++++

(Sigma, USA) for 10 min and destained in tap water for 20 min, with shaking (UMAC OMRON H7ER Orbital shaker, 28 rpm), The gel image was captured by using an image scanner Typhoon 9410 (Amersham Biosciences). The DNA patterns were analyzed by using a pattern analysis software package, Gel Compar II, Version 4.5 (Applied Math, Belgium). Pearson product-moment correlation coefficient was used to calculate similarities between patterns and a dendogram was obtained by means of unweighted pair group method with arithmetic average (UPGMA).

16S rRNA gene sequence analysis

16S rRNA gene sequencing was carried out using the methods reported by Ruiz-Garćia et al. (2005). The 16S rRNA gene was amplified by PCR with universal bacterial primers, 16S-27 F (5'-AGAGTTTGATCATGGCTCAG-3') annealed at positions 8 to 27

and 16S-1488 R (5'-CGGTTACCTGTTAGGACTTCACC-3') annealed at positions 1511-1488 (*E. coli* numbering) according to Brosius et al. (1978). Amplification reaction was carried out in 25 µl volume, using the Takara *Ex Taq* DNA polymerase and buffer system (Takara Mirus Bio Corporation, Madison, WI). The final PCR mixture comprised 1X *Ex Taq* buffer (with 1.5 mM $MgCl_2$), a 200 µM concentration of each deoxynucleoside triphosphate, an 0.2 µM concentration of each primer, 1 unit of *Ex Taq* DNA polymerase, and 50 ng of template DNA. Amplification was carried out in a thermocycler (GeneAmp PCR System 2400, PE Biosystems, Foster, California, U.S.A.) with the following cycling program: initial denaturation at 94 °C for 5 min followed by 30 cycles of 94 °C for 30 s, annealing at 55 °C for 15 s, and extension at 72 °C for 90 s, and a final extension step at 72 °C for 5 min. The PCR product was purified using QIAquick-PCR purification kit (Qiagen). The double-stranded DNA was sequenced with an ABI PRISM BigDye Terminator v3.1 Cycle Sequencing Kit (Applied Biosystems, Foster,

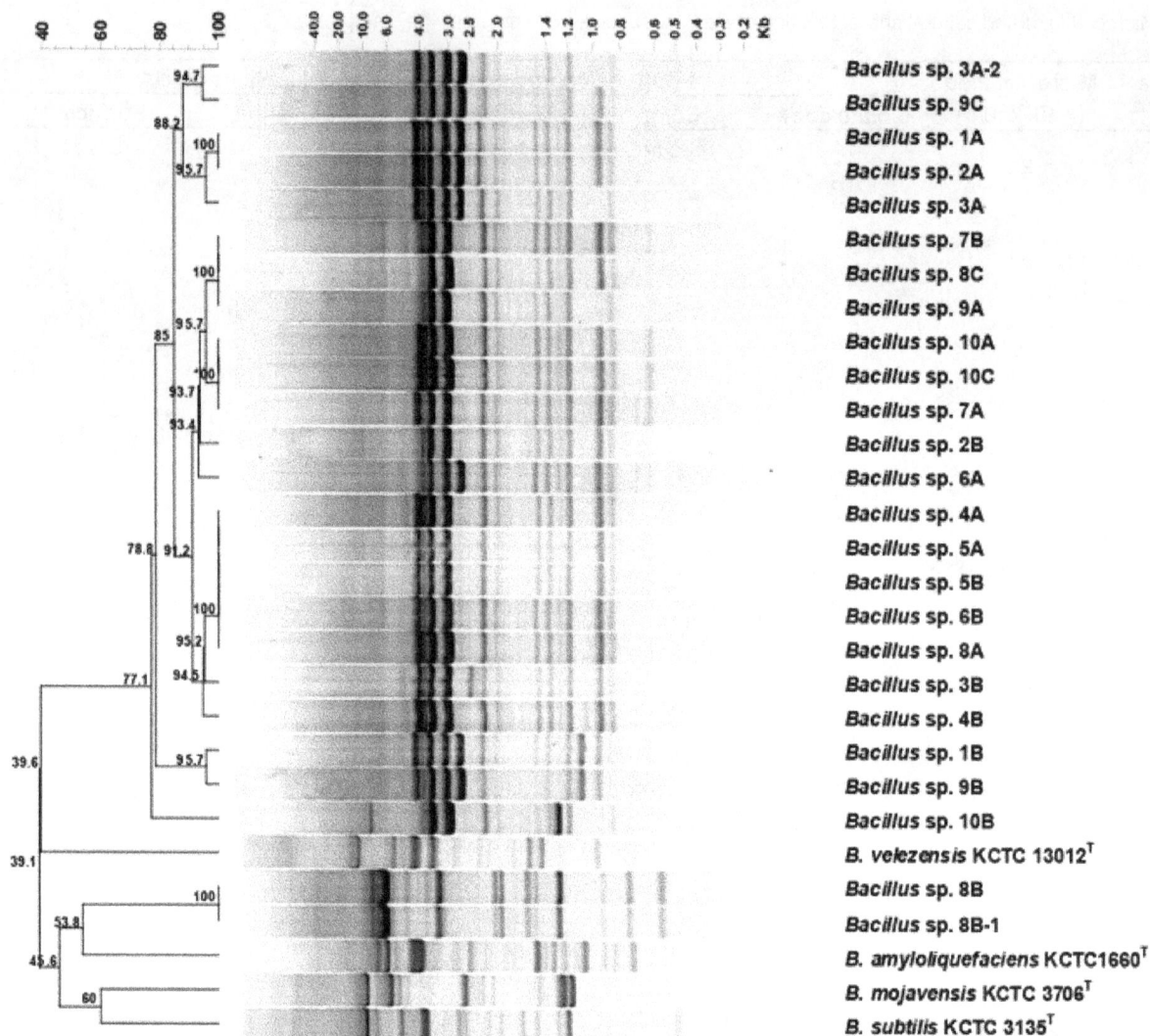

Figure 1. llustrations of the cluster analysis and (GTG)₅-PCR genomic fingerprint of *Bacillus* sp. isolated from Iru and related species of the genus *Bacillus* species. Dendrogram was based on the Dice coefficient of similarity (weighted) and obtained with the UPGMA clustering algorithm.

California, U.S.A.) according to the manufacturer's instruction, by the use of the following four primers; 16S-27 F (5'-AGAGTTTGATCATGGCTCAG-3') annealed at positions 8 to 27, 16S-421 R (5'-CGGATCGTAAAGCTCTGTTG-3') annealed at positions 401 to 421, and 16S-1488 R (5'-CGGTTACCTGTTAGGACTTCACC-3') annealed at positions 1511-1488. The PCR products were sequenced with an ABI PRISM 377 Genetic Analyzer (Applied Biosystems, Foster, California, U.S.A.). The 16S rRNA gene sequences were aligned along with the selected sequences obtained from the GenBank/EMBL/DDBJ databases by using the program CLUSTAL_X (version 1.81) (Thompson et al., 1997). Gaps and ambiguous bases were eliminated from the calculations. The distance matrices for the aligned sequences were calculated by the two-parameter method of Kimura (1980). A phylogenetic tree was constructed by the neighbor-joining method (Saitou and Nei, 1987) with a program MEGA (version 2.1) (Kumar et al., 2001). The confidence values of individual branches in the phylogenetic tree were determined by using the bootstrap analysis of Felsenstein (1985) based on 1000 samplings.

RESULTS AND DISCUSSION

Table 2 shows the microbial load and the cultural characteristics of the isolates obtained from 10 commercial 'iru' samples. All the *Bacillus isolates* were Gram- positive, spore-forming rods, catalase positive, and most were non-motile. 'Iru' produced by commercial vendors vary considerably in sensory qualities. Odunfa and Adewuyi (1985b) reported that strains of *Bacillus* involved in fermentation could influence the quality of the product. Many strains were obtained from 'iru' samples (Odunfa and Oyewole, 1986); which varied in growth rate and extracellular proteinase, amylase, polygalacturonase, galactanase and sucrase production in NB medium (Aderibigbe and Odunfa, 1990). Species of *Bacillus* have been reported to be involved in the fermentation of other plant seeds in the production of natto, 'thua-nao',' ugba',

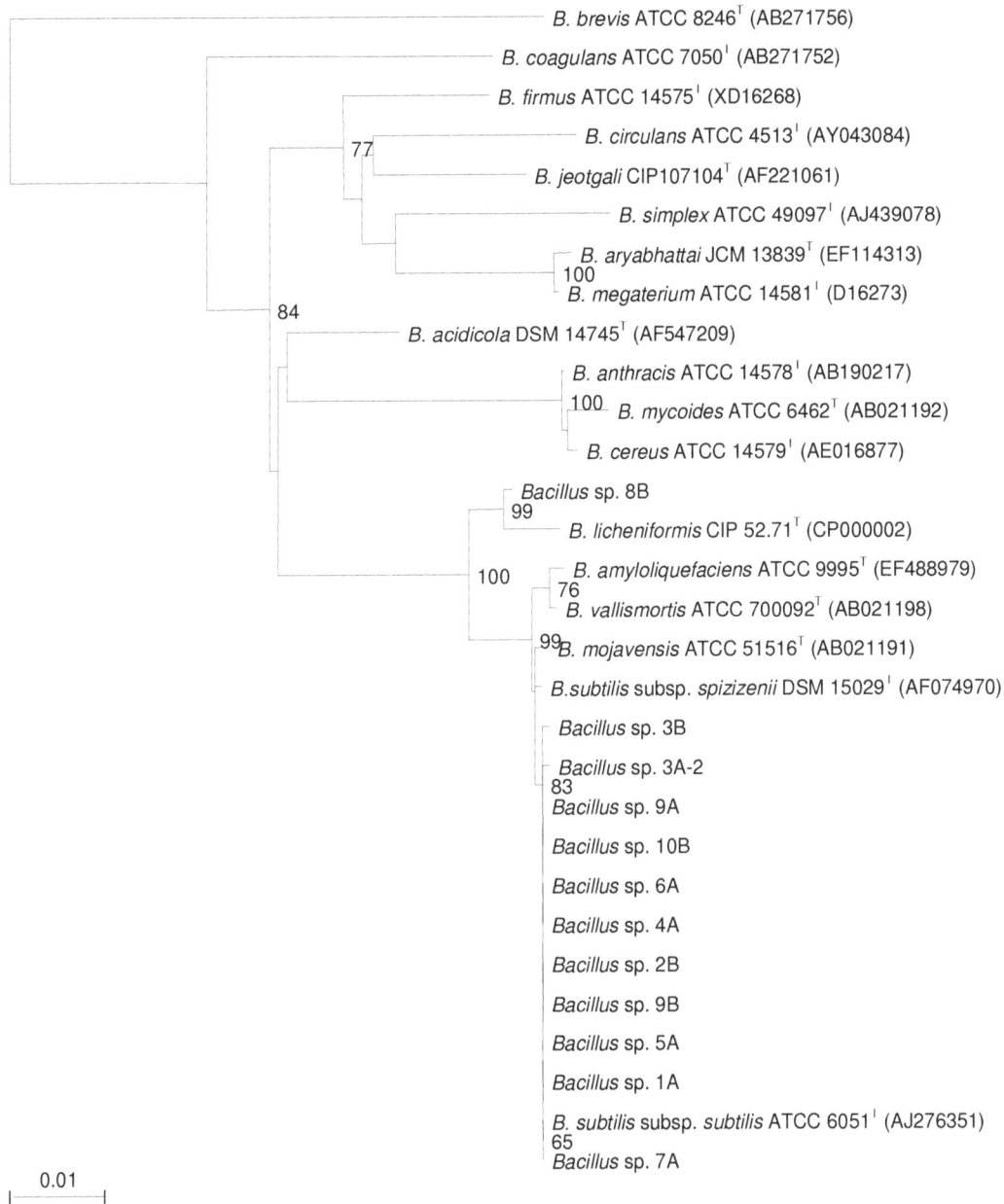

Figure 2. Phylogenic relationships of representative *Bacillus* strains obtained from 'iru' and related taxa based on 16S rRNA gene sequence analysis. The branching pattern was generated by the neighbor-joining method. Bootstrap values (expressed as percentages of 1000 replications) greater than 60% are shown at the branch points. Bar, 0.01 substitutions per nucleotide position.

and 'ogiri' (Odunfa, 1981b; Beuchat., 1997; Kiuchi and Watanabe, 2004).

Taxonomic grouping of *Bacillus* species isolates

The relationship between the *Bacillus* species was determined by the rep-PCR fingerprinting method, a useful technique for determining inter- and intra-species relatedness (Versalovic et al., 1994; Gevers et al., 2001).

The fingerprints obtained with primer (GTG)$_5$ are shown in Figure 1. The (GTG)$_5$ patterns resulted in the delineation at Pearson's correlation coefficient below 95% indicating that they were different genotypically and probably belong to the different species. Thus, on the basis of similarities in bands of the DNA dendogram, the isolates were divided into 12 groups (Table 3). Nick et al. (1999) and Rademaker et al. (2000) have compared rep-PCR genomic fingerprint analysis with DNA–DNA relatedness, they suggested that the two techniques yield

Table 3. Grouping of the *Bacillus* isolates from 'iru' on the basis of (GTG)$_5$ fingerprint.

Group	Isolate code	Group representative
I	3A-2, 9C	3A-2
II	1A, 2A, 3A	1A
III	7B, 8C, 9A	9A
IV	7A, 10A, 10C	7A
V	2B	2B
VI	6A	6A
VII	4A, 5A, 5B, 6B, 8A	4A, 5A
VIII	3B	3B
IX	4B	4B
X	1B, 9B	9B
XI	10B	10B
XII	8B, 8B-1	8B

results that are in close agreement. Heyrman et al. (2003) reported that rep-PCR fingerprinting can be used as a genomic screening method to differentiate at the species level and to select representatives for DNA–DNA reassociation experiments.

Figure 2 shows the phylogenic relationships of the *Bacillus* species from 'iru', with other type strains based on the 16S rRNA gene sequences. The 16S rRNA gene sequence similarities obtained showed that 3A-2, 1A, 9A, 7A, 2B, 6A, 4A, 5A, 3B, 4B, 9B and 10B (representative of Group I to XI, respectively) were most closely related to *B. subtilis subsp. subtilis* ATCC 6051^T with 99.5 to 100% similarity, while 8B (Group XII) was most closely related to *B. licheniformis* CIP 52.71^T (99.3%). Oguntoyinbo et al. (2004) reported variation in phenotypic and technological properties of 7 strains of *Bacillus* species involved in production of 'okpehe', with respect to production of extracellular enzymes, polyglutamate and bacteriocin. Further experiments are required to elucidate the differences in physiological properties of these isolates.

Conclusion

The 25 *Bacillus* species isolated from 10 commercial samples of 'iru' were grouped into 12 on basis of DNA fingerprinting. However, 16S rRNA gene sequencing has shown that they are closely related to *B. subtilis and B. licheniformis*.

ACKNOWLEDGEMENTS

EYA wishes to acknowledge the TWAS-UNESCO Associateship Scheme and National Centre for Genetic Engineering and Biotechnology (BIOTEC), Thailand for the financial support and opportunity of using the facilities at the Food Biotechnology Unit for this research.

REFERENCES

Aderibigbe EY, Odunfa SA (1990). Growth and extracellular enzyme production by strains of *Bacillus* species isolated from fermenting African locust bean, iru. J. Appl. Bacteriol., 69: 662-671.

Aderiye BI, Laleye SA (2003). Relevance of fermented food products in Southwest Nigeria. Plant Foods Hum. Nutr., 58: 1-16.

Azokpota P, Hounhouigan DJ, Nago MC, Jakobsen M (2006). Esterase and protease activities of *Bacillus* spp. from afitin, iru and sonru; three African locust bean (*Parkia biglobosa*) condiments from Benin. Afr. J. Biotechnol., 5(3): 265-272.

Beuchat LR (1997). Traditional Fermented Foods. In: Food Microbiology Fundamentals and Frontiers. (Eds.) MP Doyle, LR Beuchat and TJ Montville. American Society for Microbiology, Washington D. C., pp. 629-648.

Brosius J, Palmer ML, Kennedy PJ, Noller HF (1978). Complete nucleotide sequence of a 16S ribosomal RNA gene from Escherichia coli. Biochemistry, 75: 4801-4805.

Felsenstein J (1985). Confidence limits on phylogenies: an approach using the bootstrap. Evolution, 39: 783-791.

Gevers D, Huys G, Swings J (2001). Applicability of rep- PCR fingerprinting for identification of *Lactobacillus* species. FEMS Microbiol. Lett., 205 (1): 31-36.

Hamouda T, Shih AY, Baker Jr. JR (2002). A rapid staining technique for the detection of the initiation of germination of bacterial spores. Lett. Appl. Microbiol., 34: 86-90.

Heyrman J, Balcaen A, Rodriguez-Diaz M, Logan NA, Swings J, De Vos P (2003). Bacillus decolorationis sp. nov., isolated from biodeteriorated parts of the mural paintings at the Servilia tomb (Roman necropolis of Carmona, Spain) and the Saint-Catherine chapel (Castle Herberstein, Austria). Int. J. Syst. Evol. Microbiol., 53(2): 459-463.

Kimura M (1980). A simple method for estimating evolutionary rates of base substitutions through comparative studies of nucleotide sequences. J. Mol. Evol., 16: 111-120.

Kiuchi K, Watanabe S (2004). Industrialization of Japanese Natto. In: Industrialization of Indigenous Fermented Foods. (Ed.) KH Steinkraus. Marcel Dekker Inc., New York, pp. 193-246.

Kumar S, Tamura K, Jakobson IB, Nei M (2001). MEGA 2: Molecular evolution analysis software. Bioinformatics, 17: 1244-1245.

Latunde-Dada GO (1995). Fermented foods and cottage industries in Nigeria. J. Food Sci. 20: 1-33. www.unu.edu/unupress/V184e/ch4.htm

Nick G, Jusilla M, Hoste B, Niemi RM, Kaijalainen S, de Lajudie R, Gillis M, de Bruijn FJ, Lindström K (1999). Rhizobia isolated from root nodules of tropical leguminous trees characterized using DNA-DNA dot-blot hybridization and rep-PCR genomic fingerprinting. Syst. Appl. Microbiol., 22: 287-299.

Odunfa SA (1981a). Micro-organisms associated with fermentation of

African Locust bean (*Parkia filicoidea*) during iru preparation. J. Plant Foods, 3: 245-250.

Odunfa SA (1981b). Microbiology and amino acid composition of 'ogiri', a condiment from fermented melon seeds. Die Nahrung, 25: 811-816.

Odunfa SA, Adewuyi EY (1985a). Optimisation of process conditions for the fermentation of African locust bean (*Parkia biglobosa*). I. Effect of time, temperature and humidity. Chem. Technol. Food Microbiol., 9: 6 – 10.

Odunfa SA, Adewuyi EY (1985b). Optimisation of process conditions for the fermentation of African locust bean (*Parkia biglobosa*). II. Effect of starter cultures. Chem. Microbiol. Technol. Lebensm, 9: 118-122.

Odunfa SA, Oyewole OB (1986). Identification of *Bacillus* species from 'iru', a fermented African locust bean product. J. Basic Microbiol., 26: 101-108.

Oguntoyinbo FA, Sanni AI, Franz CMAP, Holzapfel WH (2004). Phenotypic diverssity and technological properties of *Bacillus subtilis* species isolated from okphehe, a traditional fermented condiment. World J. Microbiol. Biotechnol., 23(3): 401-410.

Osho AI, Mabekoje OO, Bello OO (2010). Comparative study on the microbial load of Gari, Eluboisu and Iru in Nigeria. African J. Food Sci., 4(10): 646-649.

Ouoba LII, Cantor MD, Diawara B, Traoré AS, Jakobsen M (2003). Degradation of African locust bean oil by *Bacillus subtilis* and *Bacillus pumilus* isolated from *soumbala*, a fermented African locust bean condiment. J. Appl. Microbiol., 95: 868-873.

Plengvidhya V, Breidt Jr. F, Fleming HP (2004). Use of RAPD-PCR as a method to follow the progress of starter cultures in sauerkraut fermentation. Int. J. Food Microbiol., 93: 287-296.

Rademaker JLW, Hoste B, Louws FJ, Kersters K, Swings J, Vauterin L, Vauterin P, de Brujin FJ (2000). Comparison of AFLP and rep-PCR genomic fingerprinting with DNA-DNA homology studies: *Xanthomonas* as a model system. Int. J. Syst. Evol. Microbiol., 50: 665-677.

Ruiz-Garćia C, Béjar V, Martínez-Checa F, Llamas I, Quesada E (2005). *Bacillus velezensis* sp. nov., a surfactant-producing bacterium isolated from the river Velez in Malaga, southern Spain. Int. J. Syst. Evol. Microbiol., 55: 191-195.

Saitou N, Nei M (1987). The neighbor-joining method: a new method for reconstructing phylogenetic trees. Mol. Biol. Evol., 4: 406-425.

Sumpavapol P, Tongyonk L, Tanasupawat S, Chokesajjawatee N, Luxananil P, Visessanguan W (2010). *Bacillus siamensis* sp. nov., isolated from salted crab (poo-khem) in Thailand. Int. J. Syst. Evol. Microbiol., 47: 289–298.

Terlabie NN, Sakyi-Dawson E, Amoa- Awua WK (2006). The comparative ability of four isolates of *Bacillus subtilis* to ferment soybeans into dawadawa. Int. Food Microbiol., 106: 145-152.

Thompson JD, Gibson TJ, Plewniak K, Jeanmougin F, Higgins DG (1997). The CLUSTAL_X Windows interface: flexible strategies for multiple sequence alignments aided by quality analysis tools. Nucleic Acids Res., 25: 4876-4882.

Versalovic J, Schneider M, de Brujin FJ, Lupski JR (1994). Genomic fingerprinting of bacteria using repetitive sequence-based polymerase chain reaction. Methods Mol. Cell Biol., 5: 25-40.

19

Comparative studies on properties of amylases extracted from kilned and unkilned malted sorghum and corn

OYEWOLE O. I.[1]* and AGBOOLA F. K.[2]

[1]Department of Biochemistry, Osun State University, Osogbo, Nigeria.
[2]Department of Biochemistry, Obafemi Awolowo University, Ile-Ife, Nigeria.

This study investigated the activities of α-amylase, β-amylase and glucoamylase extracted from kilned and unkilned sorghum and corn. Dry grains of sorghum and corn were obtained, steeped and allowed to undergo malting at room temperature for 48 h. Part of the malted starch was kilned by taken into an oven at 50°C for 24 h while the other part was derootted by hand before further processing. α-amylase, β-amylase and glucoamylase were extracted from the kilned and unkilned malted cereals and assayed. Results obtained showed that sorghum is richer in amylases than corn. The activities of the three enzymes were higher in unkilned malt than kilned malt which indicates an appreciable loss in enzymes activities during kilning. The cereals are also rich in glucoamylase compared to the other two enzymes. All the enzymes have appreciable glucose yield on maltose substrate. These results demonstrated that sorghum and corn are good sources of amylases which are the basic enzymes required for hydrolysis of starch to glucose in many industrial processes most especially in brewing.

Key words: Amylase, sorghum, corn, kilned malt, unkilned malt.

INTRODUCTION

Amylases are the enzymes responsible for breaking down amylose (starch). There are three types of amylase namely: α-amylase, β-amylase and glucoamylase. α-amylase (endo-1, 4-α-D-glucan glucohydrolase, EC 3.2.1.1) is an extra cellular enzyme that randomly cleave the 1,4-α-D-glucosidic linkages between adjacent glucose units in the linear amylase chain. This endozyme split the substrate in the interior of the molecules. β-amylase (β-1, 4-glucan maltohydrolase, EC 3.2.1.2) is an exoacting enzyme that cleaves non-reducing chain ends of amylase, amylopectin and glycogen molecules. It hydrolyses alternate glycosidic linkages yielding glycoside linkages maltose. Glucoamylase (exo-1, 4-α-D-glucan glucano-hydrolase, EC 3.2.1.3) hydrolyses single glucose units from the non-reducing ends of amylase and amylopectin in a stepwise manner. Biologically active amylases extracted from plants source are among the most important enzymes and are of great significance in present-day biotechnology. They could be potentially useful in the pharmaceutical and fine-chemical industries if enzymes with suitable properties could be prepared. Interestingly, the first enzyme produced industrially was amylase from a fungal source in 1894 which was used as a pharmaceutical aid for the treatment of digestive disorders (Manners and Marshall, 1969). With the advent of new frontiers in biotechnology, the spectrum of amylase application has widened in many other fields such as clinical, medicinal and analytical chemistry as well as their widespread application in starch saccharification and in the textile, food, brewing and distilling industries (Alli et al., 1998).

Sorghum (*Sorghum bicolor*) is a cereal crop cultivated in warmer climates worldwide and utilized for grain, fiber and fodder (Hulse et al., 1980). Corn or maize (*Zea mays*) constitute a staple food in many regions of the world and is used in the preparation of corn flakes, corn meal, porridge, corn bread and other baked products

*Corresponding author. E-mail: ioluoye@yahoo.com.

Table 1. Amylases activities from maltose curve.

Enzyme	Sorghum		Corn	
	Unkilned	Kilned	Unkilned	Kilned
α-amylase (mg maltose/ml/min)	1.20±0.07[a]	0.44±0.02[b]	0.52±0.03[c]	0.18±0.01[d]
β-amylase (mg maltose/ml/min)	2.10±0.12[a]	0.63±0.05[b]	0.94±0.05[c]	0.41±0.02[d]
Glucoamylase (mg maltose/ml/min)	2.78±0.38[a]	0.94±0.04[b]	1.22±0.05[c]	0.75±0.03[d]

Each value is a mean of 5 determinations ±SD. Values with different alphabetical superscript ([a, b, c, d]) along a row are statistically different at $P<0.05$.

(EtokAkpan, 1988). Starch from maize can also be made into plastics, fabrics adhesives and many other chemical products.

MATERIALS AND METHODS

Collection of plant material

Dry grains of sorghum (*S. bicolor*) were obtained from Derivative Company Ltd. Lagos, while corn (*Z. mays*) was obtained from Ora Gada Market in Ogbomoso, Nigeria. The grains were screened to remove broken seed and other impurity.

Chemicals/reagents

Dinitrosalicylic acid, sodium hydrogen phosphate, ammonium sulphate, calcium chloride, soluble starch, disodium hydrogen phosphate, ethanol, phosphoric acid, bovine serum albumin (BSA), coomasie brilliant blue G-25, sodium potassium tartarate, anhydrous sodium carbonate, maltose, sodium hydroxide, sodium sulphate, cupric sulphate, sodium acetate and glacial acetic acid were obtained from British Drug House (BDH), Poole, England.

Equipments

Laboratory equipment used includes AB205 Meltler Toledo weighing balance made in Switzerland, LEC Grant Incubator along with Golden lamp made in USA. Double Beam UV spectrophotometer made in Cambridge, England and centrifuge model 800 D micro field instrument made in England with maximum speed of 4000 rpm.

Steeping and germination of grain samples

200 g of the grains were steeped for 48 h by soaking in distilled water at room temperature. The water was changed every 8 h to prevent microbial growth. After 48 h, the water was drained from the sample and the damp seeds spread out in a malting chamber to germinate for 48 h at room temperature. The germinated grains were divided into two parts. The first part was taken into an oven at 50°C for 24 h. This is a process called kilning while the other part (unkilned) was derooted by hand before further processing.

Extraction of amylases from kilned and unkilned malt

5g of the malt was homogenized with 3 volume of the respective homogenization buffer (0.1 M acetate buffer, pH 5.5 for α-amylase, 0.1 M phosphate buffer, pH 6.0 for β-amylase and 0.5 M acetate buffer, pH 4.5 for glucoamylase) and poured in a clean beaker. They were kept in the refrigerator for 1 h with intermittent stirring every 10 min. This was followed by centrifugation at 6000 rpm for 20 min to remove the debris. It was then filtered with two layers of cheese clothes to remove fat deposits (Anon, 1986). Assay for protein and enzyme activities were then carried out.

Enzyme assay

The method of EtokAkpan (1988) was employed in the determination of enzyme activities. Crude preparations of each enzyme were diluted 2:10. The reaction mixture contained 0.5 ml of 1% starch solution, 0.48 ml distilled water and 0.02 ml of the enzyme solution. 1 ml of freshly prepared colour reagent was added to the reaction mixture and boiled for 5 min in cold running water. The absorbance was taken at 470 nm. The procedure was repeated for all amylases using respective colour reagent. Maltose concentration was determined as described by Fix and Fix (1997) and protein was measured by the Bradford method (Metwally, 1998).

Statistical analysis

Data obtained were reported as mean of 5 replicates ±SD. Statistical significance was determined by using one way analysis of variance (ANOVA) followed by Duncan Multiple Range Test and differences were considered significant at $P<0.05$.

RESULTS

Tables 1, 2 and 3 show the activities of the amylases in kilned and unkilned malt extrapolated from protein, maltose and glucose standard curves respectively. From the results, sorghum is richer in amylases than corn. The results also revealed that activities of the three enzymes are higher in unkilned malt than in kilned malt with all the substrates. It can also be seen in the tables that glucoamylase has the highest activities among the three enzymes followed by α-amylase in both kilned and unkilned malt. The activities of the enzymes with glucose substrate are the highest compared to other substrate that is maltose and protein as can be seen in the tables.

DISCUSSION

From the results obtained for enzymes activities using

Table 2. Amylases activities from protein curve.

Enzyme	Sorghum		Corn	
	Unkilned	Kilned	Unkilned	Kilned
α-amylase (mg protein/ml/min)	0.072 ± 0.003^a	0.039 ± 0.002^b	0.049 ± 0.002^c	0.030 ± 0.001^d
β-amylase (mg protein/ml/min)	0.084 ± 0.004^a	0.041 ± 0.001^b	0.060 ± 0.004^c	0.038 ± 0.002^b
Glucoamylase (mg protein/ml/min)	0.160 ± 0.009^a	0.068 ± 0.005^b	0.133 ± 0.008^c	0.074 ± 0.004^d

Each value is a mean of 5 determinations ±SD. Values with different alphabetical superscript ([a, b, c, d]) along a row are statistically different at $P<0.05$.

Table 3. Amylases activities from glucose curve.

Enzyme	Sorghum		Corn	
	Unkilned	Kilned	Unkilned	Kilned
α-amylase (mg glucose/ml/min)	1.64 ± 0.11^a	0.72 ± 0.06^b	1.17 ± 0.08^c	0.52 ± 0.04^d
β-amylase (mg glucose/ml/min)	2.23 ± 0.13^a	1.14 ± 0.08^b	1.96 ± 0.14^c	0.98 ± 0.05^d
Glucoamylase (mg glucose/ml/min)	9.90 ± 2.42^a	5.65 ± 1.60^b	5.20 ± 1.48^b	2.20 ± 0.14^c

Each value is a mean of 5 determinations ±SD. Values with different alphabetical superscript ([a, b, c, d]) along a row are statistically different at $P<0.05$.

different substrates, all the three enzymes have higher activities in unkilned malt compared to kilned malt. This might be due to loss of enzyme activities during kilning. Kilning involve heat treatment which can have adverse effect on enzyme activities. Enzymes are protein and when subjected to high temperature beyond the optimum, they are denatured due to loss in conformation (Yun and Matheson, 1990). It is very important that maximum activity of the enzymes is retained during the preparation as enzymes for industrial use are sold on the basis of overall activity (Pandey, 1992). The reduction can also be as a result of inactivation by the buffer solution employed during the preparation of the kilned cereals. Amylases from sorghum and corn have been reported to be reversibly unfolded by chemical denaturants (EtokAkpan and Palmer, 1990). Enzyme inactivation can be caused by heat, proteolysis, sub-optimal pH, oxidation, denaturants, irreversible inhibitors and loss of cofactors or coenzymes (Palmer et al., 1989). Proteolytic activities can also be responsible for the observed reduction in activities in kilned malt. Proteolysis is most likely to occur in the early stages of extraction and purification when the proteases responsible for protein turnover in living cells are still present. In their native conformations, enzymes have highly structured domains which are resistant to attack by proteases because many of the peptide bonds are mechanically inaccessible and because many proteases are highly specific. Therefore, it is important to keep enzyme preparations cold to maintain their native conformation and slow any protease action that may occur (Aniche and Palmer, 1990). The key to maintaining enzyme activity is the maintenance of their conformation so as to prevent unfolding, aggregation and changes in

the covalent structure (Glennie and Wright, 1986). From the results obtained, it can also be seen that kilned and unkilned sorghum and corn are richer in glucoamylase than the other. This is followed by β-amylase.

It has been reported that glucoamylase and β-amylase are usually of plant origin while α-amylase is obtained mostly in microorganism (EtokAkpan and Palmer, 1990). The reason might also be because glucoamylase has higher optimum temperature (60°C) than β-amylase (55°C) and α-amylase (45°C) meaning that it can withstand heat treatment than the other two enzymes (EtokAkpan, 1988). The differences in activities recorded for the three enzymes with the substrates arise probably because they have different cleaving points on the substrate. Alpha amylases cleave maltose units internally bringing little change in sweetness and large decrease in viscosity. β-amylase is an exoacting enzyme that cleave maltose units from the ends to bring large change in sweetness and little decrease in viscosity (Sen et al., 1997). Glucoamylase hydrolyses single glucose units from the non-reducing ends of amylose and amylopectin in a stepwise manner bringing large change in sweetness and little decrease in viscosity (Metwally, 1998).

Conclusion

Results obtained in this study showed that sorghum and corn are good sources of alpha-amylase, beta-amylase and glucoamylase. The enzymes are present in appreciable quantities in these cereals and can be extracted for industrial use. It is also evident from this study that appreciable amount of these enzymes were lost during the process of kilning which involve drying. Therefore, heat processing should be avoided in order to obtain maximum yield of amylases from this plant source. Further kinetic study is required on the extracted enzymes to fully understand their properties before embarking on large scale production of the enzymes from this source.

REFERENCES

Alli AL, Ogbonna CL, Rahman AT (1998). Hydrolysis of certain Nigerian cereal starches using crude fungal amylase. Nig. J. Biotech., 9: 24-36.

Aniche GN, Palmer GH (1990). Development of amylolytic activities in sorghum malt and barley malt. J. Instit. Brewing, 96: 377-379.

EtokAkpan OU (1988). Biochemical Studies of the Malting of Sorghum and Barley. PhD Thesis. Heriot-Watt University, Edinburgh. P. 245.

EtokAkpan OU, Palmer GH (1990). Comparative studies of the development of endosperm-degrading enzymes in malting sorghum and barley. World J. Microbiol. Biotechnol., 6: 408-417.

Fix GJ, Fix LA (1997). An analysis of brewing techniques. Brewers Publications Boulder Company. Pp. 180-192.

Glennie CW, Wright AW (1986). Dextrins in Sorghum beer. J. Instit. Brewing, 92: 384-386.

Hulse JH , Laing EM, Pearson OE (1980). Sorgum and millet: their composition and nutritive value. Academic Press London.

Manners DJ, Marshall JJ (1969). Studies on carbohydrate metabolizing enzymes. Part XXIII. The β-glucanase system of malted barley. J. Instit. Brewing, 75:550-561.

Metwally M (1998). Glucoamylase production in continuous culture of aspergillus niger with special emphasis on growth parameters. World J. Microbiol. Biochem., 14: 113-114.

Palmer GH, EtokAkpan OU, Igyor MA (1989). Sorghum as brewing material. J. Appl. Microbiol. Biotechnol., 5: 265-275.

Pandey A (1992). in Industrial Biotechnology (Malik V.S., Sridhar P. eds.) IBH and Oxford Publishing Co., New Delhi. Pp. 525-537.

Sen M, Thevanat C, Prioul JL (1997). Simultaneous spectrophotometric determination of amylose and amylopectin in starch from maize kernel by multiwavelength analysis. J. Cereal Sci., 26: 211-221.

Yun SH, Matheson NK (1990). Structural changes during development in the amylose and amylopectin fraction separated by precipitation with concanavalin A of starches from maize genotypes. Carbohyd. Res., 270:85-101.

20

Decolorization of anthroquinone based dye Vat Red 10 by *Pseudomonas desmolyticum* NCIM 2112 and *Galactomyces geotrichum* MTCC 1360

Archana A. Gurav[1], Jai S. Ghosh[2]* and Girish S. Kulkarni[1]

[1]Department of Technology, Shivaji University, Kolhapur 416004, India.
[2]Department of Microbiology, Shivaji University, Kolhapur 416004, India.

Wastewater, from the textile and other dyestuff industries containing synthetic dyes, require prior treatment to prevent groundwater contamination. The microbial decolorization and degradation of these dyes play a pivotal role in this aspect. *Pseudomonas desmolyticum* NCIM 2112 and *Galactomyces geotrichum* MTCC 1360 could bring about oxidative degradation resulting in decolorization of this water insoluble Vat Red 10 (Novatic red 3B) dye at pH 9, at 25°C. The decolorization was measured as the decrease in absorbance maxima at 530 nm. The end product of degradation and decolorization was 2, 6-Di isopropyl Naphthalene (2, 6-DIPN), which is an important plant growth factor.

Key words: Vat dyes, vat red 10(Novatic Red3B), dye decolorization, textile, *Pseudomonas desmolyticum*, *Galactomyces geotrichum*, wastewater, 2, 6-DIPN.

INTRODUCTION

The discharge of large amount of toxic waste in the environment is a consequence of rapid industrialization and urbanization. One of such pollutants are the textile dyes which are very chemically stable as these have to be fast to prevent easy loss of color during washing and fading on exposure to sunlight. Large numbers of synthetic and chemically different dyes are used for various industrial applications and significant proportion appears in the form of wastewater which ultimately finds their way in the environment. The textile industry is one such industry which discharges a large proportion of these pollutants. Their presence in an environment like a water body, leads to reduction in sunlight penetrations resulting in decrease in photosynthetic activity and thus reduced dissolved oxygen content in water bodies. Depending on the class of the dyes, their loss in waste waters can range from 2% of the original concentration for basic dyes to as high as 50% for reactive dyes (O'Neill et al., 1999; Tan et al., 2000; Boer et al., 2004).

Amongst these, azo and vat dyes represent the largest and most versatile class of synthetic dye (Keharia et al., 2004). Around 10,000 different dyes with annual production of more than 7×10^5 metric tones worldwide are commercially available for various applications (McMullan et al., 2001).

Dyes on the basis of their chromophore group are classified as azo, anthraquinone, nitro, nitroso, triphenylmethane, xanthene, acridine, thiazole, sulfur, indigoid, dyes etc. Dye concentrations that are used for processing vats are typically around 1,000 mg^{-1} (Ince and Tezcanli, 1999). In order to prevent such toxic effects which has an adverse effect on the natural biodiversity, it is essential to free the environment of these coloring materials. Many physical and chemical methods including adsorption, coagulation, precipitation, filtration, and oxidation have been used for the treatment of dye-contaminated effluents. These methods, however, may generate a significant amount of sludge or may easily cause secondary pollution due to excessive chemical usage. Moreover, their municipal treatment costs are high. Various wood-rotting fungi were able to decolorize azo dyes using peroxidases or laccases. Therefore, it may be economical to develop alternative means of dye

*Corresponding author. E-mail: ghoshjai@gmail.com.

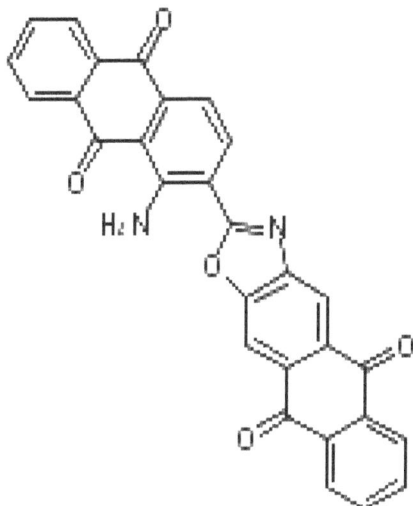

Figure 1. Chemical structure of Vat Red 10.

decolorization, such as bioremediation due to its reputation as an environmentally friendly acceptable treatment technology. The sequential anaerobic treatment followed by aerobic bacterial degradation system has proved to be efficient in the degradation of these dyes.

Microbial decolorization and degradation is an environmental friendly and cost effective process (Verma and Madamwar, 2003). Several reports revealed the existence of a wide variety of microorganisms capable of decolorizing a wide range of dyes (Banat et al., 1996). The effectiveness of microbial treatment depends on the survival, adaptability and activity of the selected organisms (Paszecazynki et al., 1992). Anthraquinone dyes like Vat Red 10 (Novatic Red 3B), which are also classified as oxazole derivatives, are resistant to degradation due to their fused aromatic structure, which remain colored for a long time (Banat et al., 1996). The molecular structure of this dye is as given in Figure 1. It has a molecular weight of 473 Da (Venkatraman, 1971). These dyes are carcinogenic and mutagenic (Itoh et al. 1996) to humans. Decolorization of anthraquinone dyes has received much attention due to their recalcitrant nature (Laszlo, 1995).

This study investigates microbial decolorization of Vat Red 10 dye by *Pseudomonas desmolyticum* NCIM 2112 and *Galactomyces geotrichum* MTCC 1360 at high pH because these dyes are soluble in water only at alkaline pH. These can be reduced, in the presence of a reducing agent in an alkaline medium; forming a water-soluble leuko-compound. The most commonly used reducing agent in vat dyeing is sodium dithionite. However, one of the major drawbacks of this industrial dyeing process is the high amount of sulfate, sulfite, and thiosulfate in the wastewater and of the toxic sulfide formed subsequently leading to bio-corrosion of the wastewater pipelines.

MATERIALS AND METHODS

Microorganisms and culture medium

Pure cultures of *P. desmolyticum* NCIM 2112 and *G. geotrichum* MTCC 1360 were maintained on solid mineral base medium having the following composition (NaNO$_3$ 0.3%, K$_2$HPO$_4$ 0.1%, MgSO$_4$ 0.05%, KCl 0.05%, yeast extract 0.02% and agar 2.5%) with 1% glucose. The culture was adapted to grow at pH 9 in the same medium at 25°C. The media used in this study were liquid mineral base medium having the same composition as aforestated:

$$\text{Decolorization rate \%} = \frac{A-B}{A} \times 100$$

A - Initial absorbance; B - Observed absorbance.

Dyes and chemicals

The commonly used vat dye for cotton dyeing- Vat Red 10 or Novatic Red 3B was used in this experiment. The other chemicals used were of analytical grade and highest purity.

Decolorization experiment

The flasks containing medium were sterilized by autoclaving. One millilitre of microbial suspension containing 120x10^6 cells was inoculated into 100 ml of the aforementioned liquid medium with glucose and containing Vat Red 10 dye at 0.01%. The incubation was carried out on a rotary shaker 120 rpm at 25°C for 23 days. At every 7 days interval, the flasks were checked for reduction in color by comparing with control set of experiments where no bacterial culture was added. The decolorization was measured as the decrease in absorbance maxima at 530 nm.

Statistical analysis

Results obtained were the mean of three or more determinants. Analysis of the variants was carried out on all data at P< 0.05 using Graph Pad software (Graph Pad Instat version 3.00, Graph Pad software, San Diego, CA, USA).

RESULTS AND DISCUSSION

Vat Red 10 is a vat dye which is soluble only at alkaline pH of 9. Therefore, to bring about microbiological decolorization, the organisms like *G. geotrichum* MTCC 1360 and *P. desmolyticum* NCIM 2112 were first adapted to grow at pH 9, as both these organisms were only reported to grow at pH 7. The adaptation studies were carried out in a stepwise manner grown in presence of 1% glucose. It was observed that when the adapted strain of *G. geotrichum* were grown in the presence of the dye and 1% glucose, 45% of the dye was decolorized in 23 days, of which 35% decolorization was in the first 7 days, as shown in Figure 2. It can be seen from the figure that the rate of decolorization was high during the initial period and it declined later.

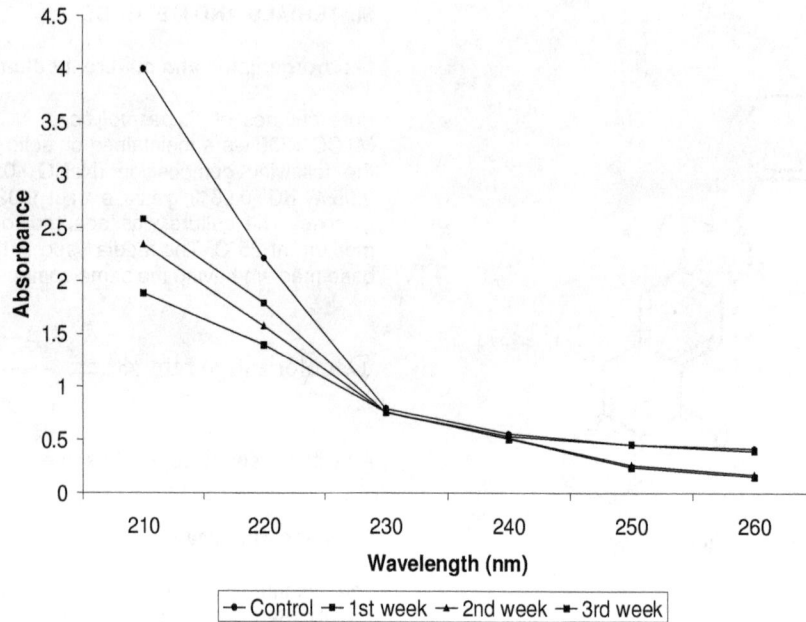

Figure 2. Decolorization of anthroquinone dye Vat Red 10 by *Galactomyces geotrichum* grown in minimal liquid base media containing 0.01% glucose. Observations were taken at regular time intervals that is after 7 days at 530 nm on UV-vis spectroscope.

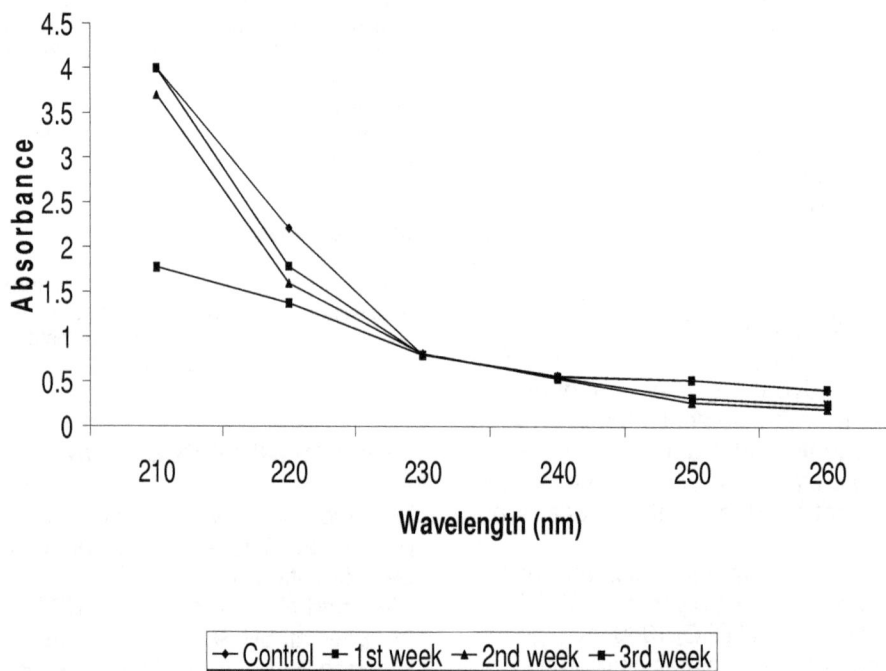

Figure 3. Decolorization of anthroquinone dye Vat Red 10 by *Pseudomonas desmolyticum* grown in minimal liquid base media containing 0.01% glucose Observations were taken place at regular time intervals that is after 7 days.

Similarly, *P. desmolyticum* NCIM 2112, in presence of glucose, could bring about 55.5% decolorization in 23 days, as shown in Figure 3. The decolorization process was aerobic as the experiments were conducted in shake flask conditions. The proposed mechanism (Figure 4) of decolorization were as per the intermediates like

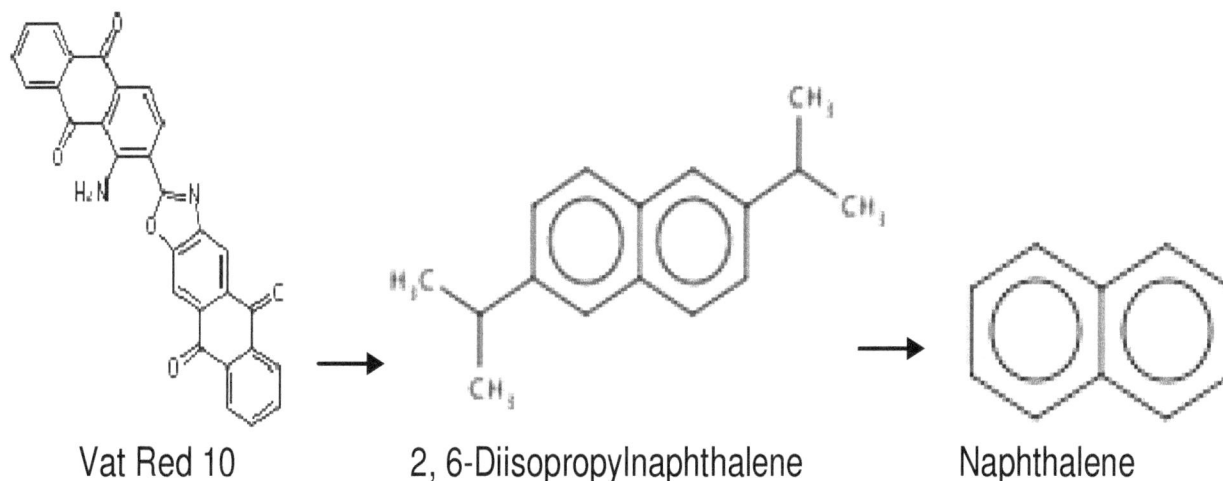

Figure 4. Proposed pathway for the decolorization of Vat Red 10 by *Galactomyces geotrichum*.

diisopropylnaphthalene and naphthalene, which were identified by GCMS as per for "diisopropylnaphthalene" (Brzozowski et al., 2007) "Naphthalene" (www.epa.gov/region01/eco/airtox/fs/naphthalene.html 2007). The dye could have been detoxified either by bioaccumulation or by biodegradation (Knapp and Newby, 1995; Sani and Banerjee, 1999). The results clearly pointed out to the fact that biodegradation was the sole means of decolorization and thus detoxification. If there was bioaccumulation, then the cells of *P. desmolyticum* and *G. geotrichum* should have shown the presence of the dye either on their outside or in the cytoplasm. This was not observed from the absorption maxima studies. Similar results were observed with *Aeromonas hydrophila* DN322 and the concerned dyes were crystal violet, malachite green, reactive red etc. (Suizhou et al., 2006).

However, crystal violet was detoxified by *Aeromonas* spp B5 by adsorption on its surface (Nobuki et al., 2000). *Aspergillus niger SA1* could detoxify anthraquinone dye – Drimarene Blue by biodegradation (Muhammad et al., 2010), *Coriolus versicolor* could degrade and detoxify the anthraquinone dye Pigment violet 12 (Itoh et al., 1998), *Trametes versicolor* could degrade 2 carpet anthraquinone dyes (Ramsay and Goode, 2004). This being a first report of its kind of microbial detoxification of such a recalcitrant like Vat Red 10 which is an anthraquinone – oxazole dye, by these 2 organisms belonging to genera *Pseudomonas* and *Galactomyces*. The end product of the degradation being a useful compound – Di-isopropyl naphthalene (2, 6-DIPN), which is a well known plant growth regulator, totally nontoxic to the environment. It is even hypothesized that this will further undergo demethylation to yield naphthalene. Therefore, biodegradation of Vat red 10 as per this investigation not only helps to protect the environment but also produces substances which are of agronomic importance.

ACKNOWLEDGEMENT

The authors are grateful to the Department of Environmental Science, Department of Technology and Department of Microbiology, Shivaji University, Kolhapur, for providing the necessary facilities towards completion of this work.

REFERENCES

Banat IM, Nigam P, Singh D, Marchant R (1996). Microbial decolorization of textile dye containing effluents. Rev. Bioresour Technol., 58: 212-227.

Boer CG, Obici L, Souze CC, Piralta RM (2004). Decolorization of synthetic dyes by solid state culture of *Lentinula (Lentinus) edodes* producing manganese peroxidase as the main lignolytic enzyme. Bioresour. Technol., 94: 107-112.

Brzozowski R, Skupinski W, Jamróz MH, Skarzyn M, Otwinowska H (2002). Isolation and identification of diisopropylnaphthalene isomers in the alkylation products of naphthalene. J. Chromatogr. A, 946: 221-227.

Ince H, Tezcanli G (1999). Treatability of textile dye-batch effluents by advanced oxidation: preparation for reuse. Water Sci. Technol., 40: 183-190.

Itoh K, Kitade Y, Yatome C (1998). Oxidative Biodegradation of an Anthraquinone Dye,Pigment Violet 12, by *Coriolus versicolor*. Bull. Environ. Cont. Toxicol., 60: 786-790.

Keharia H, Patel H, Madamwar D (2004). Decolorization screening of synthetic dyes by anaerobic methanogenic sludge using a Bacth decoloration assay. World J. Microbiol. Biotechnol., 20: 365-370.

Knapp JS, Newbym PS (1995). The microbial decolorization of an industrial effluent containing a diazo linked chromophore. Water Res., 29: 1807-1809.

Laszlo JA (1995). Electrolyte effect on hydrolysed reactive dye binding to quaternized cellulose. Textile Chem. Colorist, 27:25-27.

McMullan G, Needhan C, Connedy A, Kirby N, Robinson T, Nigam P, Banat IM, Smyth WF (2001). Microbial decolorization and degradation of textile dye. Appl. Environ. Microbiol., 56: 81-87.

Muhammad FS, Saadia A, Naeem A, Ghumro PB, Ahmed S (2010). Biotreatment of anthraquinone dye Dicromarene Blue, K2RL. Afr. J. Environ. Sci. Technol., 4: 45-50.

Nobuki H, Hazako K, Kazatoski U (2000). Isolation and Characterization of *Aeromonas* sp. B-5 capable of decolorizing various dyes. Biosci. Bioeng., 90: 570-573.

O'Neill C, Hawkes FR, Loureno ND, Pinheirio HM, Delel W (1999). Color in textile effluent source, measurement discharge content and simulation: A rev. J. Chem. Technol. Biotechnol., 74: 1009-1018.

Paszecazynki A, Pasti-Grigsby M, Gorzceynshi S, Crawfod R, Crawfod DL (1992). Mineralization of sulfonated azo dyes and sulfonilic acid by *Phyanaerochaete chrysosporium* and *S. chromofureus*. Appl. Environ. Microbiol., 58: 3598-3604.

Ramsay JA, Goode C (2004). Decoloration of a carpet dye effluent using *Trametes versicolor*. Biotechnol. Lett., 26: 197-201.

Sani R, Banerjee V (1999). Decolorization of triphenylmethane dyes and textile dye stuff by *Kurthia* spp. Enzyme Microbiol. Technol., 24: 433-437.

Suizhou R, Guo J, Zeng G, Sun G (2006). Decolorization of triphenylmethane, azo and anthraquinone dyes by a newly isolated *Aeromonas hydrophyla* strain. Appl. Microbiol. Biotechnol., 72: 1316-1321.

Tan NCG, Borg SA, Slendas P, Smitelskaya A, Lottirga G Fride JA (2000). Degradation of azo dye Mordant Yellow 10 in a sequential bioaugmented anaerobic bioreactor. Water Sci. Technol., 42: 337-344.

Venkatraman K (1971). The Chemistry of Synthetic Dyes, Academic Press, New York, pp. 170-171.

Verma P, Madamwar D (2003). Decolorization of synthetic dyes by a new isolated strain of *Serratia marcescens*. World J. Microbiol. Biotechnol., 19: 615-618.

www.epa.gov/region01/eco/airtox/fs/naphthalene.html (2007).

21

Effect of common feed enzymes on nutrient utilization of monogastric animals

Bimrew Asmare

Department of Animal Production and Technology, Bahir Dar University, Ethiopia.

Some nutrients in livestock feed may not be fully digested by animals own digestive enzymes, and hence important nutrients are unavailable to the animal. Although, supplementation of enzymes to farm animals has shown to increase the digestibility of poorly digested diets to a much greater extent. By targeting specific anti-nutrients in certain feed ingredients, feed enzymes allow especially pigs and poultry to extract more nutrients from the feed and so improve feed efficiency. Enzymes are most commonly used when the dietary ingredients contain relatively higher amounts of fiber. The classification of enzymes is usually according to the substrates they act upon and the classification can be enzymes that break down fiber, proteins, starch and phytate. Appropriate use of exogenous enzymes in feeds requires strategic reductions in dietary energy and nutrient content, as well as careful choice of feed ingredients to capture economic benefits of the various enzymes. The efficacy of enzymes will vary depending on ingredients because nutrient and energy release caused by enzyme supplementation will depend on the structure of the feedstuff itself. It is important to continue the effort to understand the use and limitations of matrix values of enzymes, which, if inappropriately applied, will result in depressed performance because of inadequacy of diets or will lead to wastage of resources.

Key words: Carbohydrate, monogastric, feed enzymes, phytase, nutrient utilization.

INTRODUCTION

Not all compounds in animal feed are broken down by animals' own digestive enzymes, and so some potential nutrients are unavailable to the animal (McDonald et al., 2010). To alleviate this problem, in the 1950s, pioneering scientists added enzymes called amylases and proteases to the diets of various farm animals and observed benefits in productivity. Such kinds of exogenous enzymes are produced commercially from microbes, fungi and yeasts in highly controlled conditions in fermentation plants (Fuller, 2004). Their main uses are in the detergent and food industries but significant quantities are manufactured for use in animal diets. As feed additives, enzymes are mainly used in the diets of non-ruminants but are also added to ruminant diets (Fuller, 2004). Among monogastric animals, pigs and poultry are important beneficiaries these days from exogenous enzyme supplementation diets, and are even used extensively for the latter (McDonald et al., 2010).

Feed enzymes help fundamentally to improve the efficiency of meat and egg production by changing the nutritional profile of feed ingredients (Bedford and Partridge, 2010). Enzymes have clearly been demonstrated to increase the digestibility of poorly digested diets to a much greater extent than well digested diets (Scott et al., 1998). Enzymes are most commonly used when the dietary ingredients contain relatively higher amounts of fiber (Bedford, 2000). For example, the various forms of fiber in the pig's diet will not be well digested by the pig; as a result, a large portion of the fiber in the diet passes through the small intestine intact, and the only breakdown that can occur is through fermentation by bacteria and yeast in the cecum and large intestine. The disruption of the cell matrix of fibrous feedstuffs by exogenous enzymes can lead to easy access of the endogenous proteolytic and cellulolytic enzymes to digest the entrapped protein and carbohydrates. This will consequently reduce the feed cost in animal production. However, the effects of exogenous enzymes can be variable and it depends on a large number of factors such as the age of the animal and the quality and type of diet (Bedford, 2000). Further, feed enzymes allow the feed producer greater flexibility in the type of raw materials that can confidently be used in feed formulation. Another application of enzymes is to break down the phytate molecule that binds phosphorus and some other mineral elements in plant based feedstuffs (McDonald et al., 2004; Fuller, 2004). Based on the fact that a significant portion of phosphorus in the diet of the pig is bound to phytate, it will not be well digested by the pig.

Another area of feed enzyme application in farm animal diet is to supplement the enzyme complement of young animals, in which the rate of endogenous enzyme production may be limiting. The effect of commercially available enzymes on the feeding value of major ingredients is often based on their effect in young chicks less than two weeks of age has been demonstrated by Meng and Solminski (2005). This is further explained by the fact that in newly hatched chicks, the enterocyte is poorly developed, limiting the bird's digestion and absorption abilities (Lji et al., 2001a). During this maturation period, the gut lacks the competency to fully digest feedstuffs and absorb smaller molecules because of a lack of brush-border enzymes, inadequate maintenance of absorptive mechanisms, and low surface area caused by immature villus height (Van Leeuwen et al., 2004). As the gastrointestinal tract develops, it is able to take advantage of the effects of fibrolytic enzymes. Before this, however, the pancreatic enzymes needed to initiate digestion in the intestinal lumen are limited in both volume and activity (Noy and Sklan, 1995). Thus, they may be unable to utilize substrates made available by a fibrolytic enzyme. Early-weaned pigs have limited amylase, protease and lipase activity, and enhancement of the extent of digestion of nutrients would improve performance and reduce the incidence of the diarrhoea

that results from undigested nutrients reaching the hind gut and being fermented by bacteria.

In a study with finishing pigs, Zhang and Kornegay (1999) reported that the digestibility of all amino acids except proline and glycine increased linearly as phytase supplementation increased. Enzymes are essential for the breakdown of cell-wall carbohydrates to release the sugars necessary for the growth of the lactic acid bacteria. Commercial hemicellulase and cellulase enzyme cocktails are now available and can improve the fermentation process considerably (Hooper et al., 1989). However, prices of these products preclude their viability for farm level application, especially in developing countries. Supplementation of a wheat by-product diet with cellulase increased the ileal digestibility of non-starch polysaccharides from 0.192 to 0.359 and crude protein from 0.65 to 0.71 (McDonald et al., 2010). The diversity of enzyme activities within commercially available enzyme preparations is probably advantageous, in that a single product can target a wide variety of substrates. Enzymes are categorized according to the substrates they act upon (Bedford and Partridge, 2010). Currently, in animal nutrition, the types of enzymes used are those that break down fibre, proteins, starch and phytate.

The objective of this paper was to review the characteristic feed enzymes and their roles in mono-gastric nutrition

COMMON FEED ENZYMES ON NUTRIENT UTILIZATION AND ANIMAL PERFORMANCE

Carbohydrases

Studies reported that carbohydrase supplementation improved the digestibility of dry matter (Nortey et al., 2007), organic matter (Li et al., 1996), and energy (Yin et al., 2000) in monogastric animal nutrition. Other studies also reported an improved result on the digestibility of amino acids due to carbohydrase-supplemented wheat- (Vahjen et al., 2007) and barley-based diets (Li et al., 1996). However, observations of increased digestibility of fiber or non starch polysacchirde components with carbohydrase supplementation emphasize the importance of release of inaccessible nutrients in enhancing amino acid digestibility.

Generally, the importance of enhanced digestibility after carbohydrase supplementation should be considered in evaluating the role of carbohydrases in enhancing nutrient utilization.

In many poultry studies, carbohydrase supplementation has been shown to improve energy utilization in corn-soybean meal diets (Rutherfurd et al., 2007; Yang et al., 2010). Others noted no improvement in energy utilization in response to carbohydrases (Olukosi et al., 2007b). In diets with cereal grains containing greater quantity of non starch polysaccharides, carbohydrase supplementation also improved energy utilization (MacLeod et al., 2008). The studies show that carbohydrases often improve energy value of diet or feed ingredients containing increased concentration of non starch polysaccharides. The differences in the effect of the enzymes on energy of feedstuffs or diet may relate to the amount of substrate for the enzyme or availability of energy from the ingredient itself, or both. Similarly, in Adeola et al. (2008) study, carbohydrases improved metabolizable energy in diets with reduced metabolizable energy but not in diets with higher metabolizable energy. There are also reports of improvement in dry matter utilization (Yang et al., 2010), fat (Boguhn and Rodehutscord, 2010), starch (Meng and Slominski, 2005) and minerals (Olukosi et al., 2008b) in response to carbohydrase supplementation. The responses to enzyme supplementation are feedstuff-, diet- and enzyme-dependent. Generally, the feedstuffs with greater amount of non starch polysaccharides, intuitively, respond to a greater extent to carbohydrase supplementation.

There is association between extent of digesta-viscosity reduction and improvement in protein digestibility (Palander et al., 2005). Many others have also reported improvement in N (Yang et al., 2010) and amino acid digestibility (Boguhn and Rodehutscord, 2010) in response to carbohydrase supplementation. However, Rutherfurd et al. (2007) noted that reduction of endogenous loss by carbohydrases is secondary to improvement in protein hydrolysis. Clearly, there is need to understand why specific amino acids respond to a greater extent and how this can be used to increase the benefit from carbohydrase supplementation.

It is important to note that improvement in nutrient digestibility does not explain all the effects of carbohydrase supplementation on performance. This is demonstrated in Barrera et al. (2004), in which xylanase supplementation to a low amino acid wheat-based diet only marginally improved performance, whereas supplementation of crystalline amino acid improved growth. In the same study, xylanase supplementation improved amino acid digestibility by an average of 11%.

In corn-soybean meal-based diets, Tahir et al. (2005) observed that cellulase, hemicellulase and their combination increased body weight gain without any effect on feed intake in broilers but Cowieson and Ravindran (2008b) observed both increased body weight gain and feed intake in response to supplementation with a mixture of xylanase, amylase and protease. Similarly, Olukosi et al. (2007a) reported a dose-related increase in body weight gain, feed intake and feed efficiency in broilers receiving wheat and rye-based diets with xylanase supplementation. In other studies, there were no responses to supplementation of carbohydrases (Olukosi and Adeola, 2008). According to the authors, part of the differences observed in the various studies can be due to the extent of nutrient density reduction in the control diets.

In comparison with broilers, the effect of non starch polysaccharides-degrading enzymes was smaller for layer pullets (Karimi et al., 2007), although the enzyme reduced digesta viscosity to the same extent in the different chicks. Some studies reported no effects of non starch polysaccharides-degrading enzymes on egg production (Hampson et al., 2002), but at the same time, reported positive effects on specific response criteria related to the quality of eggs produced (Jaroni et al., 1999). Due to variations in treatments and enzyme activities used, it is difficult to make across-study comparisons for the effects of the enzymes.

Effect of carbohydrases on non starch polysaccharides

The main reason for the use of carbohydrases is to hydrolyze complex carbohydrates that non-ruminant animals are unable to hydrolyze by themselves. Some of these compounds are present as part of the cell wall, thus shielding substrates from contact with the digestive enzymes, or as part of cell content where their presence may interfere with digestion and absorption by their chemical nature. Nitrayová et al. (2009) reported improved ileal disappearance of non starch polysaccharides in diets containing 96% rye for weanling swine; there was a 740% improvement in disappearance for xylose and a 144% improvement in disappearance for total non starch polysaccharides when xylanase was added at the rate of up to 200 mg/kg. These data indicate that non starch polysaccharides removal is one of the critical roles of carbohydrases when added to diets containing non starch polysaccharides. It seems that reduction in digesta viscosity may be one of the most important benefits of carbohydrase supplementation (Vahjen et al., 2007). Several studies have shown that xylanase attacks the arabinoxylan backbone, causing a decrease in the degree of polymerization (Courtin and Delcour, 2002) and thus liberate oligomers. The importance of this hydrolysis is that the direct link between digesta viscosity and animal performance has been demonstrated in several studies (Zhang et al., 2000a). Adeola and Bedford (2004) demonstrated that one of the modes of action of carbohydrases is their ability to reduce non starch polysaccharides-induced digesta viscosity. High-viscosity wheat responded more (in terms of improved nutrient utilization) than low-viscosity wheat when the diets were supplemented with xylanase.

Impact of carbohydrases on energy availability

Energy digestibility in swine generally decreases with increased fiber intake (Nortey et al., 2008). Explanations for reduced energy digestibility could be increased endogenous energy loss, reduced digestibility of energy yielding fraction because of impaired nutrient absorption, reduced contact of substrates and digestive enzymes, reduced proportion of energy-yielding fractions in high-fiber feedstuffs, or reduction in feed intake because of bulkiness of high-fiber diets combined with inherent stomach capacity of the animal. Johnson and Gee (1986) observed that feeding of high- non starch polysac-charides diet to rats reduced DNA and protein contents of the brush border. This reduction could be a result of increased cell turnover rate engendered by increased cell proliferation. Therefore, if carbohydrases hydrolyze the non starch polysaccharides fractions, these effects could be reversed and energy utilization should improve. Indeed, there have been observations of increased quantity of mono- and oligosaccharides in the ileum after the use of cellulase or xylanase (van der Meulen et al., 2001) and β-glucanase (Li et al., 1996) or multi-activity carbohydrases (Kiarie *et al.*, 2007). It seems that one of the ways by which carbohydrases improve energy utilization is by shifting production of volatile fatty acid and absorption of energy-yielding monosaccharides to proximal intestine. This is supported by observations of decreased net disappearance of nutrients in the large intestine of swine receiving β-glucanase-supplemented diets (Li et al., 1996). The shift in nutrient utilization to the more proximal intestine would decrease host-microbe competition for nutrients, ensure availability of nutrients where absorption efficiency is greater, reduce fermen-tative loss, and contribute to overall improved efficiency of energy utilization. However, reduction in digesta viscosity is usually far greater than improvement in energy utilization. In Macleod et al. (2008), a small increase (<5%) in metabolizable energy of naked oats was associated with a substantial decrease (>250%) in jejunal digesta viscosity. Similarly, in Adeola and Bedford (2004), the decrease in jejunal viscosity (70%) was associated with only a small increase in metabolizable energy (4%).

Since non starch polysaccharides may reduce the capacity for absorption by reducing enzyme accessibility to substrate, it is reasonable that there are observations of increased digestibility of energy-yielding nutrients after carbohydrase supplementation. For example, Adeola and Bedford (2004) reported improved starch and fat digestibility in wheat after xylanase supplementation. Juanpere et al. (2005) and Vahjen et al. (2007) also noted improved fat and starch digestibility in response to supplementation of xylanase and β-glucanase. The improvement in fat digestibility is especially noteworthy because non starch polysaccharides are known to increase hydrolysis of bile salts (Mathlouthi et al., 2002) and hence reduce fat utilization. Meng and Slominski (2005) suggested that hydrolysis of encapsulating cell walls may be responsible for enhanced energy utilization in a corn diet but that disruption of cell matrix resulting in a release of structural protein may be responsible for

improved energy utilization in a soybean meal-based diet.

Effect of carbohydrases on nitrogen and amino acids

Nitrogen digestibility reduction by neutral detergent fiber (NDF) and acid detergent fiber (ADF) has been ascribed to increased loss of endogenous and microbial N, low availability of N in the fibers themselves, or increased excretion of N trapped in the fibers or the digesta (Stanogias and Pearcet, 1985). The increased N loss with feeding of fibrous feedstuffs resulted from endo-genous (59%) and exogenous (41%) nitrogen sources. Loss of pancreatic enzymes and bile, as well as sloughed mucosa, will result in endogenous amino acid and nitrogen losses (Schulze et al., 1995). Consequently, reduction of the endogenous and exogenous losses and increased hydrolysis of dietary protein are two possible modes of action of carbohydrases in improving nitrogen and aminoacid utilization. However, observations regar-ding carbohydrases reducing endogenous protein or amino adcid have been inconsistent. Yin et al. (2000) reported a modest decrease in endogenous amin oacid loss after carbohydrase supplementation, whereas Rutherfurd et al. (2007) did not observe such effects. These may be diet-, ingredient-, or enzyme-specific (that is, presence or absence of certain antinutrients, difference enzymes and others). Tahir et al. (2008) noted that combination of enzymes were only effective in the presence of hemicellulases, which is capable of cleaving the resistant galacturonic acid and rhamnose bonds. This was corroborated by the observation that, for soybean meal, there was a strong correlation between the amount of galacturonic acid and crude protein digestibility, indicating that the hydrolysis of pectic substances in the cell wall enhanced protein digestibility. Another indirect effect of carbohydrases on nitrogen and amino acid utilization was recently demonstrated in the Yin et al. (2010) study, in which carbohydrases enhanced starch digestibility indirectly and improved nitrogen and aminoacid digestibility and absorption. The authors observed that starches with increased amylose:amylopectin are more resistant to digestion and have decreased digestibility, especially at the proximal part of the small intestine, in the overall length of the intestine, and the decreased digestibility is closely associated with decreased AA digestibility and plasma AA concentration.

Effect of carbohydrases on mineral availability

Similar to other nutrients, Nortey et al. (2008) observed reduced digestibility of minerals in diets, in which different by-products of wheat milling were added to a wheat-soybean meal basal diet. In several studies, supplemen-tation of xylanase (Nortey et al., 2007) or xylanase, amylase, and protease (Olukosi et al., 2007b) resulted in improved P digestibility. One possible explanation is provided by an observation of the relationship between phytic acid and NSP in plants. In cereal grains and legumes, most of the P is bound in phytic acid. Frølich (1990) indicated phytic acid and NSP are both found in the aleurone layer of wheat. In many cereals, grains or fractions thereof are sequestered with phytic acid. Consequently, when carbohydrases hydrolyze their substrates, phytic acid and other minerals may be exposed to digestive enzymes. As Parkkonen et al. (1997) observed, carbohydrases may increase permeability of the aleurone layer and consequently increase the release of otherwise unavailable minerals. Therefore, the increase in mineral availability as a result of carbohydrase supplementation is an indirect response to the carbohydrase effect.

Effect of carbohydrases on gut health

Increased digesta viscosity encourages slower diffusion rate, accumulation of particulate matter for microbial adhesion, and greater flow of solids rather than liquid. These factors encourage slower shedding of microorganism and encourage the proliferation of harmful microorganisms (Vahjen et al., 1998). In newly weaned swine receiving diets that promote increased digesta viscosity, McDonald et al. (2001) observed increased shedding of enterotoxigenic Escherichia coli. Poorer gut health in high-NSP diets may also arise from alteration of the morphology of the digestive surfaces in animals receiving a high-NSP diet.

Similar observations have been made in swine and poultry. Teirlynck et al. (2009) observed greater evidence and markers of gut damage, apoptosis, increased mounting of immune defense and microbial invasion of intestinal tissues in broilers receiving wheat-rye diets when compared with those on corn-based diets. The observation by Langhout et al. (2000) that the negative effects of NSP were less pronounced in germ-free birds indicates that microorganisms play a critical role in mediating the negative effects of NSP.

Carbohydrase supplementation reverses these negative effects by increasing the proportion of lactic and organic acids, reducing ammonia production (Kiarie et al., 2007), and increasing VFA concentration (Hübener et al., 2002), which is indicative of hydrolysis fragmentation of NSP and supporting growth of beneficial bacteria. Increased proportions of lactic acid promote gut health by suppressing the growth of presumptive pathogens (Pluske et al., 2001). Hillman et al. (1995) showed that certain strains of Lactobacillus inhibit the growth of coliforms such as pathogenic E. coli. Increased colonization of the gut with Lactobacilli has been associated with xylanase supplementation of a wheat diet and reduction in digesta viscosity (Vahjen et al., 1998). In addition, xylose (possible product of exogenous and endo-

genous carbohydrase activity) has been shown to be important in preferentially enhancing the growth of beneficial bifidobacteria (He et al., 2010). There were reports of improvement in the health of poultry as a result of carbohydrase supplementation (Hampson et al., 2002). In layers, Hampson et al. (2002) observed a reduced excretion of Brachyspira intermedia in hens receiving 265 mg/kg of an enzyme product containing xylanase and protease activities, which are indicative of infection reduction as a result of improved intestinal microbial population. Consequently, there is a basis for improved gut and overall health of the animal as a result of carbohydrase supplementation.

Proteases

Proteases have been added to poultry and swine diets routinely for many years as part of enzyme admixtures containing xylanases, pectinases, glucanases, amylases and other activities (Cowieson and Adeola, 2005). In recent years, proteases have grown in profile, there are currently several stand-alone proteases available, and new mechanisms of action have been proposed. Early research on the usefulness of proteases as supplemental feed enzymes is equivocal. For example, Caine et al. (1997) treated soybean meal with Bacillus subtilis in an attempt to reduce the adverse effects of proteinaceous antinutrients in soy when fed to weaner piglets. However, the authors observed a protease-induced decrease in AA digestibility from 68.7 to 63.9%. Ghazi et al. (1997) and Rooke et al. (1998) supplemented broiler and piglet diets respectively with either an acid fungal (Aspergillus) or alkaline bacterial (Bacillus) protease. In both studies, the acid fungal protease proved effective in improving body weight gain and feed conversion, whereas the bacterial protease resulted in depressed growth and poor feed conversion. For example, Blazek (2008) observed that some proteases are capable of coagulating soy protein and the extent of this reaction is dependent both on the characteristics of the soy protein and the nature of the protease.

Gelation of soy protein, as could occur in situ in the gut of poultry and swine, may be one explanation for some of the variable responses that have been observed in the literature. These effects have been previously reported where treatment of soy protein with 3 different proteases resulted in substantial, though transient, gelation of the protein (Hrckova et al., 2002). Interestingly, incubation with the different proteases resulted in different quantities and types of free AA production, with 1 protease producing mainly His (30%), Leu (24%) and Tyr (19%), and another Arg (22%), Leu (11%), and Phe (13%). The importance of these product profiles is not clear, but generation of free AA may interact with feed intake or absorption or both.

Contrary to some observed negative responses to

exogenous protease, there have been several reports where beneficial effects were reported. Mahagna et al. (1995) found positive effects of protease (and amylase) supplementation of sorghum-based diets for broiler chicks, and this was associated with a reduction in chymotrypsin secretion by the pancreas. This apparent feedback mechanism may explain why feed conversion efficiency was improved because synthesis of endogenous protein is energetically expensive for animals. Of interest is that Mahagna et al. (1995) observed no effect of supplemental enzymes on digestibility coefficients, indicating that the mechanisms involved in feed efficiency improvement may be "net." Odetallah et al. (2003) observed improved performance of broiler starters when a corn/soy-based diet was supplemented with a keratinase from *Bacillus licheniformis*. Furthermore, the beneficial effects of protease did not persist to market weight, an observation that was later confirmed (Odetallah et al., 2005). O'Doherty and Forde (1999) found that supplementation of barley/wheat/soy-based diets for swine with a neutral protease resulted in an improvement in feed efficiency. There is potential for protease in the diets of swine and chickens.

Phytase

Phytase dephosphorylates insoluble phytic acid in grains and oilseeds into orthophosphate and inositol phosphates. In broad terms, phytases are classified as 3- and 6- phytase on the basis of the site on the phytic acid molecule of initial dephosphorylation (McDonald et al., 2010). Phytate is found in most vegetable feed ingredients at concentrations from 5 to 25 g/kg, contributing between 1.4 and 7 g/kg of phytate-P. Phytate is found in virtually all seeds at concentrations from around 5 g/kg up to well over 20 g/kg and, therefore, typical pig, poultry and fish diets will contain between 8 and 12 g of phytic acid/kg (Selle et al., 2003b).

A further contribution to the elucidation of this mechanism is Cowieson et al. (2011) where the effect of phytate, phytase, and Na on ileal endogenous flow was reported. The antinutritive effects of phytate were confirmed as were the ameliorative effects of microbial phytase. However, several interact-tions were detected between Na (0.15 or 0.25%) such that greater Na concentrations were associated with a reduced antinutritive effect of phytate and a less obvious effect of phytase. As mentioned previously, it has been shown that Na is capable of partially disrupting the detrimental effect phytate has on protein solubility, and consequently greater dietary Na concentrations may reduce the anti-nutritive effect of phytate and reduce nutrient digestibility response to phytase. These effects were complex and AA specific. Thus, in a phytate-free diet, the associated Ca requirement is only about 0.6%. However, it is also important to consider that not all esters of phytate are

similarly malignant and it is not necessary to reduce all phytate to *myo*-inositol and free phosphate to remove the antinutritive effects.

CONCLUSION

Exogenous enzymes are added to an animal's food to supplement its own digestive enzymes and to break down anti-nutritive fractions in foods. Current worldwide feed enzyme utilization in diets of swine and poultry is substantially greater than originally anticipated. The non-ruminant feed enzyme market that includes phytases, carbohydrases and proteases has generated a lot of interest in recent years. Appropriate use of exogenous enzymes in feeds requires strategic reductions in dietary energy and nutrient content, as well as careful choice of feed ingredients to capture economic benefits of the various enzymes. The efficacy of enzymes will vary depending on ingredients because nutrient and energy release caused by enzyme supplementation will depend on the structure of the feedstuff itself. It is important to continue the effort to understand the use and limitations of matrix values of enzymes, which, if inappropriately applied, will result in depressed performance because of inadequacy of diets or will lead to wastage of resources. Regardless, the use of exogenous enzymes in diets of non-ruminants continues to be promising for a variety of reasons that hinge on sustainability, economics and the environment. Future research will increase our understanding of feed enzymes and how they can be more beneficially used to further improve the efficiency of non-ruminant animal production.

Conflict of Interests

The author(s) have not declared any conflict of interests.

REFERENCES

Adeola O, Bedford MR (2004). Exogenous dietary xylanase ameliorates viscosity-induced anti-nutritional effects in wheat-based diets for White Pekin ducks (Anas platyrinchos domesticus). Br. J. Nutr. 92:87-94.
Adeola O, Shafer DJ, Nyachoti CM (2008). Nutrient and energy utilization in enzyme-supplemented starter and grower diets for White Pekin ducks. Poult. Sci. 87:255-263.
Barrera M, Cervantes M, Sauer WC, Araiza AB, Torrentera N, Cervantes (2004). Ileal amino acid digestibility and performance of growing swine fed wheat-based diets supplemented with xylanase. J. Anim. Sci. 82:1997-2003.
Bedford MR, Gary GP (2010). Enzymes in farm animal nutrition, 2ND Edition, CAB, 2010 International. London.
Bedford MR (2000). Exogenous enzymes in monogastric nutrition-Their current value and future benefits. Anim. Feed Sci. Technol. 86:1-13.
Blazek V (2008). Chemical and biochemical factors that influence the gelation of soybean protein and the yield of tofu. PhD Diss. University of Sydney, Sydney, Australia.
Boguhn J, Rodehutscord M (2010). Effects of nonstarch polysaccharide-hydrolyzing enzymes on performance and amino acid

digestibility in turkeys. Poult. Sci. 89:505-513.

Caine WR, Sauer WC, Tamminga S, Verstegen MWA, Schulze H (1997). Apparent and true ileal digestibilities of amino acids in newly-weaned piglets fed diets with protease-treated soybean meal. J. Anim. Sci. 75:2962-2969.

Courtin CM, Delcour JA (2002). Arabinoxylans and endoxylanases in wheat flour bread-making. J. Cereal Sci. 35:225-243.

Cowieson AJ, Adeola O (2005). Carbohydrases, protease, and phytase have an additive beneficial effect in nutritionally marginal diets for broiler chicks. Poult. Sci. 84:1860-1867.

Cowieson AJ, Bedford MR, Ravindran V, Selle PH (2011). Increased dietary sodium chloride concentrations reduce endogenous amino acid flow and influence the physiological response to the ingestion of phytic acid by broiler chickens. Br. Poult. Sci. In press.

Cowieson AJ, Ravindran V (2008b). Sensitivity of broiler starters to three doses of an enzyme cocktail in maize-based diets. Br. Poult. Sci. 49:340-346.

Frølich W (1990). Chelating properties of dietary fiber and phytate. The role for mineral availability. Adv. Exp. Med. Biol. 270:83-93.

Fuller MF (2004). The Encyclopedia of Farm Animal Nutrition. CABI Publishing.

Ghazi S, Rooke JA, Galbraith H, Morgan A (1997). The potential for improving soya-bean meal in diets for chicks: Treatment with different proteolytic enzymes. Br. Poult. Sci. 37:554-555.

Hampson DJ, Phillips ND, Pluske JR (2002). Dietary enzyme and zinc bacitracin reduce colonization of layer hens by the intestinal spirochaete Brachyspira intermedia. Vet. Microbiol. 86:351-360.

Hillman K, Spencer RJ, Murdoch TA, Stewart CS (1995). The effect of mixtures of Lactobacillus spp. on the survival of enterotoxigenic Escherichia coli in vitro continuous culture of porcine intestinal bacteria. Lett. Appl. Microbiol. 20:130-133.

Hrckova M, Rusnakova M, Zemanovic J (2002). Enzymatic hydrolysis of defatted soy flour by three different proteases and their effect on the functional properties of resulting protein hydrolysates. Czech J. Food Sci. 20:7-14.

Hübener K, Vahjen W, Simon O (2002). Bacterial responses to different dietary cereal types and xylanase supplementation in the intestine of broiler chicken. Arch. Anim. Nutr. 56:167-187.

Jaroni D, Scheideler SE, Beck M, Wyatt C (1999). The effect of dietary wheat middlings and enzyme supplementation. 1. Late egg production efficiency, egg yields, and egg composition in two strains of Leghorn hens. Poult. Sci. 78:841-847.

Johnson IT, Gee JM (1986). Gastrointestinal adaptation in respone to soluble non-vailable polysaccharides in the rat. Br. J. Nutr. 55:497-505.

Juanpere J, Perez-Vendrell AM, Angulo E, Brufau J (2005). Assessment of potential interactions between phytase and glycosidase enzyme supplementation on nutrient digestibility in broilers. Poult. Sci. 84:571-580.

Karimi A, Bedford M, Kamyab A, Moradi M (2007). Comparative effects of xylanase supplementation on broiler, broiler breeder and layer chick performance and feed utilization on wheat based diet. Poult. Sci. 44:322-329.

Kiarie E, Nyachoti CM, Slominski BA, Blank G (2007). Growth performance, gastrointestinal microbial activity, and nutrient digestibility in early-weaned swine fed diets containing flaxseed and carbohydrase enzyme. J. Anim. Sci. 85:2982-2993.

Langhout DJ, Schutte JB, de Jong J, Sloetjes H, Verstegen WA, Tamminga S (2000). Effect of viscosity on digestion of nutrients in conventional and germ-free chicks. Br. J. Nutr. 83:533-540.

MacLeod MG, Valentine J, Cowan A, Wade A, McNeill L, Bernard K (2008). Naked oats: Metabolisable energy yield from a range of varieties in broilers, cockerels and turkeys. Br. Poult. Sci. 49:368-377.

Mahagna M, Nir I, Larbier M, Nitsan Z (1995). Effect of age and exogenous amylase and protease on development of the digestive tract, pancreatic enzyme activities and digestibility of nutrients in young meat-type chickens. Reprod. Nutr. Dev. 35:201-212.

Mathlouthi N, Lalles JP, Lepercq P, Juste C, Larbier M (2002). Xylanase and {beta}-glucanase supplementation improve conjugated bile acid fraction in intestinal contents and increase villus size of small intestine wall in broiler chickens fed a rye-based diet. J. Anim. Sci. 80:2773-2779.

McDonald DE, Pethick DW, Mullan BP, Hampson DJ (2001). Increasing viscosity of the intestinal contents alters small intestinal structure and intestinal growth, and stimulates proliferation of enterotoxigenic Escherichia coli in newly-weaned swine. Br. J. Nutr. 86:487-498.

McDonald P, Edwards RA, Greenhalgh JFD, Morgan CA, Sinclair LA, Wilkinson RG (2010). Animal Nutrition. Pearson Books.

Meng X, Slominski BA (2005). Nutritive values of corn, soybean meal, canola meal, and peas for broiler chickens as affected by a multicarbohydrase preparation of cell wall degrading enzymes. Poult. Sci. 84:1242-1251.

Meng X, Slominski BA, Nyachoti CM, Campbell LD, Guenter W (2005). Degradation of cell wall polysaccharides by combinations of carbohydrase enzymes and their effect on nutrient utilization and broiler chicken performance. Poult. Sci. 84:37-47.

Nitrayová S, Heger J, Patráš P, Kluge H, Brož J (2009). Effect of xylanase on apparent ileal and total tract digestibility of nutrients and energy of rye in young swine. Arch. Anim. Nutr. 63:281-291.

Nortey TN, Patience JF, Sands JS, Trottier NL, Zijlstra RT (2008). Effects of xylanase supplementation on the apparent digestibility and digestible content of energy, amino acids, phosphorus, and calcium in wheat and wheat by-products from dry milling fed to grower swine. J. Anim. Sci. 86:3450-3464.

Nortey TN, Patience JF, Simmins PH, Trottier NL, Zijlstra RT (2007). Effects of individual or combined xylanase and phytase supplementation on energy, amino acid, and phosphorus digestibility and growth performance of grower swine fed wheat-based diets containing wheat millrun. J. Anim. Sci. 85:1432-1443.

Noy Y, Sklan D (1995). Digestion and absorption in the young chick. Poult. Sci. 74:366-373.

O'Doherty JV, Forde S (1999). The effect of protease and alpha-galactosidase supplementation on the nutritive value of peas for growing and finishing swine. Ir. J. Agric. Food Res. 38:217-226.

Odetallah NH, Wang JJ, Garlich JD (2005). Versazyme supplementation of broiler diets improves market growth performance. Poult. Sci. 84:858-864.

Odetallah NH, Wang JJ, Garlich JD, Shih JC (2003). Keratinase in starter diets improves growth of broiler chicks. Poult. Sci. 82:664-670.

Olukosi OA, Adeola O (2008). Whole body nutrient accretion, growth performance and total tract nutrient retention responses of broilers to supplementation of xylanase and phytase individually or in combination in wheat-soybean meal based diets. Jpn. Poult. Sci. 45:192-198.

Olukosi OA, Bedford MR, Adeola O (2007a). Xylanase in diets for growing swine and broiler chicks. Can. J. Anim. Sci. 87:227-235.

Olukosi OA, Cowieson AJ, Adeola O (2008a). Energy utilization and growth performance of broilers receiving diets supplemented with enzymes containing carbohydrase or phytase activity individually or in combination. Br. J. Nutr. 99:682-690.

Olukosi OA, Cowieson AJ, Adeola O (2007b). Age-related influence of a cocktail of xylanase, amylase, and protease or phytase individually or in combination in broilers. Poult. Sci. 86:77–86.

Olukosi OA, Cowieson AJ, Adeola O (2008b). Influence of enzyme supplementation of maize-soyabean meal diets on carcase composition, whole-body nutrient accretion and total tract nutrient retention of broilers. Br. Poult. Sci. 49:436-445.

Palander S, Näsi M, Järvinen S (2005). Effect of age of growing turkeys on digesta viscosity and nutrient digestibility of maize, wheat, barley and oats fed as such or with enzyme supplementation. Arch. Anim. Nutr. 59:191-203.

Parkkonen T, Tervila-Wilo A, Hopeakoski-Nurminen M, Morgan A, Poutanen K, Autio K (1997). Changes in wheat microstructure following in vitro digestion. Acta Agric. Scand. B. Soil and Plant Sci. 47:43-47.

Pluske JR, Kim JC, McDonald DE, Pethick DW, Hampson DJ (2001). Non-starch polysaccharides in the diets of young weaned piglets. Pages 81–112 in The Weaner Pig: Nutrition and Management. M. A. Varley and J. Wiseman, ed. CABI Publishing, Wallingford, UK.

Ravindran V, Cowieson AJ, Selle PH (2008). Influence of dietary electrolyte balance and microbial phytase on growth performance, nutrient utilization, and excreta quality of broiler chickens. Poult. Sci. 87:677-688.

Rooke JA, Slessor M, Fraser H, Thomson JR (1998). Growth performance and gut function of piglets weaned at four weeks of age and fed protease-supplemented soya-bean meal. Anim. Feed Sci. Technol. 70:175-190.

performance in broilers fed a corn-soybean meal diet. Anim. Sci. J. 76:559-565.

Rutherfurd SM, Chung TK, Moughan PJ (2007). The effect of a commercial enzyme preparation on apparent metabolizable energy, the true ileal amino acid digestibility, and endogenous ileal lysine losses in broiler chickens. Poult. Sci. 86:665-672.

Schulze H, van Leeuwen P, Verstegen MW, van den Berg JW (1995). Dietary level and source of neutral detergent fiber and ileal endogenous nitrogen flow in swine. J. Anim. Sci. 73:441-448.

Scott TA, Kampen R, Silversides FG (2001). The effect of adding phosphorus, phytase enzyme, and calcium on the performance of layers fed corn-based diets. Poult. Sci. 80:183-190.

Selle PH, Walker AR, Bryden WL (2003b). Total and phytate-phosphorus contents and phytase activity of Australian-sourced feed ingredients for swine and poultry. Aust. J. Exp. Agric. 43:475-479.

Stanogias G, Pearcet GR (1985). The digestion of fibre by swine. Br. J. Nutr. 53:513-530.

Tahir M, Saleh F, Amjed M, Ohtsuka A, Hayashi K (2005). Synergistic effect of cellulase and hemicellulase on nutrient utilization and

Tahir M, Saleh F, Ohtsuka A, Hayashi K (2008). An effective combination of arbohydrases that enables reduction of dietary protein in broilers: Importance of hemicellulase. Poult. Sci. 87:713-718.

Teirlynck E, Bjerrum L, Eeckhaut V, Huygebaert G, Pasmans F, Haesebrouck F, Dewulf J, Ducatelle R, Van Immerseel F (2009). The cereal type in feed influences gut wall morphology and intestinal immune cell infiltration in broiler chickens. Br. J. Nutr. 102:1453-1461.

Vahjen W, Gläser K, Schäfer K, Simon O (1998). Influence of xylanase-supplemented feed on the development of selected bacterial groups in the intestinal tract of broiler chicks. J. Agric. Sci. 130:489-500.

Vahjen W, Osswald T, Schäfer K, Simon O (2007). Comparison of a xylanase and a complex of non starch polysaccharide-degrading enzymes with regard to performance and bacterial metabolism in weaned piglets. Arch. Anim. Nutr. 61:90-102.

Yang ZB, Yang WR, Jiang SZ, Zhang GG, Zhang QQ, Siow KC (2010). Effects of a thermotolerant multi-enzyme product on nutrient and energy utilization of broilers fed mash or crumbled corn-soybean meal diets. J. Appl. Poult. Res. 19:38-45.

Yin F, Zhang Z, Huang J, Yin Y (2010). Digestion rate of dietary starch affects systemic circulation of amino acids in weaned swine. Br. J. Nutr. 103:1404-1412.

Yin YL, McEvoy JDG, Schulze H, Hennig U, Souffrant WB, McCracken KJ (2000). Apparent digestibility (ileal and overall) of nutrients and endogenous nitrogen losses in growing swine fed wheat (var. Soissons) or its by-products without or with xylanase supplementation. Livest. Prod. Sci. 62:119-132.

Zhang Z, Marquardt RR, Guenter W (2000a). Evaluating the efficacy of enzyme preparations and predicting the performance of leghorn chicks fed rye-based diets with a dietary viscosity assay. Poult. Sci. 79:1158-1167.

22

Solid lipid nanoparticles as new drug delivery system

Abbasalipourkabir, R.[1]* Salehzadeh, A.[1] and Rasedee Abdullah[2]

[1]Department of Biochemistry, Faculty of Medicine, Hamedan University of medical Science,
65178-3-8736 Hamedan, Iran.
[2]Universiti Putra Malaysia (UPM), Malaysia.

The main challenge in cancer chemotherapy is toxic side-effects induced by chemotherapeutic drugs. Thus, alternative methods of drug administration like appropriate drug carrier system is needed to overcome this problem. The main objective of new drug delivery systems is to improve the anti-tumor efficacy of drug and reduce their toxic effects on normal tissues. Recently Solid Lipid Nanoparticles (SLN) as colloidal particulate drug delivery system have received much attention from drug development researchers. The use of solid lipid nanoparticle opens up new perspectives for the formulation of poorly soluble drugs. The SLN is a very complex system with some advantages and disadvantages over other colloidal carrier systems. Tamoxifen, an antiestrogen molecule and strong hydrophobic drug is used as a chemotherapy drug against breast cancers. When tamoxifen encapsulated within solid lipid nanoparticles, it is like free TAM display antitumoral activity against human breast cancer cells. The biological availability of drug is not affected when incorporated into SLN. Therefore SLN could be applied as a drug delivery system for cancer treatments.

Key words: Drug delivery systems, solid lipid nanoparticles.

INTRODUCTION

Breast cancer is one of the most important health concerns of the modern society (Ferlay et al., 2007). Worldwide, it is estimated that over one million new cases of breast cancer were diagnosed every year and more than 400,000 will die from the breast cancer (Coughlin and Ekwueme, 2009). The life-time risk in women contracting breast cancers was estimated to be 1 in 8, which is the highest among all forms of cancers (DevCan5.2, 2004). Although, the mortality rates from breast cancers have decreased in most developed countries because more frequent mammographic screening and extensive use of tamoxifen, it still remains the second highest in women (Clark, 2008). The main options for breast cancer treatment include surgery, radiation therapy and chemotherapy (Mirshahidi and Abraham, 2004). Surgical procedures usually lead to significant morbidity such as lymph edema, muscle wasting, neuropathy and chronic pain (Paci et al., 1996).

Radiation therapy is useful for cancer which is more localized, but it also carry a number of acute and chronic side- effects such as nausea, diarrhea, pain and fatigue (Ewesuedo and Ratain, 2003). Endocrine therapy may be used as a supplementary treatment; this method of therapy is applied to specific group of patients, for example women after menopause with hormone-responsive disease (Gradishar, 2005). In hormone-sensitive cancer patients receive chemotherapy with cytotoxic drugs. The cytotoxic drugs treat cancers by causing cell death or growth arrest. Effective cancer chemotherapy is able either to shrink a tumor or to help destroy cancer cells (Ewesuedo and Ratain, 2003).

A number of obstacles such as drug toxicity, possible undesirable drug interactions and various forms of drug resistance have to be overcome to achieve effective chemotherapy (Cardosa et al., 2009). Drug resistance is a general problem in the chemotherapy of several cancers including breast cancers (Wong et al., 2006). Failures in treatment of cancers are common. Development of new drugs is also slow to progress. Among the reasons contributing to this are weak absorption, high rate of metabolism and elimination of

---

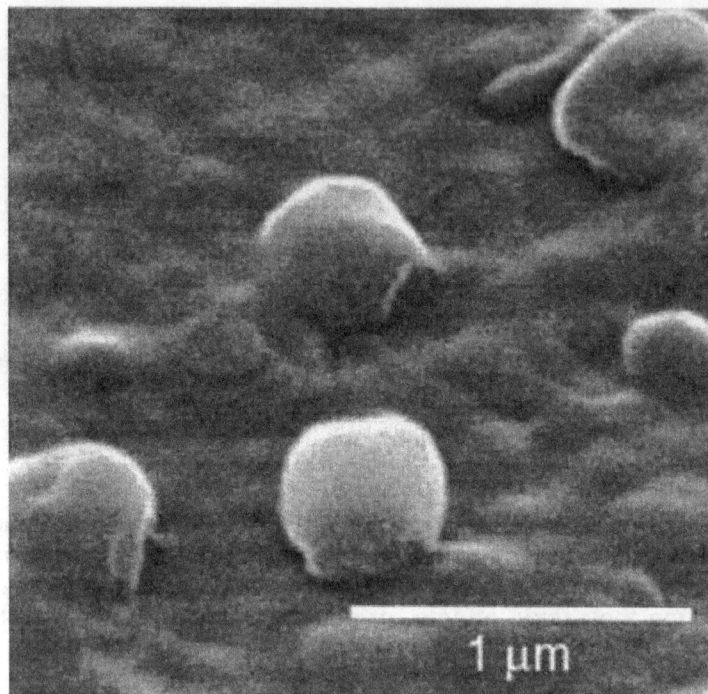

Figure 1. Electron microscopy picture of solid lipid nanoparticles made from Compritol® stabilized with poloxamer 188, diameter 400 nm (Müller et al., 2000).

drugs per oral administration resulting in less or variable concentrations in blood, poor drug solubility, unpredictable bioavailability of oral drugs due to food and tissue toxicity (Sipos et al., 1997). Thus, alternative methods of drug administration like appropriate drug carrier system is needed to overcome this problem. Targeting of unhealthy tissues and organs of the body is one of the important challenges of the drug delivery systems (Kayser et al., 2005). Depending on the route of administration, the size of drug carriers may range from a few nanometers (colloidal carriers), to micrometers (microparticles) and to several millimeters (implants). Among these carriers, nanoparticles had shown great promise for parenteral application of chemotherapeutic drugs (Mehnert et al., 2001). Nanoparticles seem to show promise as a drug targeting systems supplying drug to target tissues at the right time (Kayser et al., 2005). Nanotechnology has found application in drug delivery because with drug-loading of nanoparticles, delivery becomes effective and more specific (Brannon-Peppas and Blanchette, 2004).

Particulate colloidal delivery systems such as liposomes, nanospheres and nanocapsules have been developed and are now being studied as new carriers for drugs and vaccines (Brigger et al., 2002; Koping-Hoggard et al., 2005). These new delivery systems can be powerful, particularly in intravenous administration because they can target the macrophages of the reticuloendothelial system. The targeting of macrophages using these systems is due to the opsonization of the particles and recognition by the macrophages. Among the applications of these systems are in treatment of malignancies and infectious diseases (Youssef et al., 1988). The main objective of new drug delivery systems is to improve the anti-tumor efficacy of drug and reduce their toxic effects on normal tissues. Nanoparticle is expected to be able to diminish toxicity of chemotherapy drug. Nanoparticles based on lipids that are solid at room temperature, namely solid lipid nanoparticle (SLN) using physiological well-tolerable lipids have potentially wide application (Figure 1) (Müller et al., 2000).

Solid lipid nanoparticles (SLNs)

Beginning from early 1990s, pharmaceutical researches began to shift towards production of nanoparticles from solid lipids which were named the solid lipid nanoparticles (SLN) or liposomes or nanospheres (Siekmann and Westesen, 1994). The SLN is a drug delivery system that loads lipophilic or chemically unstable drugs. This system for drug delivery has all the advantages of other developed delivery systems that are physical stability, protection of encapsulated labile drugs from degradation, controlled-release and high tolerability. Among the advantages of SLN are high potential for management of

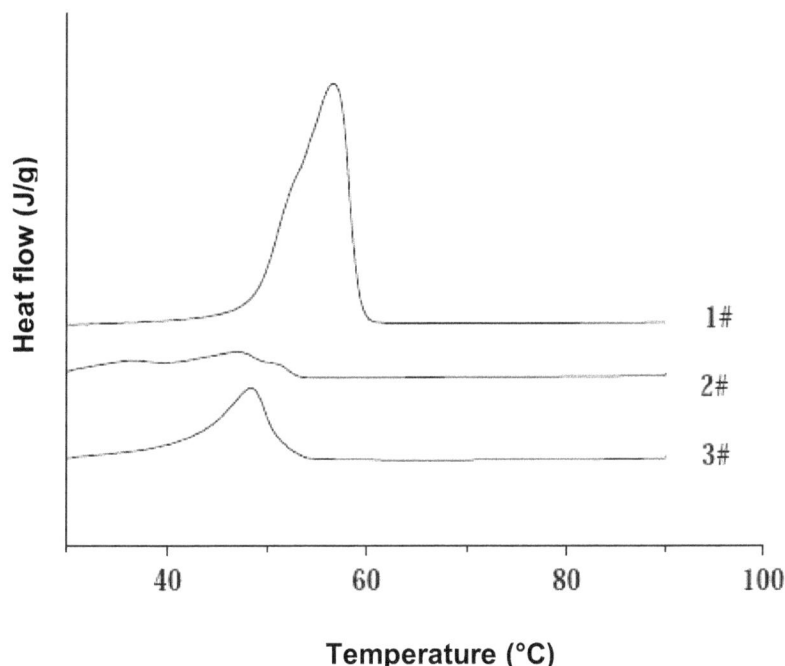

Figure 2. DSC scan of lyophilization SLNs powder heating from 301 to 901°C at a rate of 5 k/min. 1#-Bulk matrix material, 2#-Placebo SLNs, 3#-drug loaded SLNs powder containing 2 mg/ml model drug (Hou et al., 2003).

drug release and drug targeting, high stability for drug loading and high capacity for drug payload. This delivery system makes possible the encapsulation of lipophilic and hydrophilic drugs without the toxic effect of the carriers. SLN also avoids the use organic solvents and has potential for large scale production. One of the major advantages of SLNs as a drug carrier is the high rate of drug entrapment efficiency (EE) that can result in long-term physical stability of the drug. The crystalline structure of the lipid phase of SLN that represents its chemical nature plays an important role in determining whether or not a guest molecule can be strongly encapsulated within delivery system. If the lipid structure of the nanoparticle delivery system is composed of a high crystalline structure with perfect lattice, the incorporated drug would be expelled. In other words, the existence of imperfect crystalline structure or lattice in the lipid phase of SLN makes it possible for stronger drug incorporation. Therefore, the lipids with lesser ordered arrangements in the chemical structure of SLN would be useful for high-rate drug loading.

This observation is supported by another study using differential scanning calorimetric (DSC) and x-ray diffraction (XRD) measurements which show that most of the SLNs have less ordered arrangement crystalline structure favoring increasing capacity drug-loading. Hou et al. (2003) showed when the bulk matrix material is turned into SLNs, the melting point was depressed no matter whether it was drug-free or not (Figure 2). The

decrease of the onset and maximum temperature could be attributed to the small size effect and explained by the Thomson equation. Also the presence of the reflections in X-ray diffraction pattern of the SLN compared to the bulk materials (Figure 3) confirm that lipid within nanoparticles are arranged in low ordered crystalline structure. It was clear that in the drug-free SLNs, the amorphous state would contribute to the higher drug loading capacity. However, several disadvantages are associated with SLNs to include particle growth, particle aggregation, unpredictable gelation tendency, polymorphic transitions, burst drug release and inherently low incorporation capacities due to the crystalline structure of the solid lipid (Mehnert and Mäder, 2001).

PRODUCTION OF SOLID LIPID NANOPARTICLES

Ingredients

The ingredients used for production of SLN include solid lipids, emulsifiers and water. The lipids used cover a wide range from triglycerides (for example tristearin), partial glycerides (for example imwitor) and fatty acids (for example stearic acid), to steroids (for example cholesterol) and waxes (for example cetyl palmitate). Many emulsifiers of varying charges and molecular weights can potentially be used to stabilize the lipid dispersion. If a combination of certain emulsifiers were

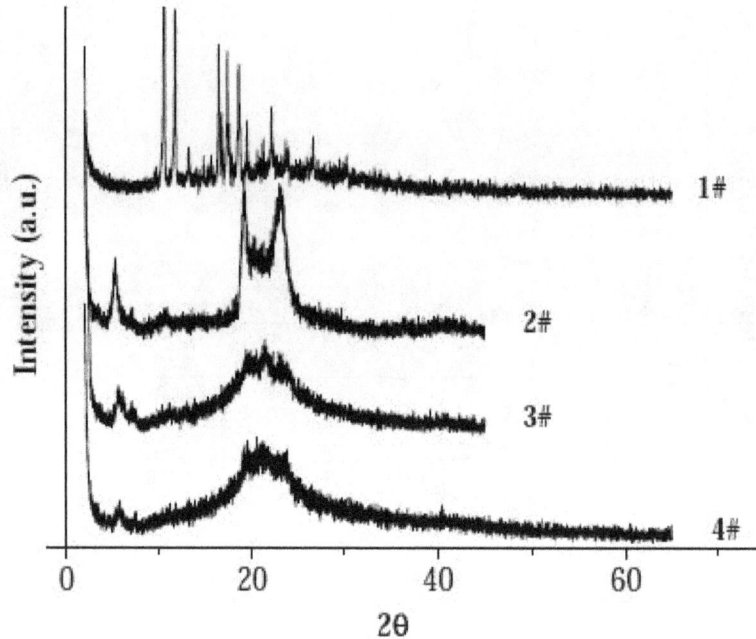

Figure 3. X-ray diffraction pattern. 1#-Pure mifepristone (model drug), 2#-bulk material, 3#-Placebo SLNs, 4#-lyophilization SLNs powder containing 2 mg/ml model drug (Hou et al., 2003).

used together, particles aggregation may be prevented or minimized. Physiological lipids which are used for production of SLN will reduce acute and chronic toxicity of chemical components (Abbasalipourkabir et al., 2011a). Emulsifiers are generally toxic and rarely used for parenteral administrations. Therefore the use of this ingredient in production of SLN should take into consideration their toxic effects and route of administration. It has been shown that formulations with a mixture of surfactants would produce a more stable SLN with lower particle size than formulations with only one surfactant. Table 1 shows ingredients that are generally used in the production of SLN.

Lyophilization is the best way to increase chemical and physical stability for long-term storage of SLN and for improvement of SLN-incorporation into pellets, tablets and capsules (Mehnert and Mäder, 2001).

Technique

The most common production technique of SLNs are high-pressure homogenization (HPH) of lipids in the fluid phase, precipitation of microemulsions, high-shear homogenization combined with ultrasound, solvent emulsification/evaporation and microemulsion techniques. The HPH is the predominant production method because it is easy to handle and scale-up. Scale up methods has been developed to produce SLN in commercial quantities (Copland et al., 2005; He et al.,

2007). High pressure homogenization (HPH) is a relevant method for the preparation of SLN and can be conducted above or below room temperature (hot or cold HPH technique) (Schwarz et al., 1994). Homogenization provides cavitations forces which break down the particles into smaller sizes (Schwarz and Mehnert, 1997). Harivardhan et al. (2006) found homogenization at 15,000 psi for 3 cycles resulted in smaller sized nanoparticles. High pressure leads to increase particle size. It is postulated that at high homogenization pressure, the particles may have coalesced as a result of the high kinetic energy of the particles which impaired the homogenization process.

Increasing the kinetic energy increases particle collision and consequently resulting in coagulation. High particle collisions also distort the surfactant film coating the particle surface and hence, enhance particle aggregation (Siekmann and Westesen, 1994).

CHARACTERIZATION OF SLNS

Characterization of SLN is necessary for control of the quality of the product. The parameters in the characterization of SLN are directly involved in stability and release kinetics are particle size, particle size distribution (PI) and zeta potential, degree of crystallinity and lipid modification, co-existence of additional colloidal structures like micelles, liposomes, supercooled melts, drug nanoparticles and dynamic phenomena (Laggner, 1999). Particle size and particle size distribution are

Table 1. Lipids and emulsifiers for preparation of SLN.

Lipids	Emulsifiers/coemulsifiers
Triglycerides	
Tricaprin	Soybean lecithin
Trilaurin	Lipoid® S75, Lipoid® S100)
Trimyristin	Egg lecithin (Lipoid® E80)
Tripalmitin	Phosphatidylcholine
Tristearin	Epikuron® 170, Epikuron® 200)
Hydrogenated coco-glycerides	
	Poloxamer 188
	Poloxamer 182
(Softisan®142)	Poloxamer 407
	Poloxamine 908
Hard fat types	
Witepsol W 35	
Witepsol H 35	Tyloxapol
Witepsol 42	Polysorbate 20
Witepsol E 85	Polysorbate 60
Glyceryl monostearate (Imwitor®900)	Polysorbate 80
Glyceryl behenate (Comporitol 888®ATO)	Sodium cholate
Glyceryl palmitostearate (Precirol®ATO5)	Sodium glycocholate
Cetyl palmitate	Taurocholic acid sodium salt
Stearic acid Palmitic acid	Taurodeoxycholic acid sodium salt
Decanoic acid	Butanol
Behenic acid	Butyric acid
Acidan N12	Dioctyle sodium sulfosuccinate
	Monooctylphosphoric acid sodium

Mehnert and Mäder (2001).

important factors in the physical stability of nanodispersions (Attama and Müller-Goymann, 2007). Schubert and Müller-Goymann (2005) showed in the preparation of SLN, at least 10% emulsifier such as lecithin should be incorporated in the lipid matrix. As lecithin concentration increased, particles size decreased to 95.5 ± 13.1 nm. The presence of lecithin within the multilayer structure of the particle surfaces provides interface between oil and water (Westesen and Siekmann, 1997). Since lecithin has limited mobility, aggregation of particles that could lead to particle growth does not occur (Heiati et al., 1996). Schubert and Müller-Goymann (2005) also found high concentrations of lecithin content lead to significant increases in particle sizes. It is possible that beyond this critical concentration of lecithin, presence of other colloidal structure within the aqueous phase and/or around the particles will cause increased mobility of lecithin resulting in particle size increase (Westesen and Wehler, 1993).

A similar result obtained high concentrations of poloxamer 188 was incorporated into the SLN (Sanjula et al., 2009). This again may be attributed to the increase in particle size as well as reduction in hydrophobicity. Small particle sizes and presence of emulsifiers may retard lipid crystallization and modification changes (Mehnert and Mäder, 2001).

Incorporation of drugs in SLNs

Incorporation of the drug into a particulate carrier can protect it from degradation *in vitro* and *in vivo* (Figure 4). The release of the incorporated drug can be controlled and greater specific targeting is possible (Müller and Keck, 2004). A potent delivery system should be able to load drug with high capacity. The incorporated drugs can be accommodated between the fatty acid chains, lipid bilayers or in imperfect spaces. The drugs are usually located in the core of the particles or shell or molecularly dispersed throughout the matrix. The location of drugs in SLN is generally depending on ratio of drug to lipid and solubility of drug in lipid. The solid lipid core of SLN has the capability to increase the chemical stability of incorporated drugs (Müller and Bohm, 1998). The

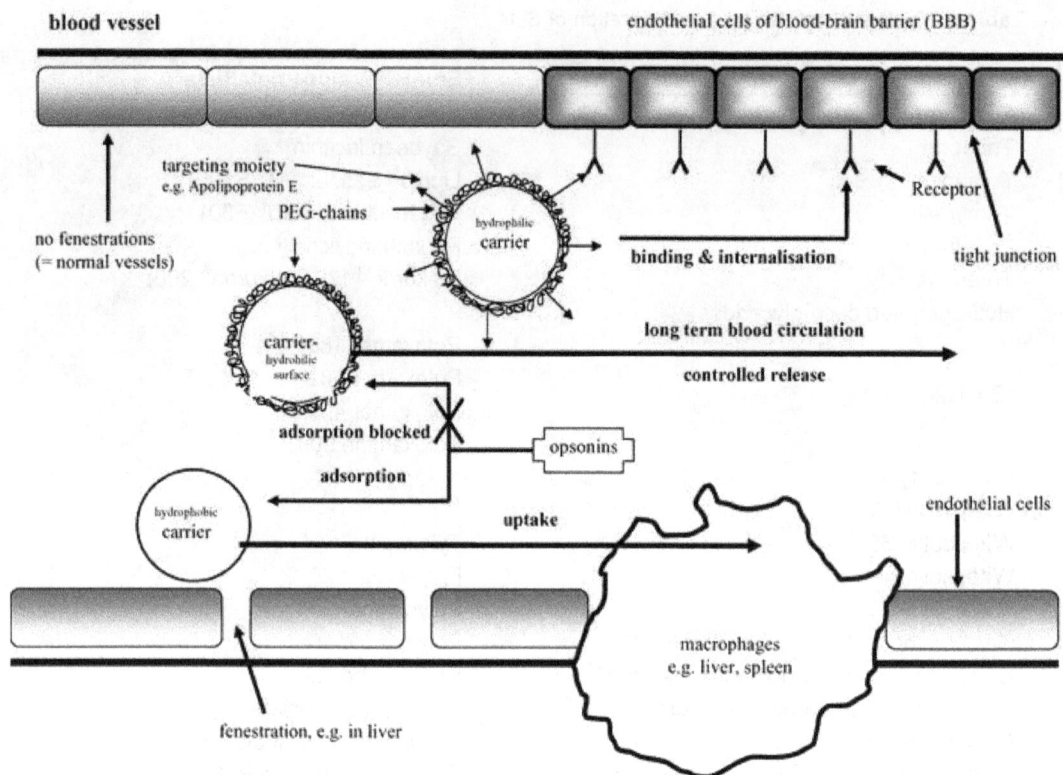

Figure 4. Approaches to achieve passive targeting, long-circulating carriers for prolonged drug release and target-specific carriers (Müller and Keck, 2004).

Drug-enriched core   Solid solution   Drug-enriched shell

Figure 5. Proposed structural models for drug loading in SLN (Mühlen et al., 1998).

proposed structural models are shown in Figure 5. Abbasalipourkabir et al. (2011b) studied the incorporation of an antiestrogen molecule and strong hydrophobic drug, tamoxifen (TAM) in palm oil SLN. They found the palm oil which consist of triglyceride mixture of natural fatty acids produce the highest entrapment efficiency (89.98 to 90.25%). Increasing drug concentration led to significant increase (P<0.05) in drug loading (DL) (Figure 6) and significant decrease (P<0.05) in entrapment efficiency (EE) (Figure 7). These effects are due to the reduced SLN dispersions and higher solubility of the drug in higher lipid concentrations.

Factors affecting loading capacity of drug in nanoparticles include solubility of drug in melted lipid, miscibility of drug and lipid melt, chemical and physical structure of solid lipid matrix and polymorphic state of lipid material. Therefore, high solubility of the drug in the lipid matrix leads to adequate loading capacity (Westesen et al., 1997). Increasing lipid matrices lead to increased EE and DL suggesting that greater solubility of the drug in the higher lipid concentrations (Harivardhan et al., 2006). Mühlen et al. (1998) reported that drug incorporation can accelerate the transformation to the stable modification in comparison to drug free particles. However, when

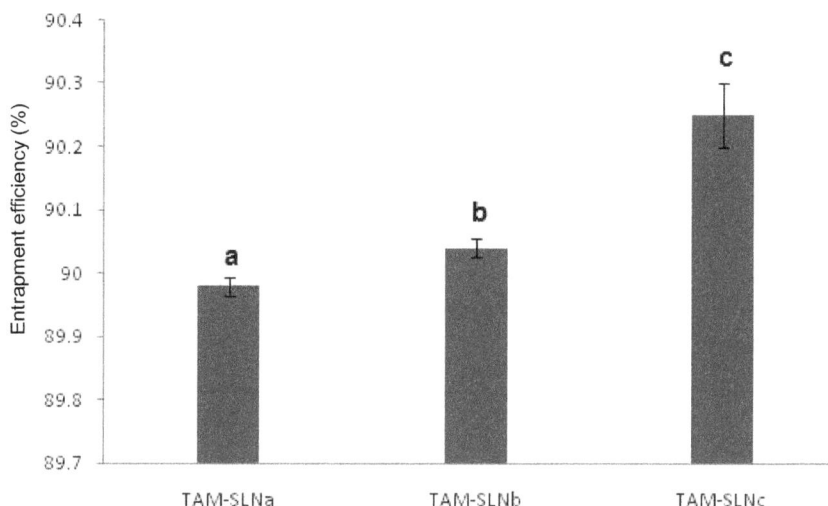

Figure 6. Entrapment efficiency of tamoxifen-loaded solid lipid nanoparticles. TAM-SLNa, TAM-SLNb and TAM-SLNc represent 1 mg TAM incorporated in 5, 10 and 20 mg SLN respectively. Means with superscripts a to c are significantly different ($p<0.05$), (n = 5) (Abbasalipourkabir et al., 2011b).

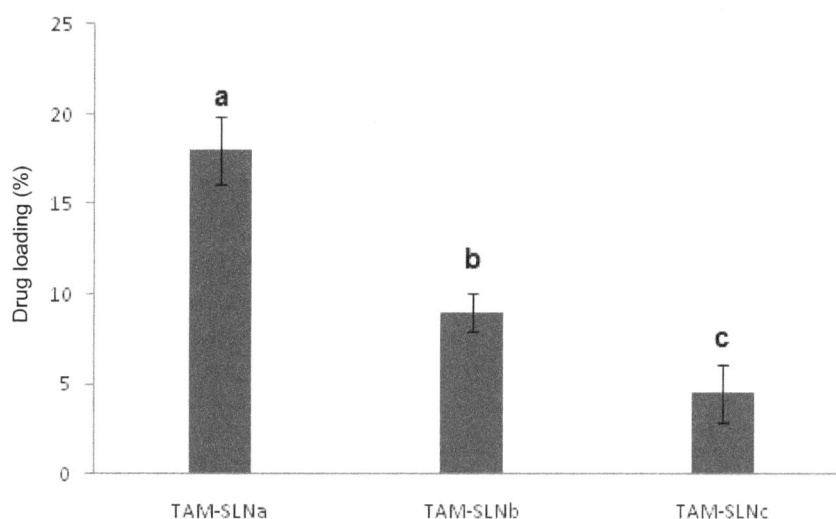

Figure 7. Drug loading of tamoxifen-loaded solid lipid nanoparticles. TAM-SLNa, TAM-SLNb and TAM-SLNc represent 1 mg TAM incorporated in 5, 10 and 20 mg SLN respectively. Means with superscripts a to c are significantly different ($p<0.05$), (n = 5) (Abbasalipourkabir et al., 2011b).

thermodynamic stability and lipid packing density increase, drug incorporation rates decrease in the order: supercooled melt, α-modification, ß–modification, β'-modification (Müller et al., 2000).

Release of drugs from SLNs

The profile of drug release depends on some parameters such as modification in lipid matrix, surfactant concentration and production conditions. The probable mechanism for drug release from SLN is that drug loading in SLN is associated with the partitioning of the drug in the water phase. During the cooling step that phase separation occurs, the lipid precipitates and partitions into liquid-lipid phase. The drug which is present at high concentrations in the outer shell of the nanoparticles crystallizes. If the drug is primarily loaded in the outer shell of the particles, a burst release mechanism may occur. Huang et al. (2008) reported that

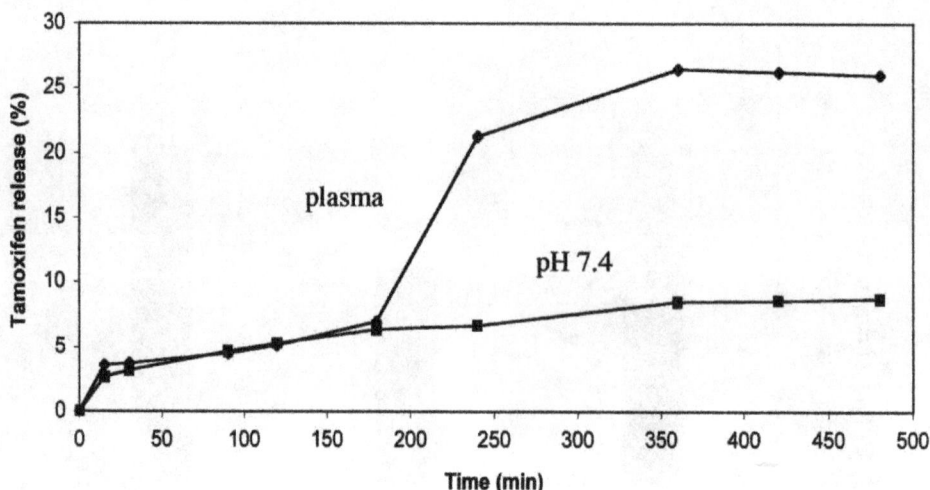

Figure 8. Release profiles of tamoxifen at pH 7.4 and in plasma from palmitic acid nanoparticles. Each value is the mean of three experiments (Fontana et al., 2005).

SLN made of 'precirol' may have the potential to serve as a delivery system for parenteral camptothecin administration because of the sustained drug release, strong cytotoxicity, limited hemolysis and good storage stability.

Memisoglu-Bilensoy et al. (2005) obtained a delay for tamoxifen citrate pre-loaded nanospheres based on 'amphiphilic h-cyclodextrin' that liberate the drug within 6 h while Fontana et al. (2005) showed the release rate of tamoxifen from SLN based on 'palmitic acid' is quite low until 3 h after the beginning of the experiment and then the successively release rates increase quickly and the amount of drug released reaches 21% after 4 h and 26% after 6 h (Figure 8) suggesting other factors such as large surface area, high diffusion coefficient (small molecular size), low matrix viscosity and short-diffusion distance of the drug are also contributory to the fast release of the drug (Mühlen et al., 1998).

Antitumor efficacy of drug-loaded SLNs

The basis of drug-loaded SLN is to obtain the necessary dose of drug at tumor location for a known period of time and reducing adverse effects on normal organs in the body (Chawla and Amiji, 2003). According to Abbasalipourkabir et al. (2010), tamoxifen-loaded SLN has similar effect on the tumor as free TAM which promotes apoptosis in the rat mammary gland tumor. The efficacy of free-TAM and TAM-loaded SLN is similar. However, the TAM-loaded SLN show a more prolonged effect suggesting that incorporation of TAM in SLN is suitable for delayed drug release in the chemotherapy of breast cancers while decreasing the hepatotoxic effects. According to Fontana et al. (2005), *in vitro* anti-proliferative activity of SLNs containing tamoxifen on

MCF-7 cell line (human breast cancer cells), maintain an antitumoral activity comparable to free drug (Figure 9). These results, demonstrate that drug activity is not reduced in the presence of nanoparticles carrier.

SUMMARY AND OUTLOOK

The use of solid lipid nanoparticle (SLN) opens up new perspectives for the formulation of poorly soluble drugs. The SLN is a very complex system with some advantages and disadvantages over other colloidal carrier systems. Our study showed that lipophilic drugs can be encapsulated into the SLN formulations. However, further research should be conducted to determine the structure and dynamics of drug-loaded SLN at molecular level, so as to further understand the mechanism of drug delivery using this system. This study may be conducted using NMR, ESR and synchrotron irradiation techniques. This will throw some light on the question whether the drug is really incorporated in the solid lipid or the lipid and drug nanosuspensions coexist in the formulation. The lipid matrix of SLN is more mobile than polylactide-co-glycolide based nanoparticles and diffusion of the drug within the lipid matrix is limited, therefore the mechanism of controlled-release of drug and factors influencing release need to be further investigated. Further study is needed to determine the interaction (adsorption, desorption processes, enzymatic degradation, agglo-meration) of SLN with biological environment. A better understanding of the plasma protein adsorption patterns and the *in vivo* fate of this drug delivery system need to be further studied to ascertain effectiveness of tissue-targeting by the drug.

The main challenge in cancer chemotherapy is toxic side-effects induced by chemotherapeutic drugs. Single

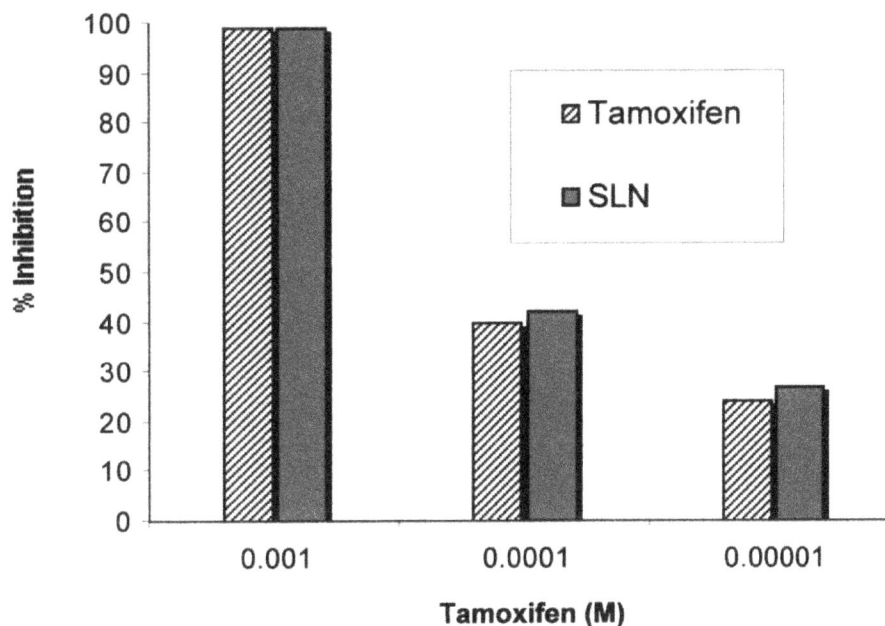

Figure 9. Antiproliferative activity of free tamoxifen and tamoxifen-loaded SLN on human breast cancer MCF-7 cells. Each value is the mean of three experiments (Fontana et al., 2005).

dose or short-time application (1 to 2 weeks) will probably cause serious health problems, but the use of biodegradable nano-sized particles for long-term or life-time therapy may produce other serious side-effects. Increasing the encapsulation efficiency of poorly water-soluble molecules will lead to the development of improved SLN formulations. In the near future, it is expected that more studies will focus on improving SLN and drug-loaded SLN formulations to increase the efficacy and reduce the side-effects of chemotherapeutic drugs for anticancer treatment. These studies should include preparation of formulations with different particle size and distributions, different matrix lipids and additional ingredients. Thus, if nanoparticulate drug delivery systems to be used effectively and routinely, the matter of toxicity of the components of nanoparticles must be addressed. Thus, SLN warrants further development before it can be used as a new drug delivery system for chemotherapy drugs in treatment of human cancers.

REFERENCES

Abbasalipourkabir R, Salehzadeh A, Rasedee A (2010). "Antitumor activity of tamoxifen loaded solid lipid nanoparticles on induced mammary tumour gland in Sprague-Dawley rats." Afr. J. Biotech., 9(43): 7337-7345.

Abbasalipourkabir R, Salehzadeh A, Rasedee A (2011a). Cytotoxicity effect of solid lipid nanoparticles on human breast cancer cell lines. Biothecnology, 10(6): 528-533.

Abbasalipourkabir R, Salehzadeh A, Rasedee A (2011). Delivering tamoxifen within solid lipid nanoparticles. Pharm. Tech., 35(4): 74-79

Attama AA, Müller-Goymann CC (2007). Investigation of surface-modified solid lipid nanocontainers formulated with a heterolipid-templated homolipid. Int. J. Pharm., 334: 179-189.

Brannon-Peppas L, Blanchette JO (2004). Nanoparticle and targeted systems for cancer therapy. Adv. Drug Deliv. Rev., 56: 1649-59

Brigger I, Dubernet C, Couvreur P (2002). Nanoparticles in cancer therapy and diagnosis. Adv. Drug Deliv. Rev., 54: 631-51

Cardosa F, Bedard PL, Winer EP, Pagani O, Senkus-Konefka E, Fallowfield LJ, Kyriakides S, Costa A, Cufer T, Albain KS (2009). International Guidelines for Management of Metastatic Breast Cancer: Combination vs Sequential Single-Agent Chemotherapy. J. National Cancer Ins., 101: 1174-1181

Chawla JS, Amiji MM (2003). Biodegradable poly (e-caprolactone) nanoparticles for tumor targeted delivery of tamoxifen. Int. J. Pharm., 249: 127–138.

Copland MJ, Rades T, Davies NM, Baird MA (2005). Lipid based particulate formulations for the delivery of antigen. Immunol. Cell Biol., 83: 97-105

Coughlin SS, Ekwueme DU (2009). Breast cancer as a global health concern. Cancer Epidemiol., 33: 315-318.

Clark MJ (2008). WITHDRAWN: Tamoxifen for early breast cancer. Cochrane Database of Systematic Rev., 8(4): CD000486.

DevCan5.2 (2004). Probability of Developing or Dying of Cancer Software, Version Statistical Research and Applications Branch, National Cancer Institute. http://srab.cancer.gov/devcan

Ewesuedo RB, Ratain MJ (2003). Principles of cancer therapeutics. In: Vokes EE, Golomb HM (eds) Oncologic therapies. Springer, Secaucus, NJ, pp. 19–66.

Ferlay J, Autier P, Boniol M, Heanue M, Colombet M, Boyle P (2007). Estimates of the cancer incidence and mortality in Europe in 2006. Ann. Oncol., 18: 581-592.

Fontana G, Maniscalco L, Schillaci D, Cavallaro G (2005). Solid Lipid Nanoparticles Containing Tamoxifen. Characterization and in vitro Antitumoral Activity. Drug Deliv., 12: 385–392.

Gradishar WJ (2005). Safety considerations of adjuvant therapy in early breast cancer in postmenopausal women. Oncol. 69: 1-9.

Harivardhan Reddy L, Vivek K, Bakshi N, Murthy RSR (2006). Tamoxifen Citrate Loaded Solid Lipid Nanoparticles (SLN^{TM}): Preparation, Characterization, In Vitro Drug Release, and Pharmacokinetic Evaluation. Pharm. Dev. Tech., 11: 167-177.

He J, Hou S, Lu W, Zhu L, Feng J (2007). Preparation, pharmacokinetics and body distribution of Silymarin-loaded Solid Lipid Nanoparticles after oral administration. J. Biomed. Nanotech., 3:195-202.

Heiati H, Philips NC, Tawashi R (1996). Evidence for phospholipid bilayers formation in solid lipid nanoparticles formulated with phospholipid and triglyceride. Pharm. Res., 13: 1406-1410.

Hou DZ, Xie CS, Huang K and Zhu CH (2003). The production and characteristics of solid lipid nanoparticles (SLNs). Biomat., 24: 1781-1785.

Huang Z-r, HUA S-c, Yang Y-I, Fang J-yl (2008) Development and evaluation of lipid nanoparticles for camptothecin delivery: a comparison of solid lipid nanoparticles, nanostructured lipid carriers, and lipid emulsion. Acta Pharmacol., 29(9): 1094–1102.

Kayser O, Iemke A, Hernandez-Trejo N (2005). The impact of nanobiotechnology on the development of new drug delivery systems. Current Pharm. Biotech., 6: 3-5.

Koping-Hoggard M, Sanchez A, Alonso MJ (2005). Nanoparticles as carriers for nasal vaccine delivery. Expert Rev. Vaccin., 4: 185-96.

Laggner P (1999). X-ray diffraction of lipids, in: Hamilton, R.J. and Cast, J. (Eds), Spectral properties of Lipids, Sheffield Academic Press, Sheffield, pp. 327-367.

Mehnert W, Mäder K (2001). Solid lipid nanoparticles Production, characterization and applications. Adv. Drug Deliv. Rev., 47: 165–196.

Memisoglu-Bilensoy E, Vural I, Bochot A, Renoir JM, Ducheneb D, Atilla Hincal A (2005). Tamoxifen citrate loaded amphiphilic β-cyclodextrin nanoparticles: In vitro characterization and cytotoxicity. J. Controlled Release, 104: 489–496.

Mirshahidi HR, Abraham J (2004). Managing early breast cancer: prognostic features guide choice of therapy. Postgrad. Med., 116: 23-34.

Mühlen AZ, Schwarz C, Mehnert W (1998). Solid lipid nanoparticles (SLN) for controlled drug delivery-Drug release and release mechanism. Eur. J. Pharm. Biopharm., 45: 149-155.

Müller RH, Bohm BHL (1998). In: Müller RH, Benita S, Bohm B editor. Emulsions and nanosuspensions for the formulation of poorly soluble drugs. Stuttgart: Medpharm Scientific, pp. 149-73.

Müller RH, Mäder K, Gohla S (2000). Solid lipid nanoparticles (SLN) for controlled drug delivery - a review of the state of the art. Eur. J. Biopharm., 50: 161-177

Müller RH, Keck CM (2004). Challenges and solutions for the delivery of biotech drugs - a review of drug nanocrystal technology and lipid nanoparticles. J. Biotech., 113: 151-170.

Paci E, Cariddi A, Barchielli A, Bianchi S, Cardona G, Distante. V, Giorgi D, Pacini P, Zappa M, Del Turco MR (1996). Long-term sequelae of breast cancer surgery. Tumori, 82: 321-4.

Sanjula B, Shah FM, Javed A, Alka A (2009). Effect of poloxamer188 on lymphatic uptake of carvedilol-loaded solid lipid nanoparticles for bioavailability enhancement. J. Drug Target., 17: 249-56.

Schubert MA, Müller-Goymann CC (2005). Characterization of surface-modified solid lipid nanoparticles (SLN): Influence of lecithin and nonionic emulsifier. Eur. J. Pharm. Biopharm., 61: 77-86.

Schwarz C, Mehnert W, Lucks JS, Müller RH (1994). Solid Lipid Nanoparticles (SLN) for controlled drug delivery. Production, Characterization and sterilization. J. Controlled Release, 30: 83-96.

Schwarz C, Mehnert W (1997). Freez-drying of drug-free and drug loaded solid lipid nanoparticles (SLN). Int. J. Pharm., 157: 171-179.

Siekmann B, Westesen K (1994). Melt-homogenized solid lipid nanoparticles stabilized by the nontonic surfactant tyloxapol. Preparation and particle size determination. Pharm. Pharmacol. Lett., 3: 194-7.

Sipos EP, Tyler B, Piantadosi S, Burger PC, Brem H (1997). Optimizing interstitial delivery of BCNU from controlled release polymers for the treatment of brain tumors, Cancer Chemo. Pharmacol., 39: 383-389.

Westesen K, Bunjes H, Koch MHJ (1997). Physicochemical characterization of lipid nanoparticles and evaluation of their drug loading capacity and sustained release potential. J. Controlled Release, 48: 223-226.

Westesen K, Siekmann B (1997). Investigation of the gel formation of phospholipid stabilized solid lipid nanoparticles. Int. J. Pharm., 151: 35-45.

Westesen K, Wehler T (1993). Characterization of submicron-sized oil-in-water emulsion. Colloids and Surface A, 78: 115-123.

Wong HL, Rauth AM, Bendayan R, Manias JL, Ramaswamy M, Liu Z, Erhan SZ, Wu XY (2006). A New Polymer-Lipid Hybrid Nanoparticle system Increases Cytotoxicity of Doxorubicin Against Multidrug-Resistaant Human Breast Cancer Cells. Pharm. Res., 23: 1574-1584.

Youssef M, Fattal E, Alonso MJ, Roblot-Treupel L, Sauzieres J, Tancre'de C (1988). Effectiveness of nanoparticle-bound ampicillin in the treatment of Listeria monocytogenes infection in athymic nude mice. Antimicrob. Agents Chemo., 32: 1204-1207.

23

Biofilm forming bacteria isolated from urinary tract infection, relation to catheterization and susceptibility to antibiotics

Nermeen Mahmoud Ahmed Abdallah[1], Shereen Bendary Elsayed[1]*, Manal Mohamed Yassin [1]Mostafa and Ghada Metwally El-gohary[2]

[1]Departments of Medical Microbiology and Immunology, Ain Shams University, Cairo 11341, Egypt.
[2]Internal medicine, Faculty of Medicine, Ain Shams University, Cairo- Egypt.

In human medicine, it has been estimated that 65% of nosocomial infections are biofilm associated, loading the health care system enormous costs. These biofilm infections are 10 to 1000 times more resistant to the effects of antimicrobial agents. This study aimed at showing the difference between patients with catheter associated urinary tract infection (CAUTI) and those with non catheter associated (UTI) in terms of the type of isolated pathogens, antibiotic susceptibility of isolated pathogens, detection of their ability to form biofilm, and comparing (antibiotic susceptibility of sessile cells) minimal biofilm eradication concentration (MBEC) and (their planktonic counterpart) minimal inhibitory concentration (MIC) for biofilm forming bacteria. The most frequently isolated micro-organisms were *Escherichia coli* (31.7%) followed by *Klebsiella* (15%); *Staphylococcus aureus*; coagulase negative *Staphylococcus* (CoNS); *Enterococcus* (11.7%); Proteus (10%); *Pseudomonas* (6.7%) and the least common was *Enterobacter* (1.7%). In the catheterized patients, 13 isolates out of thirty bacterial isolates (43.3%) were biofilm forming and 17 isolates (56.7%) were non biofilm forming, while in the non catheterized patients, 9 isolates out of thirty bacterial isolates (30%) were biofilm forming and 21 isolates (70%) were non biofilm forming. Antibiotic sensitivity of the isolated pathogens was done using disc diffusion method which showed that Imipenem and Amikacin were most effective antibiotics against gram-negative isolates while for gram positive isolates, Vancomycin and Ciprofloxacin were most effective. There was no statistical difference between the two groups regarding the isolated pathogens or the antibiotic susceptibility pattern. For the biofilm forming isolates, antibiotic susceptibility of sessile cells MBEC were tested and compared to the MIC of their planktonic counterpart. For gram negative isolates, Amikacin and Imipenem were used and for gram positive isolates, Ciprofloxacin and Vancomycin were used. The difference between MBEC and MIC for tested strain was statistically significant. Therefore, researches on easier methods for diagnosing and quantifying biofilm infection would surely help the fight against biofilm formation. Also for certain infection such as CAUTI, it is advised to test antimicrobial susceptibility in biofilm form MBEC.

Key words: Urinary tract infection, Biofilm.

INTRODUCTION

Urinary tract infection, with its diverse clinical syndromes and affected host groups, remains one of the most common but widely misunderstood and challenging infectious diseases encountered in clinical practice. Antimicrobial resistance is a leading concern and efforts should be made to ensure an appropriate duration of therapy for symptomatic infections (Drekonja and Johnson, 2008). The risk of developing urinary tract infection increases significantly with the use of indwelling devices such as catheters and urethral stents or sphincters (Foxman, 2003). Urinary tract infections

*Corresponding author. E-mail: Sherin_bendary@yahoo.com.

account for an estimated 25 to 40% of nosocomial infections and represent the most common type of these infections (Bagshaw and Laupland, 2006).

Clinical observations have established that, the microbial populations within catheter associated urinary tract infection (CAUTI) frequently develop as biofilms, directly attaching to the surface of catheters (Trautner and Darouiche, 2004). Biofilms are microbial communities of surface-attached cells embedded in a self-produced extracellular polymeric matrix (Donlan and Costerton, 2002). They can cause significant problems in many areas, both in medical settings (For example; persistent and recurrent infections, device-related infections) and in non-medical (industrial) settings (For example, biofouling in drinking water distribution systems and food processing environments) (Flemming, 2002; Fux et al., 2005). A worrying feature of biofilm-based infections is represented by the higher resistance of bacterial and fungal cells growing as biofilms to antibiotics and disinfecting chemicals as well as resisting phagocytosis and other components of the body's defence system, when compared to planktonic cells (Hoiby et al., 2010).

Traditionally, microbiologists have evaluated the efficacy of an antibiotic (AB) by measuring the minimum inhibitory concentration (MIC) and minimum bactericidal concentration (MBC). In virtually all diagnostic laboratories, these measurements are made on freely floating, planktonic, laboratory phenotypes. These assays measure only the concentration of chemotherapeutic agent required to inhibit growth or kill planktonic bacteria. For some antibiotics, the concentration required to kill sessile bacteria may be greater than a thousand times that required to kill planktonic bacteria of exactly the same strain. Thus, even well-chosen treatment based upon laboratory results often merely suppresses an infection until biofilm-associated organisms are reactivated and cause another clinical infection (Olson et al., 2002).

Therefore, this study aimed at showing the difference between patients with CAUTI and those with non catheter associated urinary tract infection (UTI) in terms of type of isolated pathogens, antibiotic susceptibility of isolated pathogens, detection of their ability of to form biofilm, and comparing antibiotic susceptibility of sessile cells minimal biofilm eradication concentration (MBEC) and their planktonic counterpart minimal inhibitory concentration MIC for biofilm forming bacteria.

MATERIALS AND METHODS

The study was performed in the Department of Medical Microbiology and Immunology of Ain Shams University and was conducted on seventy two patients divided into two groups admitted to Ain Shams University Hospitals during the period from November2009 to December 2010 to get the final number of thirty bacterial isolate per group:

Group 1: Forty patients had either an indwelling urinary catheter in place at the time of specimen collection and at least one of the following: fever (>38°C), suprapubic tenderness, costovertebral angle pain or tenderness, or the catheter removed within 48 hours prior to specimen collection and had at least one of the following: fever (>38°C), urgency, frequency, dysuria, suprapubic tenderness, or costovertebral angle pain or tenderness.

Group 2: Thirty two patients did not have an indwelling urinary catheter in place at the time of specimen collection nor within 48 hours prior to specimen collection and had at least one of the following: fever (>38°C, urgency, frequency, dysuria, suprapubic tenderness, or costovertebral angle pain or tenderness).

All samples were cultured on cysteine lactose electrolyte deficient agar (CLED) and blood agar media using 1 µl calibrated loop to detect colony forming unit and incubated aerobically at 37°C for 24 h, count more than 10^5 was considered significant. Identification of isolated organisms was done by conventional microbiological methods (Cheesebrough, 2007).

Biofilm formation test

The isolated organisms were tested for their ability to form biofilm, according to Stepanovic et al. (2007). Briefly, flat-bottomed 96-well clear polystyrene tissue culture treated microtitrplate (MP) with a lid (TPP- Switzerland) were inoculated with 200 µl of a bacterial suspension in corresponding to 0.5 McFarland (with further 1:100 dilution). After 24 h incubation at 37°C, the contents of each well were removed by decantation and each well was washed three times with 300 µl of sterile saline. The remaining attached bacteria were heat-fixed by exposing them to hot air at 60°C for 60 min in Fisher isotemp incubator, then150 µl crystal violet (2%) stain was added to each well. After 15 min, the excess stain was rinsed off by decantation, and the plate was washed., 150 µl 95% ethanol was added to each well, and after 30 min, the optical densities (OD) of stained adherent bacterial films were read using a microtiter-plate reader (Tecan Sunrise remote Austria) at 620 nm. The average OD values were calculated for all tested strains and negative controls, the cut-off value (ODc) was established. It is defined as a three standard deviations (SD) above the mean OD of the negative control: ODc=average OD of negative control + (3×SD of negative control). Final OD value of a tested strain was expressed as average OD value of the strain reduced by ODc value (OD= average OD of a strain -ODc); ODc value was calculated for each microtiter plate separately. When a negative value was obtained, it was presented as zero, while any positive value indicated biofilm production. For easier interpretation of the results, strains were divided into the following categories:

1. Non biofilm producer (0) OD ≤ODc
2. Weak biofilm producer (+ or 1) = ODc <OD ≤2×ODc,
3. Moderate biofilm producer (++ or 2) = 2×ODc <OD≤4×ODc
4. Strong biofilm producer (+++or 3), 4×ODc <OD

Antibiotic susceptibility testing for planktonic cells

Disc diffusion method

Antibiotic susceptibility of all isolated organisms was done by disc diffusion method, using Muller-Hinton (MH) agar plates. After overnight incubation results were reported and interpretation was done according to clinical and laboratory standards institute (CLSI M100-S20, M100-S20U, guidelines 2010). The antibiotics evaluated were those commonly approved for the treatment of bacterial infections National Committee for Clinical Laboratory Standards

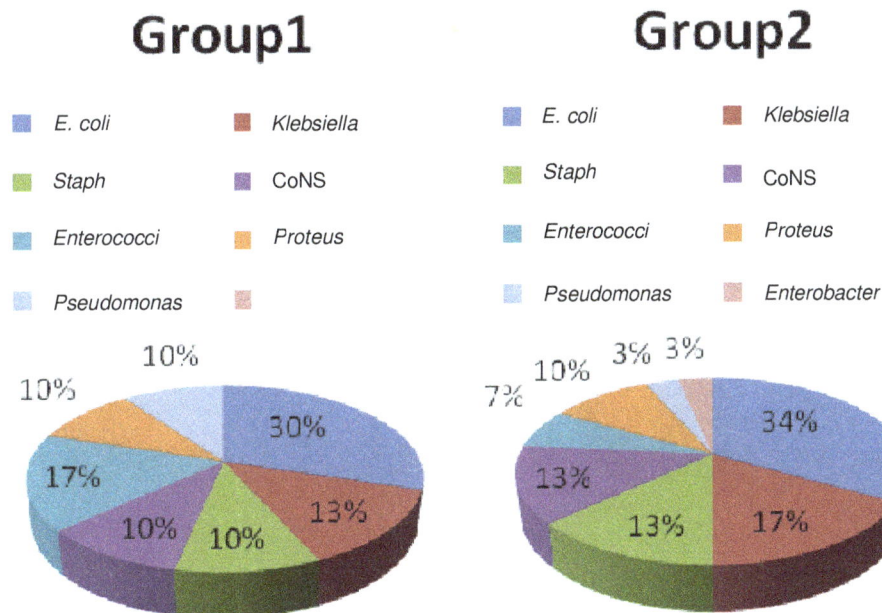

Figure 1. Distribution of isolated pathogen between the two groups.

(NCCLS).

Minimal inhibitory concentration (MIC)

This was done only for biofilm forming isolates. The antibiotics were selected as the biofilm forming isolates were commonly sensitive to it. For gram negative isolates, Amikacin and Imipenem were used, for gram positive isolates, Ciprofloxacin and Vancomycin were used. MIC was determined by broth micro dilution using 96 wells MP and results were interpreted according to CLSI, guidelines (2010).

Antibiotic susceptibility of biofilm

For the biofilm forming isolates, antibiotic susceptibility of sessile cells were tested according to (Cernohorska and Votava, 2004, Passerini de Rossi et al., 2009) and compared to the MIC of their planktonic counterpart.

Statistical analysis

Analysis of data was done by IBM computer using statistical program for social science (SPSS) version 12. Quantitative variables were expressed as mean ± standard deviation (SD). Statistical tests included Chi-square test and Spearman correlation coefficient rank test (r) which was used to rank different variables against each other either positively or inversely. Unpaired t-test was used to compare two independent groups as regards quantitative variables in parametric data (SD < 50% mean). Results were considered significant when p value was ≤ 0.05.

RESULTS

In this study, sixty bacterial isolates were obtained from fifty patients divided into two groups. The first group included 23 catheterized patients (8 males, 15 females) giving thirty bacterial isolates and the mean age of group 1 patients was 52 ± 8.8 ranging from 34 to 65. The second group included 27 non catheterized patients (9 males, 18 females) giving thirty bacterial isolates and the mean age of group 2 patients was 41.5 ± 11.5 ranging from 23 to 66. Out of the sixty isolates the most common isolated pathogens were: *Escherichia coli* (31.7%); (47.4%) in group 1, (52.6%) in group 2, followed by Klebsiella (15%) (44.4%) in group 1, (55.6%) in group 2, then *Staphylococcus aureus*, CoNS, (11.7%) (42.9%) in group 1, (57.1%) in group 2, *Enterococcus* (11.7%) (71.4%) in group 1, (28.6%) in group 2, *Proteus* (10%) (50% in group 1 and 2, *pseudomonas* (6.7%) (75%) in group1, (25%) in group 2, and lastly, *Enterobacter* (1.7%) single isolate in group 2. The difference between two groups was statistically non significant (Figure 1). Out of sixty isolates 38 isolates (63.3%) were non biofilm forming (44.7%) in group1, (55.3%) in group 2 and 22 isolates (35.0%) were biofilm forming [(59.1%) in group1, (40.9%) in group 2 (Figures 2 and 3). Out of 22 biofilm forming isolates 18(81.8%) were weak biofilm forming (50%) in group 1,2 and 4 (18.2%) were moderate biofilm forming all (100%) in group 1 but this difference was statistically non significant (Chi-square test). The highest percent of biofilm formation was detected in CoNS as (57.1%) of CoNS isolates were biofilm forming followed by *Pseudomonas* (50.0%), *Klebsiella* (44.4%), *Staph* (42.9%), *E. coli*(31.6%) and *Enterococci* (28.6%) while least biofilm forming was *Proteus*(16.7%) and *Enterobacter* (0.0%).

The most effective antibiotics against Gram-negative

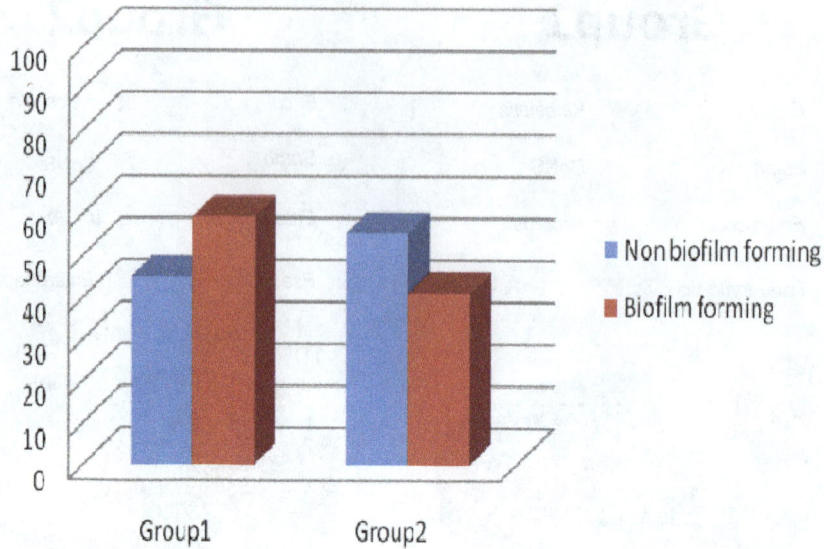

Figure 2. Shows the difference between the two groups in biofilm formation.

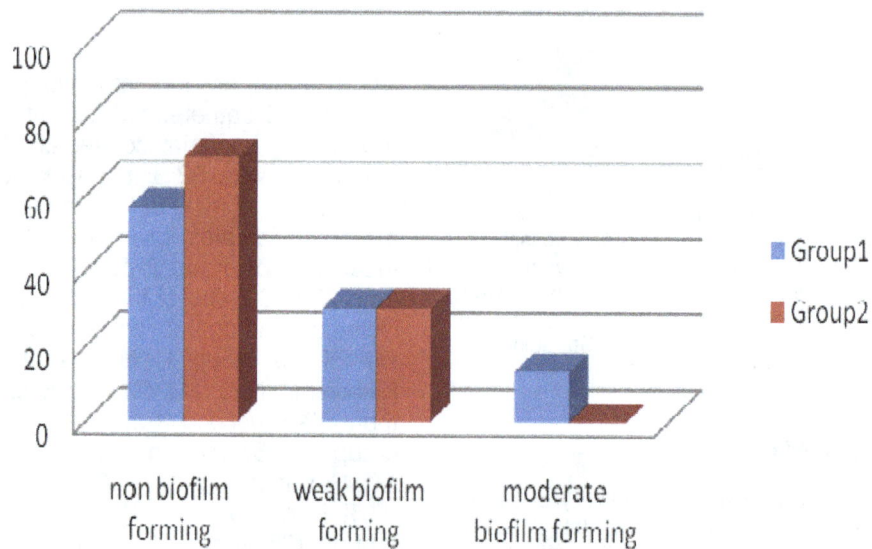

Figure 3. Difference in degree of biofilm formation between two groups.

isolates were found to be Imipenem and Amikacin and for gram positive isolates Vancomycin and Ciprofloxacin. A statistically significant difference (Unpaired t-test) existed between MIC and MBEC of Vancomycin and Ciprofloxacin exist for gram positive isolate. The isolates that were sensitive or intermediately sensitive in their MIC values became resistant in their MBEC value, and for gram negative isolates the difference between MIC and MBEC of Imipenem and Amikacin was statistically highly significant as for Imipenem isolates that were sensitive or intermediately sensitive in their MIC values became resistant in their MBEC value and for Amikacin isolates that were sensitive or intermediately sensitive in their MIC

value became resistant in their MBEC values except for two isolates as the first isolate remained sensitive, while the second became intermediately sensitive (Table 1).

DISCUSSION

In this study, the frequency of UTI was greater in women as compared to men as 66% of the patients were females and 34% were males principally owing to anatomic and physical factors. Similar results were shown by kashef et al. (2010) and Al Benwan et al. (2010). This study showed that the most common causative organism of UTI

Table 1. Difference between MIC and MBEC of biofilm forming isolates.

	Type of organism	Degree Of Biofilm Formation	Ciprofloxacin µg/ml		Vancomycin µg/ml	
			MIC	MBEC	MIC	MBEC
Gram positive isolates	S. aureus	Weak	0.5	64	2	>256
	Enterococci	Weak	2	>64	4	>256
	S. aureus	Moderate	1	8	4	>256
	CoNS	Weak	0.125	>64	4	>256
	Enterococci	Weak	1	>64	(16)	>256
	CoNS	Moderate	1	8	16	>256
	S. aureus	Moderate	1	16	1	>256
	CoNS	Weak	2	64	16	>256
	CoNS	Weak	(2)	>64	(16)	>256

			Amikacin (µg/ml)		Imipinem (µg/ml)	
			MIC	MBEC	MIC	MBEC
Gram negative isolates	E coli	Weak	4	512	1	128
	Klebsiella	Weak	2	64	1	> 128
	Klebsiella	Weak	1	64	0.5	> 128
	Proteus	Weak	2	128	2	64
	E. coli	Weak	4	128	1	> 128
	Klebsiella	Weak	2	64	2	> 128
	E. coli	Weak	2	512	0.25	> 128
	Pseudomonas	Weak	0.25	64	0.125	64
	E. coli	Weak	1	64	0.25	32
	E. coli	Weak	2	32	0.25	64
	Klebsiella	Weak	1	16	0.25	128
	Pseudomonas	Moderate	32	64	8	> 128
	E. coli	Weak	8	512	2	> 128

was E. coli (31%) followed by klebsiella (15%),Staphylococcus, Coagulase negative staphylococci CoNS, Enterococcus (11.7%) each, Proteus (3.8%), Pseudomonas (6.7%), and the least common cause being Enterobacter (3.8%). Similar results were reported by Neto et al. (2003), Bgshaw and Lupland, (2006), Savas et al. (2006) and EL-Banoby et al. (2007) who found that E. coli was the most common cause of UTI .

There is a reported difference in prevalence of various uropathogens between patients with indwelling urinary catheters and non catheterized patients. In this study, it was found that E. coli, Klebsiella, S. aureus and CoNS were more often recovered from non catheterized patients while Enterococcus spp. and Pseudomonas spp. were more often recovered from catheterized patients but due to small number of samples the difference was not statistically significant. Similar results were reported by (Savas et al., 2006, ko et al., 2008, Milan and Ivan, 2009) who found that E. coli was the most prevalent organism in both catheterized and non catheterized patient.

As regards the antibiotics used, the present study showed that the most effective antibiotics against Gram-negative isolates were found to be Imipenem and Amikacin and for gram positive isolates Vancomycin and Ciprofloxacin. This is in agreement with Savas et al. (2006) who found that the most effective antibiotics against Gram-negative bacteria were imipenem and meropenem.

Similar results were reported by EL-Banoby et al.(2007) who stated that Amino glycosides were the most common antibiotics to which the organisms were sensitive for nosocomial UTI (46.2%) followed by Monobactam (34.6%), Quinolones (30.8%), and lastly Vancomycin (3.8%). The present study showed no statistically significant difference (Unpaired t-test) between the catheterized and non catheterized UTI patients as regard antibiotic susceptibility. In disagreement with the results of the present study, Ko et al. (2008) noticed that urinary catheters increased antimicrobial resistance of Enterobacteriaceae and rare gram negative bacilli (For example, Acinetobacter spp.) to nearly all antibiotic tested.

Another contradictory results reported by Milan and Ivan ,(2009) who evaluated resistance between community-acquired urinary tract infections, nosocomialy-acquired urinary tract infections , and found that the highest level of general resistance was among isolates of nosocomialy-acquired UTI and catheter-associated UTI followed by community-acquired UTI isolates. This study found that (43.3%) of isolates of group 1 were biofilm forming while in group 2 (30%) of the isolates were biofilm forming, yet the difference between two groups was statistically non significant. Similar results were found by Watts et al. (2010) who compared the virulence properties of a collection of asymptomatic bacteriuria (ABU) and catheter associated asymptomatic bacteriuria (CA-ABU) nosocomial E. coli isolates and found that the CA-ABU strains displayed no significant difference from the ABU strains in the biofilm formation phenotype; (48%) of CA-ABU and (59%) of ABU strains formed a biofilm.

In this study, antibiotic susceptibility of planktonic cells presented as MIC was compared to and their counterpart sessile cells presented by MBEC. Amikacin and Imipinem were chosen for gram negative biofilm forming isolates and Vancomycin, Ciprofloxacin for gram positive biofilm forming isolates. This choice was based upon disc diffusion antibiotic susceptibility as we chose the common sensitive or intermediately sensitive antibiotic. A statistically significant difference in antibiotic susceptibility between planktonic populations and biofilm populations of the same organism was demonstrated in this work.

Similar results were reported by Sepandj et al. (2004) who compared the MIC and MBEC of Ampicillin, Cefazolin, Cefotaxime, Ciprofloxacin, Gentamicin and Tobramycin of eight E.coli strains, Amikacin, Aztreonam, Ceftazidime, Ciprofloxacin, Gentamicin, Imipenem, Piperacillin and Tobramycin susceptibility were also tested for eight *Pseudomona strains*, isolated from patients with peritoneal dialysis - associated peritonitis. The authors concluded that in their biofilm state, gram-negative bacteria are much less susceptible to antibiotics compared to their antibiotic susceptibility in the planktonic state.

Another agreeing results were reported by Sepandj et al. (2007) who compared the MIC and MBEC of Six samples of *Enterococcus faecalis* and five isolates of *E. faecium* and found that the *E. faecalis* isolates that were sensitive in their MIC results but demonstrated resistance in their MBEC results to Ciprofloxacin, Vancomycin and Ampicillin. Ampicillin combined with Gentamicin (Gentamicin 1 and 4 mg/ml) resulted in 0 MIC and 2 MBEC resistances. As regards *E. faecium,* the MIC results were sensitive for Vancomycin only. The MBEC results revealed uniformly poor sensitivity of *E. faecium* biofilms to antibiotics, including Ampicillin with Gentamicin combinations. Also, Passerini de Rossi et al. (2009) who compared MIC and MBEC of Levofloxacin and Ciprofloxacin for 32 *Stenotrophomonas maltophilia* isolates and found that *S. maltophilia* isolates which were sensitive to fluoroquinolones according to their MICs were highly resistant according to the MBEC values.

Another agreeing results were reported by Naves et al. (2010) who studied the *in vitro* susceptibility of seven *E. coli* biofilm producing strain in their planktonic and biofilm

associated forms to Amoxicillin, Amoxicillin/Clavulanic, Cefotaxime, Gentamycin and Ciprofloxacin and found that *E. coli* biofilms were much less sensitive than their planktonic counterparts to tested antibiotics. Lastly, Antunes et al. (2010) who compared the MIC and MBEC of Vancomycin for thirty staphylococci isolates found that all isolates presented higher MBEC than the MIC for Vancomycin. But in disagreement with the above results Spoering and Lewis (2001) who performed a comparative examination of tolerance of biofilms versus stationary and logarithmic-phase planktonic cells with four different antimicrobial agents and concluded that, at least for *Pseudomonas aeruginosa*, one of the model organisms for biofilm studies, the notion that biofilms have greater resistance than do planktonic cells is unwarranted. They suggested that tolerance to antibiotics in stationary-phase or biofilm cultures is largely dependent on the presence of persister cells.

RECOMMENDATION

Current antibiotics have classically been developed to treat infections involving planktonic bacterial populations in acute infection settings and are typically ineffective in the eradication of bacteria in biofilm- associated, persistent infections. The MBEC assay were developed for rapid and reproducible antimicrobial susceptibility testing for bacterial biofilms in the anticipation that the MBEC would be more reliable for selection of clinically effective antibiotics.

Many researches are needed to find easier methods for diagnosing and quantifying biofilm infection, to develop more specific antimicrobial agents and ideal device surfaces that would surely help the fight against biofilm formation

REFERENCES

Antunes AL, Trentin DS, Bonfanti JW, Pinto CC, Perez LR, Macedo AJ, Barth AL (2010). Application of a feasible method for determination of biofilm antimicrobial susceptibility in staphylococci. APMIS, 118(11): 873-877.

Bagshaw SM, Laupland KB (2006). Epidemiology of intensive care unit-acquired urinary tract infections. Curr. Opin Infect Dis., 19(1): 67-71.

Cernohorska L, Votava M (2004). Determination of minimal regrowth concentration (MRC) in clinical isolates of various biofilm-forming bacteria, Folia Microbiol (Praha) 49: 75-78.

Cheesebrough M (2007). District laboratory practice in tropical countries part 2 second edition pp. 62-70,105-115.

Clinical Laboratory Standards Institute (CLSI) (2010). Performance Standards for Antimicrobial Susceptibility Testing; twentieth Informational Supplement M100-S20vol.30 no.1 and Performance Standards for Antimicrobial Susceptibility Testing; twentieth Informational Supplement (June 2010 update) M100-S20-U, 3: 1.

Donlan RM, Costerton JW (2002). Biofilms: Survival mechanisms of clinically relevant microorganisms. Clin. Microbiol. Rev., 15: 167-193.

Drekonja DM, Johnson JR (2008). Urinary tract infections.Prim Care, 35(2): 345-367.

EL-Banoby MH, Ebeid SM, EL-Bedewy RS, Amrousy MA (2007). Nosocomial pneumonia and urinary tract infections in elderly patients admitted to the ICU. Egypt J. Med. Lab. Sci., 16(1): 51-56.

Foxman B (2003). Epidemiology of urinary tract infections: incidence, morbidity and economic costs. Dis Mon. 49: 53-70.

Fux CA, Costerton JW, Stewart PS, Stoodley P (2005). Survival strategies of infectious biofilms. Trends Microbiol, 13: 34-40.

Flemming HC (2002). Biofouling in water systems — cases, causes and countermeasures. Appl. Microbiol. Biotechnol., 59: 629–640.

Hoiby N, Bjarnsholt T, Givskov M, Molin S, Ciofu O (2010). Antibiotic resistance of bacterial biofilms. Int. J. Antimicrob. Agents, 35(4): 322-332.

Kashef N, Djavid GE, Shahbazi S (2010). Antimicrobial susceptibility patterns of community-acquired uropathogens in Tehran, Iran. J. Infect. Dev. Ctries. 14(4): 202-206.

Ko MC, Liu CK, Woung LC, Lee WK, Jeng HS, Lu SH, Chiang HS, Li CY (2008). Species and antimicrobial resistance of uropathogens isolated from patients with urinary catheter Tohoku J. Exp. Med., 214(4): 311-319.

Milan PB, Ivan IM (2009). Catheter-associated and nosocomial urinary tract infections: antibiotic resistance and influence on commonly used antimicrobial therapy. Int. Urol Nephrol., 41(3): 461-464.

Naves P, Del Prado G, Ponte C, Soriano F (2010). Differences in the in vitro susceptibility of planktonic and biofilm-associated Escherichia coli strains to antimicrobial agents. J. Chemother., 22(5): 312-317.

Neto JAD, da Silva LDM, Martins ACP (2003). Prevalence and bacterial susceptibility of hospital acquired urinary tract infection. Acta Cirurgica Brasileira 18 (Suppl 5).

Olson M, Ceri H, Morck D, Buret A, Read R (2002). Biofilm bacteria formation and comparative susceptibility to antibiotics Can. J. Vet. Res. 66(2): 86-92.

Passerini de Rossi B, García C, Calenda M, Vay C, Franco M (2009). Activity of levofloxacin and ciprofloxacin on biofilms and planktonic cells of Stenotrophomonas maltophilia isolates from patients with device-associated infections. Int. J. Antimicrobiol. Agents, 34(3): 260-264.

Savas L, Guvel S, Onlen Y, Savas N, Duran N (2006). Nosocomial urinary tract infections: micro-organisms, antibiotic sensitivities and risk factors. West Indian Med. J., 55(3): 188-193.

Sepandj F, Ceri H, Gibb A, Read R, Olson M (2004). Minimum inhibitory concentration (MIC) versus minimum biofilm eliminating concentration (MBEC) in evaluation of antibiotic sensitivity of gram-negative bacilli causing peritonitis. Perit. Dial. Int., 24(1): 65-67.

Sepandj F, Ceri H, Gibb A, Read R, Olson M (2007). Minimum inhibitory concentration versus minimum biofilm eliminating concentration in evaluation of antibiotic sensitivity of enterococci causing peritonitis. Perit Dial Int., 27(4): 464-465.

Spoering AL, Lewis K (2001). Biofilms and planktonic cells of Pseudomonas aeruginosa have similar resistance to killing by antimicrobials. J. Bacteriol., 183(23): 6746-6751.

Stepanovic S, Vuković D, Hola V, Di Bonaventura G, Djukić S, Cirković I, Ruzicka F (2007). Quantification of biofilm in microtiter plates overview of testing conditions and practical recommendations for assessment of biofilm production by staphylococci. APMIS, 115(8): 891-899.

Trautner BW ,and Darouiche RO (2004). Role of biofilm in catheter-associated urinary tract infection. Am. J. Infect. Control, 32: 177–118

Watts RE, Hancock V, Ong CL, Vejborg RM, Mabbett AN, Totsika M, Looke DF, Nimmo GR, Klemm P, Schembri MA(2010). Escherichia coli isolates causing asymptomatic bacteriuria in catheterized and noncatheterized individuals possess similar virulence properties. J Clin. Microbiol., 48(7): 2449-2458.

24

Effects of polysorbate-80 on liver and kidney function in broiler chicken during juvenile growth period

Khosravinia, H.[1]*, Manafi M.[2] and Rafiei Alavi, E.[3]

[1]Department of Animal Sciences, Faculty of Agricultural Sciences, Lorestan University, Khorrambad-68159, Lorestan, Iran.
[2]Department of Animal Science, Faculty of Agricultural Sciences, Malayer University, Malayer, Iran.
[3]Department of Pathology, Faculty of Medicine, Lorestan University of Medical Sciences, Khorramabad, Iran.

An experiment was conducted to examine the effects of polysorbate-80 (PS-80) on liver and kidney function in broiler chicken using 120 one-day-old Cobb 500 chicks. The birds were randomly divided into 6 groups, of 20 birds each (as one replicate) and then allocated to 1 of 6 floor pens (90×180 cm) at the stocking density of 0.08 m^2/ bird in a concrete floor, cross ventilated house. All the birds received the same corn and soybean meal based starter (1 to 21 days) and grower (15 to 28 days) diets for *ad libitum* consumption. The birds in 3 pens continually received water supplemented with either 0 (control) or 3500 ppm PS-80 throughout the experiment. Live weight gain as well as, feed conversion ratio were significantly suppressed in response to supplementation of drinking water with 3500 ppm PS-80 ($P<0.05$). Mean liver weight and liver pH did not significantly change, for the birds received treated water compared to those grown on normal water ($P>0.05$). No alteration in concentration of GPT, GOT, Urea, uric acid and creatinine in blood serum were observed in the chicks receiving PS-80-supplemented water as compared to control group ($P>0.05$). The PS-80 treated water resulted in increased levels of serum ALP ($P<0.05$). Based on these findings, persistent exposure of broiler chicks to 3500 ppm PS-80 through drinking water has a negative impact on juvenile growth performance in chicks. No indication found that such discouraging impact is caused by liver or kidney dysfunction.

Key words: Polysorbate-80, GPT, GOT, ALP, liver, kidney.

INTRODUCTION

Emerging use of poorly soluble or permeable compounds such as vaccines, herbal extracts and essential oils in poultry production, represents a reason to introduce inexpensive and non toxic vehicle formulations and emulsifiers (Gad et al., 2006). Polysorbate-80 (PS-80), also known as Tween-80, is a commercially available non-ionic surfactant (Chou et al., 2005) which originates from polyethoxylated sorbitan and oleic acid and is mainly used in food industry (Goff, 1997). PS-80 is also used as an excipient as well as, emulsifier in the manufacture of anti arrhythmis amiodarone (Path et al.,

1991), microbial media culture (Jacques et al., 1980), cosmetics and oral, parentral and topical pharmaceutical formulations (Rowe, 2009).

PS-80 is shown to be safe for few spices of laboratory animals and is well-tolerated by human (Steele et al., 2005; Roberts, 2010). Rats fed on diets supplemented with 5% PS-80 (v/v) showed no toxic effects (Oser and Oser, 1956). In the same study, the adverse effects of PS-80 on reproduction performance were pointed out where the dietary dose increased to 20%. Early reports on safety study of PS-80 in dogs and other canine species reveal that IV administration has been associated with an idiosyncratic reaction characterized by a prolonged depressor response (Krantz et al., 1951). Elder, (1984) and Masini et al. (1985) showed that this hypotensive response was caused by a marked release

---

*Corresponding author. E-mail: Khosravi_fafa@yahoo.com.

Table 1. Mean water and tween-80 intake for experimental flock in different periods from 1 to 28 days of age.

Age (day)	Water consumption (ml/bird)			ADG[1]	Tween intake (µg)	
	Control birds	Treated birds	Difference (%)		Per day	µg/g WG/day
1 - 7	990	1232	+24.00	13.25	616	46.50
8 - 14	2650	2686	+1.30	28.35	1434	50.62
15 - 21	4540	4750	+4.60	70.04	2375	33.91
22 - 28	13500	13100	-2.96	66.64	5614	82.24
1 - 28	21680	21760	-2.45	45.57	2510	55.01

Average daily weight gain.

of histamine after IV intake of PS-80. Thackaberry et al. (2010) confirmed that PS-80 at the rate of 10 mg/kg was safe and well tolerated when administered to mice, rats, dogs and cynomolgus monkeys through oral gavage over the course of 3 months administration. However, experimental results on effects of PS-80 on chicken as well as, other avian species are scarce.

Wide application of vaccines and increasing use of herbal extracts and essential oils in avian drinking water, warrant further investigation on the effects of continual administration of PS-80 in avian species. This study was undertaken to study the response of juvenile broiler chicks to intake of PS-80 through drinking water with respect to indicators of liver and kidney function

MATERIALS AND METHODS

Experimental flock

A total of 120 one-day-old Cobb 500 broiler chicks of mixed sex were provided from Zarbal hatchery, Borujerd, Iran. Up on arrival, the birds were randomly divided in 6 groups of 20 birds each and then randomly allocated to 1 of the 6 floor pens (90×180 cm). The pens were located in a cross-ventilated negative-pressure house equipped with infra red brooders. The space allowance for each bird was 0.08 m^2. Corn and soybean meal based starter (230 g/kg CP, 3100 Kcal ME/kg 1 to 21 days) and grower (191 g/kg CP, 3220 Kcal ME/kg 15 to 28 days) diets and water were provided for ad libitum consumption under a round the clock lighting regime.

Data collection

Live weight gain, feed intake, feed conversion ratio and mortality per cent were weekly determined from 1 to 28 days. At 28 days of age, two male and two female chicks per pen (those with the minimum difference from the mean pen weight for each sex) were killed and manually processed to obtain liver-to-body weight ratio and liver pH data. Blood samples (4 ml) were collected from all the killed birds to assess the serum enzymes including serum glutamate pyruvate transaminase (SGPT), serum glutamate oxaloacetate transaminase (SGOT), serum alkaline phosphatase (SALP) and certain serum biochemical parameters. Concentration of SGPT and SGOT enzymes were assayed based on IFCC method without pyridoxal phosphate (P-5-P) as described by Bermeyer et al. (1986). SALP were measured based on DGKC method (Anonymous, 1972) with slight modification as adopted by SCE (Anonymous, 1974). The blood serum samples were evaluated for concentration of urea, uric acid and creatinine using an automatic analyzer (Selects E Autoanlyzer, Sr. No. 8-7140, Vital

Company, Netherlands), following the indications for commercial kits at 25ºC.

Experimental design

A supplemented water (with 0 or 3500 ppm PS-80) × gender (male or female) factorial design was used to evaluate the effects of two main fixed factors (PS-80 level and sex) and their interaction on liver function of broiler chicks during early growth period, up to 28 d. The birds in each replicate (a pen) were of mixed sex and received drinking water supplemented with either 0 (control) or 3500 ppm PS-80 (PS-80). However, the PS-80 intake was associated with water consumption which by itself is a function of feed intake. The ultimate PS-80 intake based on water consumption and average daily gain is presented in Table 1.

Statistical analysis

Collected data were subjected to analysis of variance utilizing the General Linear Model Procedure of Statistical Analysis System (SAS Institute, 2002). Means for each fixed effect as well as, interactions were separated based on Duncan's Multiple Range Test (DMRT). Means were considered significantly different at P<0.05.

RESULTS

A summary of the results achieved for the response of the chicks in conventional productive performance parameters to continual receiving polysorbate-80 (PS-80) through drinking water is presented in Table 2. The birds with access to treated water demonstrated significantly lower average daily gain (P<0.05). Weight gain suppression for treated birds initiated from neonatal days improved by age such that it reached 117 g lower body weight at 28 days as compared to control group (Figure 1). Mean feed intake did not significantly differ for treated birds as compared to the control group (P>0.05), while feed conversion ratio was significantly increased for the birds receiving PS-80 supplemented water compared to the birds fed on normal water (P<0.05). No mortality was confirmed for the birds receiving either normal or treated water.

Supplementation of 3500 ppm PS-80 into drinking water showed no significant effect on liver weight and liver significant differences were pointed out in SGPT and SGOT enzymes between the birds receiving either normal or treated water (P>0.05). However, administration

Table 2. Effect (mean ± S.E.) of Polysorbate 80 on average daily weight gain (AVG), average daily fed intake (ADFI), feed conversion ratio (FCR) and mortality in chicks.

Variable	Tween-80 (ppm)		P-Value
	0	3500	
ADG (g; 1-28 d)	47.54±0.17[a]	45.57±0.66[b]	0.0582
ADFI (g; 1-28 d)	67.89±0.20[a]	68.42±0.36[a]	0.3609
FCR (g:g; 1-28 d)	1.43±0.01[b]	1.50±0.02[a]	0.0212
Mortality (%, 1-28 d)	0.00±0.00[a]	0.00±0.00[a]	0.3295

[a-b] Means within a raw without a common superscript differ significantly (P<0.05).

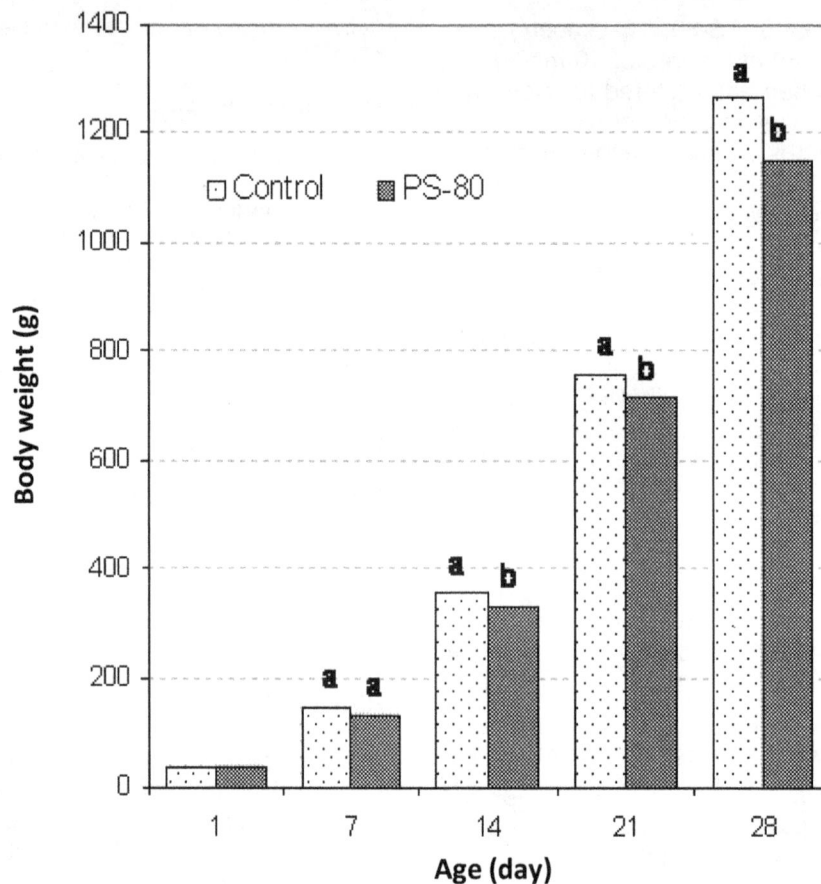

Figure 1. Change in live body weight (g) of the birds received normal water (control) and supplemented water with 3500 ppm polysorbate-80 at 1 to 28 days.

of PS-80 caused a considerable increase in ALP enzyme as compared to the normal chicks (62 vs. 45 IU/l; P=0.0984). The mean liver weight, liver pH, SGPT and SGOT were not significantly differ for the treated birds compared to control group for gender as well as gender × PS-80 interaction. The female birds generally showed lower ALP concentration in serum and their response in enhanced levels of the same enzyme with PS-80 was greater than the male chicks.

No significant differences in average uric acid, urea and creatinine concentrations (mg/dl) in serum were seen at 28 days of age between the birds received either normal or PS-80 supplemented water (P>0.05; Table 4). The mean values for all three variables with no significant differences were greater in male chicks as compared to females. There was a significant difference on uric acid level in serum for interaction between PS-80 and gender. However, it does not seem that administration of PS-80 through drinking water affects male and female birds in contrary manners. The PS-80 supplanted water resulted in lowered serum uric acid in males and females by 0.74 and 0.71 mg/dl, respectively.

Table 3. Effects (Mean ± SE) of normal and PS-80 treated water on liver percentage and pH, blood serum level for glutamate pyruvate transaminase (SGPT), serum glutamate oxaloacetate transaminase (SGOT), serum alkaline phosphatase (SALP) in broiler chicks at 28 days of age.

Factor\ level		Liver (%)	Liver pH	SGPT (IU/L)	SGOT (IU/L)	ALP (IU/L)
Polysorbate-80 (ppm)						
0		2.42±0.13[a]	6.24±0.03[a]	36.67±11.24[a]	26.18±1.74[a]	45.87±6.29[b]
3500		2.44±0.12[a]	6.27±0.04[a]	29.27±6.57[a]	25.98±1.01[a]	62.33±6.98[a]
Sex						
Female		2.42±0.13[a]	6.24±0.03[a]	36.67±11.24[a]	26.18±1.74[a]	45.87±6.29[b]
Male		2.44±0.12[a]	6.27±0.04[a]	29.27±6.57[a]	25.98±1.01[a]	62.33±6.98[a]
Sex						
Female		2.45±0.12[a]	6.24±0.03[a]	32.31±7.35[a]	27.32±1.08[a]	45.62±5.07[b]
Male		2.42±0.12[a]	6.27±0.04[a]	33.64±11.38[a]	24.60±1.67[a]	66.78±8.03[a]
PS-80 × Sex						
0	Female	2.50±0.17[a]	6.23±0.03[a]	31.11±7.22[a]	27.41±1.44[a]	46.82±6.23[b]
0	Male	2.22±0.15[a]	6.27±0.05[a]	53.33±43.33[a]	22.47±5.73[a]	67.28±9.56[a]
3500	Female	2.35±0.16[a]	6.27±0.09[a]	35.00±18.48[a]	27.13±1.67[a]	53.11±8.43[a]
3500	Male	2.49±0.16[a]	6.28±0.05[a]	26.25±4.98[a]	25.40±1.28[a]	66.28±9.26[a]
SEM		0.08	0.03	6.41	0.98	4.99
				P > F		
PS-80		0.8889	0.5275	0.5774	0.9206	0.0984
Sex		0.8547	0.5524	0.9214	0.1859	0.0407
PS-80 × Sex		0.2909	1.0000	0.2787	0.3827	1.0000

[a-b] Means within a column for each main effect and interaction without a common superscript differ significantly (P<0.05).

DISCUSSION

The present study intended to evaluate the effect of high levels of polysorbate-80 (PS-80) on liver and kidney functions in broiler chicks when it may be consumed through oily medications or additives as in the case of phytogenic preparations. A noticeable indication of hepatic damage is the leaking of cellular enzymes into the plasma (Schmidt and Schmidt, 1983), due to instability reasoned in the transport functions on hepatocytes (Pratt and Kaplan, 2000). The estimation of serum marker enzymes was reported as a useful quantitative marker for determining the extent and type of hepatocellular damage in chicken as well as, many other species exposed to toxic substances through feed (Celyk et al., 2003), water (Wu, 1997) and air (Amakiri et al., 2008). When liver cell plasma is injured, a variety of enzymes situated in the cytosol is released into the circulation, thus, causing increased enzyme levels in the serum. Constant exposure of the chicks to 3500 ppm PS-80 through drinking water resulted in no increase in liver weight (hypertrophy), no change in liver pH and no elevation in GPT and GOT marker enzymes in serum. These results considered as pathological evidences to confirm the non toxicity of PS-80 with respect to liver

function. However, the increased ALP is interesting, but may be misleading, since it was minimal and not associated with an increase in any other liver-specific enzymes.

In avian species the kidney function is well evaluated based on concentration of uric acid, urea and creatinine (Selvaraj et al., 1998). The concentration of these three biochemical components in peripheral blood have shown to be reliable indicators for nitrogen metabolism and kidney function (Zantop, 1997). Huff et al. (1988) also confirmed that significant increase in serum uric acid and creatinine levels are indicative of nephrotoxicity in broiler chicken. Among them, uric acid is of prime importance. Broiler chickens like other birds are urecotelic and eliminates 60 to 80 percent of nitrogen in the form of uric acid. Uric acid serum levels changes with protein content in feed, quantity of feed ingested (Costa et al., 1993), and water consumption and kidney health for normal filtration rate, among many other factors. In our study, no noticeable changes were pointed out in concentration of uric acid, urea and creatinine in the serum of the treated birds. It reveals that PS-80 is safe for kidney function at the high dose of 3500 ppm.

The results evidently demonstrated that exposure of a

Table 4. Effects (Mean ± SE) of normal and PS-80 treated water on blood serum level (mg/dl) of uric acid, urea and creatinine in male (M) and female (F) broiler chicks at 28 days of age.

Factor\ level		Acid uric (mg/dl)	Urea (mg/dl)	Ceratinine (mg/dl)
PS-80 (ppm)				
0		5.11±0.67[a]	2.92±0.19[a]	0.28±0.02[a]
3500		4.82±0.70[a]	3.00±0.21[a]	0.28±0.01[a]
Sex				
Female		4.12±0.49[a]	2.92±0.18[a]	0.27±0.02[a]
Male		5.96±0.79[a]	3.00±0.23[a]	0.28±0.01[a]
PS-80 × Sex				
0	F	4.64±0.60[ab]	2.89±0.20[a]	0.27±0.02[a]
0	M	6.50±2.08[a]	3.00±0.58[a]	0.30±0.00[a]
3500	F	3.93±0.47[b]	3.00±0.41[a]	0.28±0.03[a]
3500	M	5.76±0.85[ab]	3.00±0.27[a]	0.28±0.02[a]
SEM		0.48	0.01	0.14
			_P > F_	
PS-80		0.7471	0.7849	0.9988
Sex		0.0521	0.8017	0.5887
PS-80× Sex		0.1881	0.9987	0.4886

[a-b] Means within a column for each main effect and interaction without a common superscript differ significantly ($P<0.05$).

juvenile chick to PS-80 supplemented water leads to no negative consequence with respect to liver as well as, kidney metabolic pathways. However, the observations on growth and feed utilization parameters are not in favor of inclusion PS-80 to drinking water for chicks at the level of 3500 ppm. Such adverse impacts might rise in commercial broiler practices whereas PS-80 contained fat soluble substances such as phytogenic additives are given to birds via water or feed as medication or growth promoters. These evidences give an idea about presumable occurrence of an antagonistic impact of PS-80 on digestion or absorption process of one or more nutrient in gastrointestinal tract of chicks. Interpretation of the data gathered in this study offer no evidences to sustain the idea mentioned. Evaluation of digestibility for different nutrients with the diets containing PS-80 may approve the idea. From the economic standpoint, these results are rather significant as the minor differences multiply by huge number of birds in a commercial chicken house.

In conclusion, results of the present study demonstrate that persistent exposure of broiler chicks to 3500 ppm PS-80 through drinking water suppress gain weight and feed conversion ratio in chicks during the early growth periods. However, no evidences were found that such negative impacts are arising from either liver or kidney dysfunction.

ACKNOWLEDGEMENT

The authors are thankful to Dr. Ali Salehnia, Director of Lorestan Medicinal Plants Laboratory, Khorraman, for financial assistance.

REFERENCES

Amakiri AO, Monsi A, Teme SC, Ede PN, Owen OJ, Ngodigha EM (2008). Air quality and micro-meterological monitoring of gaseous pollutants/flame emissions from burning crude petroleum in poultry house. Toxicological Environ. Chem., 91: 225 – 232.

Anonymous (1972). Determination of serum alkaline phosphatase. German Society for Clinical Chemistry. Z. Klin. Chem. Klin. Biochem., 10: 281.

Anonymous (1974). Assessment of serum alkaline phosphatase. Committee on Enzymes of the Scandinavian Society for Clinical Chemistry and Clinical Physiology. Scand. J. Clin. Lab. Invest., 33: 291.

Bermeyer HU, Horder M, Rej R (1986). IFCC method for aspartat aminotrasferase. International Federation of Clinical Chemistry (IFCC) Scientific Committee. Analytical section: approved recommendation on IFCC methods for the measurement of catalytic concentration of enzymes. Part 2. J. Clin. Chem. Clin. Biochem., 24: 497-510.

Celyk K, Denly M, Savas T (2003). Reduction of toxic effects of Aflatoxin B1 by using baker yeast (Saccharomyces cerevisiae) in growing broiler chicks dieta. R. Bras. Zootec., 32: 615-619.

Chou DK, Krishnamurthy R, Randolph TW, Carpenter JF, Manning MC (2005). Effects of Tween 20 and Tween 80 on the stability of Albutropin during agitation. J. Pharm Sci., 94: 1368-1381.

Costa ND, McDonald DE, Swan RA (1993). Age-related changes in plasma biochemical values of farmed emus (Dromaius novaehollandie). Australian Vet., J. 70: 341-344.

Elder RL, Ed (1984). Final report on the safty assessment of polysorbates 20, 21, 40, 60, 61, 65, 80, 81, and 85. Int. J. Toxicol., 3: 1-82.

Gad SC, Cassidy CD, Aubert N, Spainhour B, Robbe H (2006). Nonclinical vehicle use in studies by multiple routes in multiple species. Int. J. Toxicol., 25: 499-521.

Goff HD (1997). Colloidal aspects of ice cream, A review. Int. Dairy J., 7: 363-373.

Huff, WE, Kubena, LF, and Harvey, RB (1988). Progression of ochratoxicosis in broiler chickens. Poultry Science, 67(8), 1139-1146.

Jacques NA, Hardy L, Knox KW, Wicken AJ (1980). Effect of Tween 80 on the morphology and physiology of Lactobacillus salivarious.strain IV CL-37 growth in a chemostat under glucose limitation. J. Gen. Microbiol., 119: 195-201.

Krantz JC, Culver PJ, Carr CJ, Jones CM (1951). Sugar alcohols, XXVIII. Toxicologic, pharmacodynamics and clinical observations on Tween 80. Bull. Sch. Med. Univ. Md. 36:48-56.

Masini E, Planchenault J, Pezziardi F, Gautier P, Gagnol JP (1985). Histamine- releasing properties of polysorbate 80 in vitro and in vivo: Correlation with its hypotensive action in the dog. Agents Actions,16: 470-477.

Oser BL, Oser M (1956). Nutritional studies on rats of diets containing high levels of partial ester emulsifiers. II. Reproduction and lactation. J. Nutr., 60: 489-505.

Path GJ, Dai XZ, Schwartz JS (1991). Effects of Amiodarone with and without polysorbate-80 on myocardial oxygen consumption and coronary blood flow during treadmill exercise in the dog. J. Cardiovasc. Pharmacol., 18: 11-16.

Pratt, DS, Kaplan MM (2000). Evaluation of abnormal liver-enzyme results in asymptomatic patients. N. Engl. J. Med. 4:1266–1271.

Robert CL (2010). Translocation of Crohn's disease Esherichia coli across M-cells: contrasting effects of soluble plant fibers and emulsifiers. Gut, 59: 1331-1339.

Rowe RC (2009). In Hanbook of pharmaceutical Excipiants, 6th ed., pharmaceutical Press, Washington, DC.

SAS Institute (2001). SAS/STAT® Guide for personal computers. Version 9.1 Edition. SAS Institute, Inc., Cary, NC.

Schmidt E and Schmidt FW (1983). *Glutamate* dehydrogenase, In: Methods of Enzymatic Analysis, 3th Edn. By U. Bergmeyer (Ed), Academic Press, New York, pp. 216-217.

Selvaraj P Thangavel A, Nanjappan K (1998). Plasma biochemical profile of broiler chickens. Indian Vet., J. 75: 1026-1027.

Steele RH, Limaye S, Cleland B, Chow J, Suranyi MG (2005). Hypersensitivity reactions to the Polysorbate contained in recombinant erythropoietin and darbepoietin. Nephrology, 10: 317-320.

Thackaberry EA, Kepytek S, Sherrett P,Trouba K, McIntryre B (2010). Comprehensive investigation of hydroxypropyl methylcellulose, propylene glycol, Polysorbate 80 and hydroxypropyl-beta-cyclodextrin for use in general toxicology studies. Toxicol. Sci., 117: 485-492.

Wu W (1997). Counteracting Fusarium proliferatum toxicity in broiler chicks by supplementing drinking water with Poultry Aid Plus. Poult. Sci., 76: 463-468.

Zantop DW (1997). Biochemistries. Pages115–129 in Avian Medicine: Principles and Applications. B. W. Ritchie, G. J. Harrison, and L. R. Harrison, ed. Wingers Publishing Inc., Lake Worth, FL.

25

Production, purification and characterization of a thermostable alkaline serine protease from *Bacillus lichniformis NMS-1*

C. D. Mathew and R. M. S. Gunathilaka

Department of Biochemistry and Molecular Biology, Faculty of Medicine, University of Colombo, Sri Lanka.

Alkaline proteases are used in food industry, leather tanning and processing industry, preparation of pharmaceuticals and also in the fiber industry. An alkaline serine protease producing strain was isolated using soil sample from a natural hot water spring in Sri Lanka. It was identified based on morphological, biochemical and 16s rRNA identifications as *Bacillus licheniformis* NMS-1. The extacellular protease enzyme was purified by two steps procedure involving ammonium sulfate precipitation followed by DEAE-Sephadex A-25 gel chromatography. The purification gave a 56 fold increase of the specific activity with a yield of 16%. The optimal pH and optimal temperature of the protease were pH 9 and 60°C, respectively. The protease was relatively stable between 20– 80°C. The enzyme was stable within the pH values of 8 – 12. The K_m and V_{max} values calculated from Lineweaver – Burk plot were 2.7×10^{-3} mg/ml and 263 mU/mg. Among the protease inhibitors that were tested, PMSF completely inhibited the enzyme activity indicating that the protease is a serine protease. The enzyme retained more than 50% of its activity after 60 min incubation at 60°C. The major protease types used commercially are heat stable alkaline proteases. Alkaline serine proteases are enzymes that cleave peptide bonds in protein in which serine serves as the nucleophilic amino acid at the enzyme active site. Properties of this protease have shown it's suitability for industrial applications such as detergent industry.

Key words: Alkaline protease, purification, characterization, *Bacillus licheniformis* NMS-1, thermophilic, serine protease.

INTRODUCTION

Proteases count for nearly 65% of the world enzyme market (Rao et al., 1998). Commercial proteases are mostly produced from the bacteria and it is reported that 35% of total microbial enzymes used for detergent industry are the proteases from bacterial sources (Ferrero et al., 1996). Proteases are commonly classified according to their optimum pH as acidic protease, neutral protease and alkaline protease. There have been extensive researches on functions of acidic and alkaline proteases. Bacterial alkaline proteases generally have an optimum pH of 10 and optimum temperature at 60°C. The genus *Bacillus* has been studied in considerable depth and the ability of *Bacillus* strains to produce and secrete large quantities (20-25 g/l) of extracellular enzyme has placed them among the most importance industrial enzyme producers (Schallmey et al., 2004). These properties make them suitable for use in the enzyme industry. Alkaline proteases are used in food industry, leather tanning and processing industry, preparation of pharmaceuticals and also in the fiber industry (Van-kessel et al., 1991; Ming chu et al., 1992). Therefore research is done on large scale production of alkaline proteases. Thermophilic bacteria from hot water springs produce thermostable enzymes. Proteases produced from thermophilic bacteria are used in a range of commercial applications (Sonnleitner and Fiechter, 1983; Rehman et al., 1994; Rao et al., 1998; Adams and Kelly, 1998; Zeikus et al., 1998; Singh et al., 2001). Enzymes isolated from these organisms are not only thermostable and active at high temperature but are also often resistant to and active in presence of organic solvents and detergents. Thermophiles such as *Bacillus licheniformis* (Ferrero et al., 1996) and *Bacillus clausii* (Kumar et al., 2004), have been studied for their ability in producing thermostable proteases. In recent years, there has been a great amount of research and development effort focusing with the aim of obtaining high yields of alkaline protease in the fermentation medium (Varela et al., 1996; De Coninck et al., 2000; Puri et al., 2002). Serine proteases are characterized by the presence of a serine group in their active site, found in virus, bacteria, and eukaryotes; this class comprises two distinct families, the chymotrypsin family which include the mammalian enzyme such as chymotrypsin, trypsin, elastase, and the sublilisin family include the bacterial enzymes such as subtilisin although the three dimensional structure is different in the two families, they possess similar active site geometry and catalytic mechanism. Serine proteases exhibit different substrate specificities. Serine alkaline proteases are produced by several bacteria, molds, yeast, and fungi; optimum pH of serine alkaline protease is around 10 (Ibrahim et al., 2011; Jayakumar et al., 2012). They are inhibited by diisopropyl-fluorophosphate (DFP) or a potato protease inhibitor but not by tosyl L-lysine chloromethyl ketone (TLCK). Their molecular masses are in the range of 15 to 30 kDa (Rao et al., 1998).

In this paper we report the purification and characterization of a serine protease from *B. licheniformis* NMS 1.

MATERIALS AND METHODS

Isolation of bacterial strain using enrichment media

B. licheniformis strain NMS 1 was isolated from a soil sample collected from Nelumwewa hot water spring, in Polonnaruwa, Sri Lanka. The bacterial strain was grown in an enrichment medium containing 0.5% (w/v) yeast extract, 1.0% (w/v) peptone, 0.5 g/l glucose, 0.4 g/l Na_2HPO_4, 0.085 g/l Na_2CO_3, 0.0 2g/l $ZnSO_4$, 0.02 g/l $MgSO_4$, 0.02 g/l $CaCl_2$ and pH 7.2. Cultures were incubated in conical flask at 37, 55 and 60°C for 48 h in an orbital shaker at 150 rpm and then centrifuged at 15000 g for 30 min at of 4°C (Kubota, 6500). To determine the growth of the microorganism, absorbance was measured at 600 nm. Streak plate isolation procedure and the dilution plate technique were used in isolation of bacterial colonies. The supernatant was assayed for proteolytic activity. Screening test was performed for all bacterial strains isolated using a plate technique to detect the production of protease using skimmed milk as the substrate.

Study of the colony morphology of bacterial species

The bacterial colonies isolated were streaked on MacConky and Nutrient agar. Plates and slants of MacConky agar and nutrient agar then were incubated for 24 hat 37°C. Morphological characteristics of colonies appeared on MacConky agar and nutrient agar of the bacteria were studied. Isolated bacterial species NMS1 was the only bacterial species which produced clear zone in skim milk media at 50°C. It was selected for further studies. This bacterial species was identified by morphological characteristics, biochemical test and 16s r RNA analysis.

PCR amplification and sequencing of 16S rRNA identification

Total DNA was isolated from small colony of the NMS1 bacterial sample and 5 µl of extracted DNA was subjected to the polymerase chain reaction (PCR) using 27F/800R and 518F/1492R. Amplified DNA was subjected to DNA sequencing using 518F and 800R primers and the obtained DNA sequences were compared with already existing DNA sequences in Gene bank. Bacterial universal 16s rRNA primers were 27_forward 5'-AGAGTTTGATCCTGGCTCAG-3' 1492_reverse 5'-TACGGTTACCTTGTTACGACTT-3' PCR products were sequenced in Genetech laboratory, Sri Lanka, forward and reverse primers were aligned using ABI Auto assembler software, and the over lapping consensus sequence was compared with sequence in the NCBI data base using FASTA3 sequence homology searches.

Selection of the optimum temperature for protease production

To determine optimum temperature for protease production, enrichment culture medium was used with casein in place of peptone. 10 ml of initial cultures were inoculated into 150 ml of culture media at different temperatures separately. The cultures were incubated at 37, 50, 55, and 60°C in an orbital shaker at 150 rpm for 48 h. Optical densities at 600 nm, pH, protease activity, and cell number were estimated in culture media at different temperatures at 4 hour intervals.

Purification

Purification was done at 4°C. The bacterial strain NMS1 grown in a fresh 300 ml culture medium for 32 h at 50°C was harvested by centrifugation at 15000 g for 20 min. The supernatant was precipitated in ammonium sulphate solution of 90% saturation. Saturated solution was centrifuged at 15'000 g for 30 min. The pellet obtained was dissolved in 50 mM phosphate buffer at 7.5. The solution was dialysed in the same buffer for 16 h before applying to a DEAE-Sephadex A-25 column (2.6×30 cm). The protease was eluted with a 1:1 gradient of NaCl (0.1 to 0.7 N NaCl in 50 mM pH 7.5 phosphate buffer) at a flow rate of 0.8 ml/min. The fractions were assayed for protease activity. The protein content was determined using Lowry's method (Lowry et al., 1951) using bovine serum albumin as the standard.

Protease assay

The protease activity was determined using the method described by Cupp – Enyard (2008) with slight modifications. The enzyme activity was calculated using a tyrosine standard graph. Activity is defined as the amount of enzyme that hydrolyzed one micromole (µmol) of casein per minute.1 ml of each culture supernatant was mixed with equal amount of Tris-HCl buffer and was mixed thoroughly.1 ml of this mixture was incubated with 5 ml of 0.65% casein for 10 min at 37°C. The reaction was terminated by addition of 5 ml of 110 mM TCA to each tube. The blanks were prepared by incubating the buffer without supernatant. After 30 minutes of incubation each test solution was centrifuged and filtered. Then 2 ml of each filtrate and blank was added to separate vials and 5 ml of 0. 5 M Na_2CO_3 added to each tube and immediately added 1 ml Folin- Ciocalteus reagent was added. After incubating at 37°C for 30 min, absorbance was measured at 660 nm.

Polyacrylamide gel electrophoresis

Polyacrylamide gel electrophoresis (PAGE) was carried out according to the method described by Davis (1964) using Shandon apparatus.

Characterization of protease

All assays were done in triplicate.

Effect of temperature on protease activity

To determine the optimum temperature, protease activity was measured using 5 ml of 0.65% casein solution in 50 mM pH 7.2 phosphate buffer solutions incubated at 10 to 60°C, respectively. 1 ml of partially purified enzyme was added to each tube and assayed for protease activity

Stability of enzyme at different temperature

5 ml of the purified protease was incubated at temperatures ranging from 10 to 98°C for 1 h. 1 ml of each enzyme solution was added to tubes containing 5 ml of 0.65% casein in 50 mM phosphate buffer (pH 7.5). The protease assay was done under standard assay conditions.

Effect of pH on protease activity

Protease activity was measured at 29°C in the following buffer systems: HCl-KCl buffer solution of pH 2, citrate buffer solution of

pH 3, 4, 5, 6 and phosphate buffer of pH-6.5, 7, 7.5, 8, 9 bicarbonate buffer of pH 10, 11, 12 to measure the optimum pH. The protease assay was performed as described under standard assay conditions.

Stability of enzyme at different pH

To test the pH stability of protease HCl-KCl buffer solution of pH- 2, citrate buffer solution of pH -3, 4, 5, 6 and phosphate buffer of pH-6.5, 7, 7.5, 8, 9 bicarbonate buffer of pH 10, 11 were prepared. 5 ml of each buffer solution and 1 ml of partially purified enzyme were incubated at 29°C over a period of 1 h. The protease assay was performed as described under standard assay conditions.

Effect of substrate concentration on enzyme activity

The protease assay was performed varying the substrate concentration and the K_m and V_{max} values were calculated using Lineweaver-Burk plot.

Determination of protease type

To determine the type of protease, the purified protease was pre-treated for 30 min at 37°C and 5 mM PMSF residual enzyme activity was measured under standard assay conditions.

RESULTS AND DISCUSSION

Three bacterial strains were isolated based on their morphology. The extracellular protease producing ability was screened by streaking bacterial strains on skimmed milk agar and incubation for 24 h at 55°C. It has been reported that *B. licheniformis* produces very narrow zone of hydrolysis on casein agar despite being a very good producer of proteases in sub merged cultures (Mao et al., 1992) but *B. licheniformis* NMS 1 produces a very prominent and distinct zone of hydrolysis. *B. licheniformis* NMS 1 grows in the temperature range of 30 to 60°C. After 28 h incubation at 50°C *B. licheniformis* NMS 1 strain produced a maximum protease activity of 200 mU/ml but at lower temperatures like 30°C a lower production was observed. These results are compatible with findings of Atalo and Gashe (1993) and Johnvesly et al. (2002).

Razik et al. (1994) isolated a *Bacillus stearothermophilus* AP-4 with a optimum protease production at 55°C. Mabrouk et al. (1999) found optimum temperature as 37°C for *B. lichniformis* 21415 after 5 days of incubation. Rehman et al. (2005) also observed optimum temperature for enzyme production in *P. aeruginosa* strain K as 37°C. NMS1 Colonies were convex, large, and sticky, off white/yellow in colour and had rough edges, and were extremely mucoid and had a characteristic smell. This bacterial strain was used for the further studies. The bacterium was Gram positive, rod shaped, with a bulge

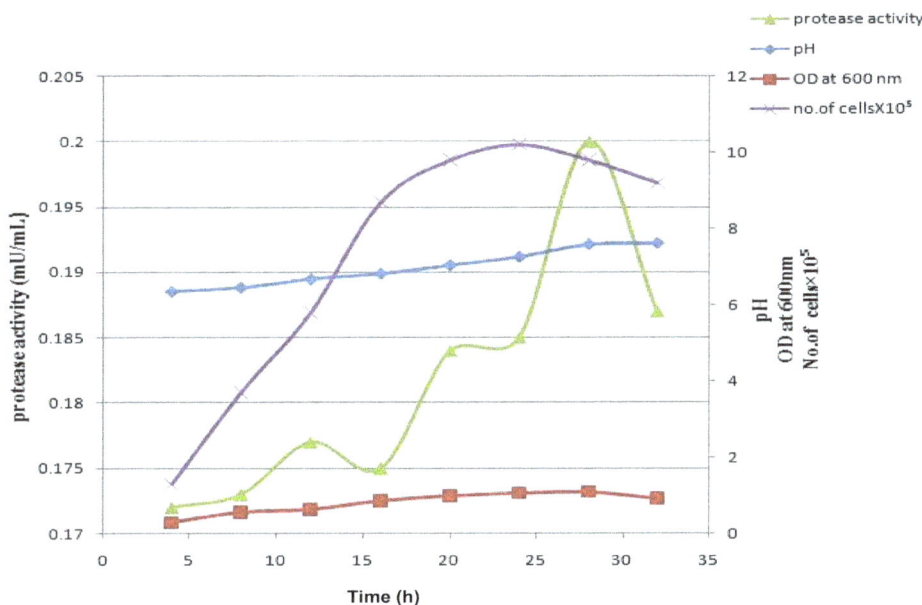

Figure 1. Extracellular protease production by *Bacillus licheniformis* NMS-1 grown at 50°C.

Table 1. Purification of extracellular proteases from *Bacillus licheniformis* NMS1

Purification stage	Total volume (ml)	Total activity (mU)	Total protein (mg)	Specific activity (mU/mg)	Recovery (%)	Purification fold
Initial crude Supernatant	500	45110	829	54	100	1
0-90% ammonium sulphate fractionated pellet	100	22240	49	453	49	8
DEAE ion exchange chromatography	30	7380	2.4	3049	16	56

central spore. These features were observed by light microscopy. The bacteria were highly motile as observed by the hanging drop method. The isolate could ferment various types of sugars as shown in biochemical test results. At the late stationary phase the organism produced spores abundantly and this showed a relationship with protease production. This isolated organism grows at temperatures ranging 30 to 60°C and pH of 4 to 11 and up to 10% of NaCl.

PCR amplification and sequencing of 16S rRNA identification

99% homology sequence was shown with *B. licheniformis*. NCBI BLAST result shows a significant relation with isolates NMS1 with other *Bacillus* in NCBI with a 99% identity and query recovery. Based on biochemical,

morphological tests and 16S Rrna, identification of isolated bacteria was found to be *B. licheniformis*- NMS1.

Optimization of the temperature for production of extra cellular protease

The maximum protease production was observed at 50°C as 200 mU/ml. After 48 h of incubation at 50°C (Figure 1).

Purification of the protease enzyme

The purification of enzyme by ammonium sulphate precipitation and ion exchange chromatography gave a specific activity increase of 54 mU/mg (Table 1). From DEAE sephadex A-25 gel chromatography, two enzymes peak were obtained indicating the presence of isoenzymes (Figure 2). Only one peak was pooled and

Ion exchange chromatography

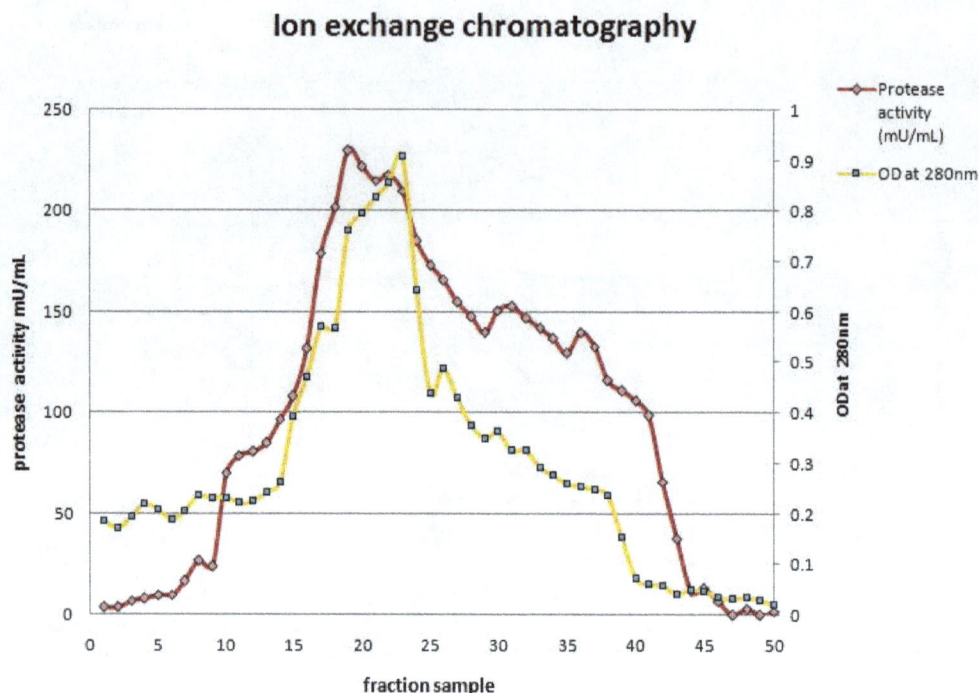

Figure 2. Ion exchange chromatography on DEAE sephadex A-25 of the crude extract of protease from *Bacillus licheniformis* NMS 1.

used for further studies. The number of folds purification was 56 folds with yield of 16%. The specific activity was 3049 mU/mg. Two protein bands were observed in polyacrylamide gel electrophoresis indicating that the enzyme has been partially purified. Rai and Mukharjee (2009) reported that a bacterium may produce arrays of extracellular protease isoenzymes for its survival and growth in particular habitat. Only a limited number of studies have been done on isoenzymes of alkaline proteases produced by Bacillus genus. Among these studies, Mala and Srividya (2010) reported isolating two different isoenzymes with molecular weights of 66 and 18 KDa from a Bacillus species. Tekin et al. (2012) reported similar results, isolating alkaline proteases from *Bacillus cohnii* APT5.

Characterization

Effect of temperature on enzyme activity and stability

Enzyme activity increases from 20 to 60°C (Figure 3). The results show that it has stability between 20 to 80°C after 1 hof incubation (Figure 4). Therefore this enzyme can be classified as a thermostable enzyme. Gupta et al. (2008) found optimum temperature for protease from *Virgibacillus pantpthenticus* as 50°C with a retained

activity of 82%, when incubated at this temperature for 1 h. Comparison of these results of El-Hawary and Ibrahim (1992) and Nilegaokare et al. (2002) concluded that optimum temperature of proteolytic activity frequently exceed the optimum temperature for enzyme production. It was also suggested that the stability of protease enzyme could be due to their genetic adaptability to carry out their biological activities at a higher temperature (Gaure et al., 1989; Whittle and Bloomfield, 1999; Kanekar et al., 2002)

Effect of pH on enzyme activity and stability

The maximum activity was observed at pH 9 (Figure 5) and enzyme was stable in the pH range 4-12 (Figure 6). These types of protease are very unique and are widely used in detergent industry. Prescott et al. (1995) observed Wai 21a protease stable over the pH range 2-5. Kumar (2005) observed that *Bacillus pumilus* protease had pH optimum for hydrolysis of casein from 10.5-12 with optimum activity at 11.5. He reported that the enzyme was stable over a broad pH range of 6-12 for 4 hand a pH range of 6-11 for 24 h at 30°C. Related results were also given by Johnvesly and Naik (2001) who reported a pH optimum of 11 for protease enzyme from *Bacillus* species JB-99 and a high activity in the range of pH 8-12.

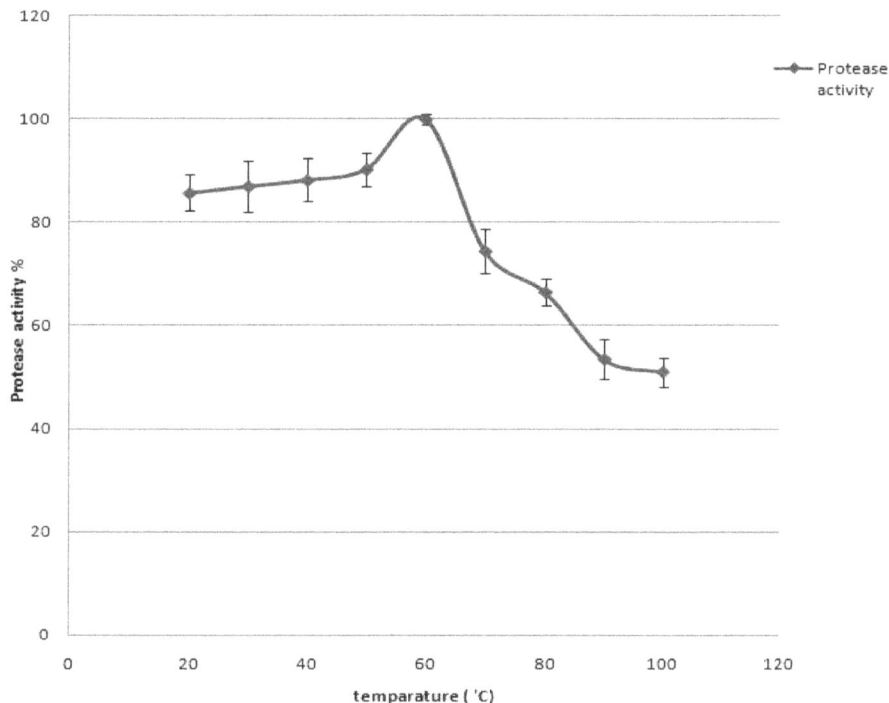

Figure 3. Effect of temperature on activity of purified extracellular thermostable protease.

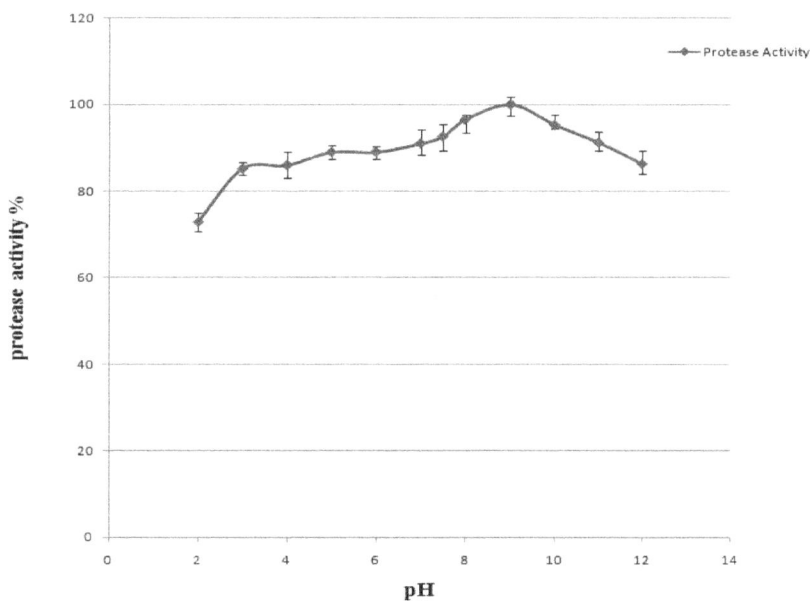

Figure 4. Effect of pH on activity of purified extracellular thermostable protease.

Calculation of K_m and V_{max}

The results show K_m value as 2.7×10^{-3} mg/ml and V_{max} as 263 mU/mg (Figure 7). K_m value of *B. licheniformis* NMS

1 protease was lower than those of Bacillus species TKU004 metallo-protease (2.98 mg/ml) (Wang et al., 2006) and TKU2007 protease (0.13 mg/ml) (Wang and Yah, 2006), *P aeruginosa* Psa A protease (2.69 mg/ml)

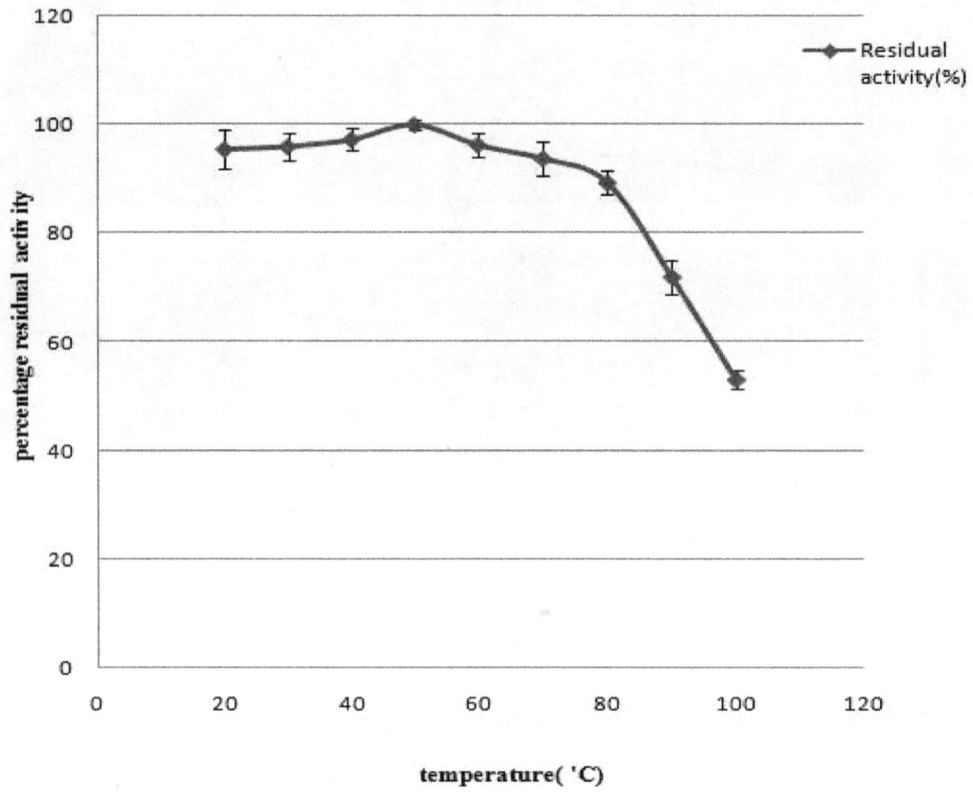

Figure 5. Effect of temperature on the stability of purified extracellular thermostable enzyme.

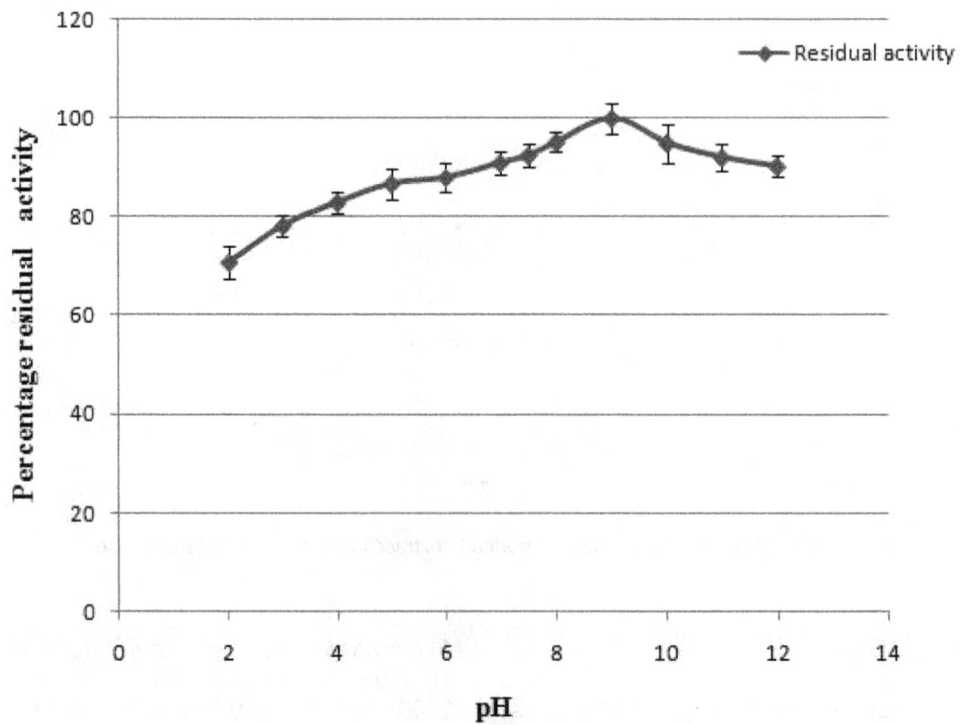

Figure 6. Effect of pH on the stability of purified extracellular thermostable enzyme.

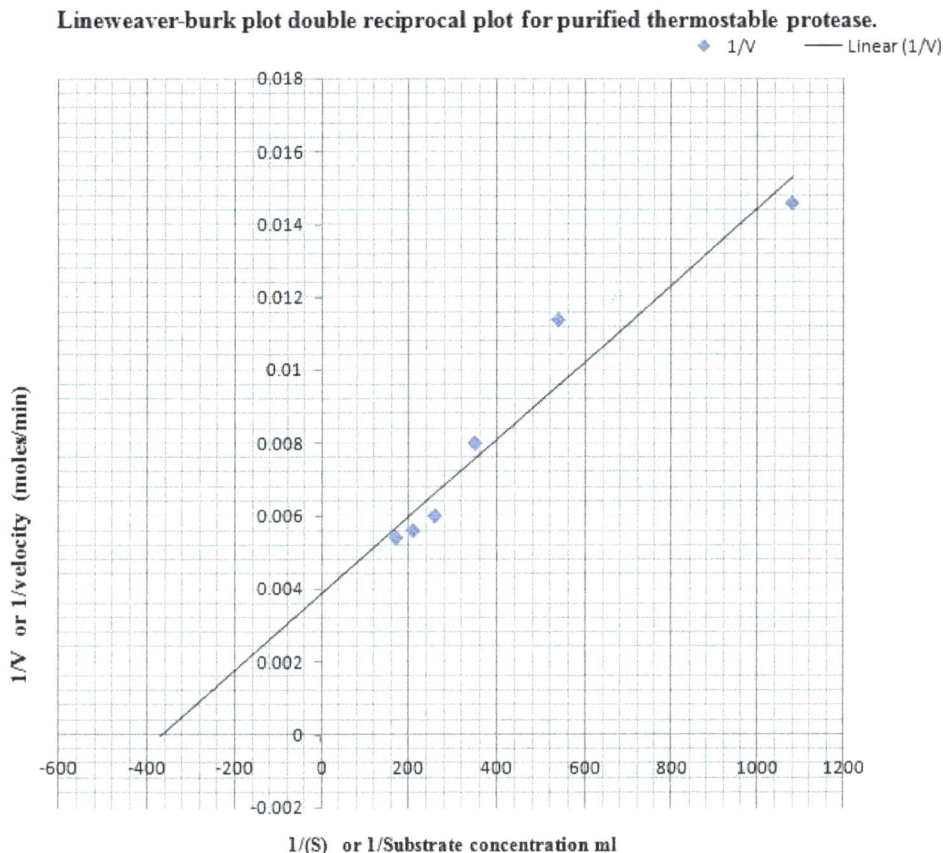

Figure 7. Lineweaver- Burk double reciprocal plot for purified protease enzyme.

(Gupta et al., 2005a). The V_{max} of *B. licheniformis* NMS 1 was higher than that of TKU007 (0.86 U/ml) (Wang and Yeh, 2006), 0.14 U/ml of *Bacillus* species TKU004 metallo-protease (Wang et al., 2006) 1.26 U/ml of *Bacillus cereus* (Sierecka, 1998) and 3.03 U/ml of *P. aeruginosa* Psa A (Gupta et al., 2005a).

Determination of protease type

Inhibition by phenylmethylsulfonyl fluoride (PMSF) indicated NMS 1 protease to be a serine protease. Such reports have been shown earlier for bacillus species (Dhandapani, 1994) PMSF causes Sulphonation of the serine residues residing in the active site of the protease and has been reported to result in the complete loss of enzyme activity (Beynon and Bond, 2001).

Conclusion

Optimum temperature for production of *B. licheniformis* NMS 1 extracellular protease was at 50°C. The enzyme was stable at 20 to 60°C with maximum protease activity at 60°C. The enzyme showed a high activity between pH values 8 - 12 with optimum pH at 9. Enzyme was stable between pH values 4 - 12. K_m and V_{max} values were 2.7×10^{-3} and 263 mU/mg. Characteristics of enzymes show that it is an alkaline serine protease suitable for use in the detergent industry.

Conflict of interests

The authors did not declare any conflict of interest.

REFERENCES

Adams MWW, Kelly RM (1998). Finding and using thermophilic enzymes. Trends Biotechnol. 16:329-332.

Atalo K, Gashe BA (1993). Protease production by a thermophilic *Bacillus* species(P-001A) which degrades various kinds of fibrous proteins. Biotechnol. Lett. pp.1151-1156.

Beynon R, Bond JB (2001). Proteolytic Enzymes, A Practical Approach, IRL Press, New York. pp. 113.

Cupp-Enyard C (2008). Sigma's Non-specific Protease Activity Assay - Casein as a Substrate. J. Vis. Exp. 19:899.

Davis M (1998). Making a living at the extremes. Trends Biotechnol. 16:102-104.

Dhandapani R, Vijayaragavan R (1994). Production of thermophilic extracellular alkaline protease by *Bacillus stearothermophilus* Ap-4. World J. Microbiol. Biotechnol. 10:33-35.

El-Hawary FI, Ibrahim II (1992). Comparative study on protease of three thermophilic *Bacilli*. Zagazig J. Agric. Res. 19:777-787.

Gaure R, Yadav J, Panday L (1989). Thermostability of Extracellular protease enzyme produced by *Spicaria fusispora*. Hindustan Antibiot. Bull. 31:36-37.

Gupta A, Joseph B, Mani A, Thomas G (2008). Biosynthesis and properties of an extracellular thermostable serine alkaline protease from *Virgibacillus pantothenticus*. World J. Microbiol. Biotechnol. 24:237-243.

Gupta A, Roy I, Khare SK, Gupta MN (2005a). Purification and characterization of a solvent stable protease from *Psedomoas aeruginosa* PseA. J. Chromatogr. 1069:155-161.

Ibrahim KS, Muniyandi J, Karutha Pandian S (2011). Purification and characterization of manganese-dependent alkaline serine protease from Bacillus pumilus TMS55. J. Microbiol. Biotechnol. 21(1):20-27.

Jayakumar R, Jayashree S, Annapurna B, Seshadri S (2012). Characterization of thermostable serine alkaline protease from an alkaliphilic strain Bacillus pumilus MCAS8 and its applications. Appl. Biochem. Biotechnol. 168(7):1849-66.

Joo HS , Choi JW (2012) Purification and characterization of a novel alkaline protease from Bacillus horikoshii. J. Microbiol. Biotechnol. 22(1):58-68.

Johnvesly B, Manjunath BR, Naik GR (2002). Pigeon pea waste as a novel, inexpensive, substrate for production of a thermostable alkaline protease from thermophilic *Bacillus sp*. JB-99, Bioresour. Technol. 82:61-64.

Johnvesly B, Niak GR (2001). Studies on production of thermostable alkaline protease from thermophilic and alkalophilic *Bacillus* sp.JB-99 in a chemically defined medium. Proc. Biochem. 57:139-144.

Kanekar P, Nilegaonkar S, Sarnaik S, Kelkar A (2002). Optimization of protease activity of alkaliphilic bacteria isolated from an alkaline lake in india. Bioresour. Technol. 85:87-93.

Kumar CG, Joo HS, Koo YM, Paik SR, Chang CS (2004). Thermostable alkaline protease from a novel marine holo alkalophilic *Bacillus clausii* isolate. World J. Microbio. Biotech. 20:351-357.

Lowry OH, Rosebrough NJ, Farr AL, Randall RJ (1951). Protein measurement with the Folin-Phenol reagents. J. Biol. Chem. 193: 265-275.

Mala M, Srividya S (2010). Partial purification and properties of a laundry detergent compatible alkaline protease from a newly isolated *Bacillus* species Y. Indian J. Microbiol. 50:309-317.

Mabrouk SS, Hashem AM ,El-Shayeb NMA, Ismail AMS and Fattah AFA (1999). optimization of alkaline protease productivity by *Bacillus licheniformis* ATCC 21415. Bioresour. Technol. 69:155-159.

Mao W, Pan R, Freedman D (1992). High production of alkaline protease by *Bacillus licheniformis* in a fed batch fermentation using synthetic medium. J. Ind. Microbiol. 11(1):1-6.

Ming CI, Lee C, Tsu-Shun L (1992). Production of alkaline protease in batch cultures of *Bacillus subtilis* ATCC 14416. J. Enzym. Microbiol. Technol. 14: 755-761.

Nilegaokare S, Kanekar P, Sarnaik S, Kelkay A (2002). production Isolation and characterization of extracellular protease of an alkaliphilic strain of *Arthrobacter ramosus*,MCM B-351 isolated from the alkaline lake of loner India. World J. Microbiol. Biotechnol.18:785-789.

Prescott M, Peek K, Daniel RM (1995). Characterization of thermostable pepstatin insensitive acid protease from a *Bacillus* species. Int. J. Biochem. Cell. Biol. 27(7):724-739.

Rai SK, Mukherjee AK (2009). Ecological significance and some biotechnological application of an organic solvent stable alkaline serine protease from *Bacillus subtilis* strain DM-04. Bioresour. Technol. 100:2642-2645.

Rao MB, Tanksale AM, Ghatge MS, Deshpande VV (1998). Molecular and biotechnological aspect of Microbial proteases. Microbiol. Mol. Biol. Rev. 62(3):597-635.

Razak NA, Samad MYA, Basri M, Yunus WMZW, Ampon K, Salleh AB (1994). Thermostable extracellular protease of *Bacillus stearothermophilus* factors affecting its production. World J. Microbiol. Biotechnol. 10:260-263.

Rehman RNZA, Razak CN, Ampon K, Basri M, Yunus WMZW, Salleh AB (1994). Purification and characterization of a heatstable alkaline protease from *Bacillus stearothermophilus* F1. Appl. Microbiol. Biotechnol. 40:822-827.

Sierecka JK (1998). Purification and partial characterization of a neutral protease from a virulent strain of *Bacillus cereus*. Int. J. Biochem. Cell Biol. 30:579-595.

Singh J, Vohra RM, Sahoo DK (2001). Purification and characterization of two extracellular alkaline proteases from a newly isolated obligate alkalophilic *Bacillus sphaericus*. J. Ind. Microbiol. Biotechnol. 26:387-393

Sonnleitner B, Fiechter A (1983). Advantages of using thermophiles in biotechnological processes: expectations and reality. Trends Biotechnol. 1:74-80.

Tekin N, Cihan AÇ, Takaç ZS, Tüzün CY, Tunç K, Çökmüş C (2012). Alkaline protease production of Bacillus cohnii APT5Turkish J. Bio. 36(4):430-440.

Van Kessel KPM, Van Strijp JAG, Verhoeff J (1991). Inactivation of recombinant human tumor necrosis factor-α by proteolytic enzymes released from stimulated human neutrophils. J. Immunol. 147:3862-3868.

Wang SL, Kao TY, Wang CL, Yen YH, Chern MK, Chen YH (2006). A solvent stable metallo protease produced by *Bacillus sp*. TKU004 and its application in the deproteinization of squid pen for β-chitin preparation. Enzym. Micro. Technol. 39:724-731.

Wang SL, Yeh PY (2006). Production of a surfactant-and solvent stable alkaliphilic protease by bioconversion of shrimp shell wastes fermented by *Bacillus subtilis* TKU007. Process Biochem. 41:1545-1552.

Whittle G, Bloomfield GA, McDonagh MB, Katz ME, Cheetham BF (1997). Analysis of sequences flanking the vap regions of *Dichelobacter nodosus*: evidence for multiple integration events, a killer system, and a new genetic element. Microbiol. 143(2):553-562.

Zeikus JG, Vieille C, Savchenco A (1998). Thermozymes: Biotechnology and structure function relationship. Extremophiles, 2:179-183.

26

Lead toxicity, oxidative damage and health implications. A review

Sadhana Sharma, Veena Sharma[*], Ritu Paliwal, Pracheta

Department of Bioscience and Biotechnology, Banasthali University, Banasthali- 304022, Rajasthan, India.

The toxicity of Lead was recognized centuries ago, and it continued to pose serious threat to the health of children as well as adults. This review presents an overview of the current knowledge of toxic effects of Lead induced oxidative damage and also suggests some possible measures which could reduce the toxic effects of the metal. This paper examines the effects of Lead in blood, soft tissues, haematopoietic system and the antioxidant defense system. On the other hand, data also indicated that lead is an essential element at low dietary intakes. Its deficiency was shown to depress growth, disturb iron metabolism, alter activities of some enzymes and disturb the metabolism of cholesterol, phospholipids and bile acids. It was found that lead toxicity is significant but a preventable health problem. Furthermore, work is needed to find the effective and safe intervention for lowering the lead exposure at the general population level.

Key words: Lead toxicity, oxidative damage, haematological, antioxidant.

INTRODUCTION

Lead is ubiquitous, and the most common environmental pollutant naturally present in the earth's crust in small concentrations, for centuries it has been mined and disseminated throughout the environment from where it has gradually become incorporated into the structural tissue of plants, animals and humans (Pracheta et al., 2009). However, both occupational and environmental exposure has made lead a serious problem in many developing and industrializing countries (Yucebilgic et al., 2003). It has many undesired effects, including neurological (Senapati et al., 2001; Soltaninejad et al., 2003; Bellinger, 2008; Sharma et al., 2011), behavioural

(Moreira et al., 2001; De Marca, 2005; Adeniyi et al., 2008), immunological (Razani-Boroujerdi et al., 1999; Bunn et al., 2001; Rosenberg et al., 2007), renal (Lockitch, 1993; Vargas et al., 2003; Rastogi, 2008; Sharma et al., 2011c), hepatic (Lockitch, 1993; Patra et al., 2001; Sharma et al., 2011b), cardiovascular system and haematological dysfunctions (Mousa et al., 2002; Adeniyi et al., 2008). Lead pollution can also cause irreversible encephalopathy, seizure, coma and even death. Fatigue, memory loss, high blood pressure, nephropathy, gastrointestinal disturbances, weight loss and immuno-suppression are other common toxic effects of lead exposure in animals. Prenatal exposure to metal may also cause birth defects, miscarriage and underdeveloped babies (Ehle and Mckee, 1990; Pracheta et al., 2009).

*Corresponding author. E-mail: drvshs@gmail.com.

Abbreviations: **ACP**, Acid phosphatase; **ALA**, aminolevulinic acid; **ALAD**, δ-aminolevulinic acid dehydratase; **ALP**, alkaline phosphatase; **ALT**, alanine transaminase; **AST**, aspartate transaminase; **CAT**, catalase; **GPX**, glutathione peroxidise; **GSH**, reduced glutathione; **GST**, glutathione-S-transferase; **HDL**, high density lipoprotein; **LDL**, low density lipoprotein; **LPO**, lipid peroxidation; **ROS**, reactive oxygen species; **SOD**, superoxide dismutase.

EFFECTS OF LEAD AND ITS POTENTIAL HEALTH EFFECTS

Lead is a poison that affects virtually every system in the body. Children are more vulnerable to lead exposure than adults because of the frequency of pica, hand-to-mouth activity, and a higher rate of intestinal absorption and retention. The most deleterious effects of lead are on erythropoiesis, soft tissues, kidney function, and the central nervous system (ATSDR, 1993).

Effect of Lead in soft tissue

Levels of lead in soft tissue appear to be relatively constant during life, despite a fairly high turnover rate (Barry, 1975). Lead is stored in almost all soft tissues (Doyle and Younger, 1984); autopsy studies show that liver is the largest repository of soft tissue lead (33%), followed by kidney cortex and medulla, pancreas, ovary, spleen, prostate, adrenal gland, brain, fat, testis, heart, and skeletal muscle (Senapati et al., 2001; Adeniyi et al., 2008; Sharma et al., 2011b, c; Sharma et al., 2011). Dose related accumulation of most lead in heart and kidney in new born rat pups was reported by Singh et al. (1976). It is observed that chronic oral administration of low doses of lead results in accumulation particularly in bone, kidney and skeletal muscle in most animal species (NRC, 1972). Lead of 1700µg/dl after 35 days of exposure resulted in a testicular Pb concentration of 0.8 µg/dl in Sprague- Dawley rats. Thoreux- Manlay et al. (1995), reached testicular lead concentrations of 2.0, 1.6, 2.6, and 4.3 µg/dl with Pb B of 56, 91, 196, and 332 µg/dl respectively after 30 days of exposure. The liver contains numerous proteins to which Pb may bind. One of these proteins is metallothionein (Hamer, 1986), which has a high affinity for lead in vitro (Waalkes et al., 1984), although lead can induce metallothionein production in liver (Waalkes and Klaasen, 1985; Ikebuchi et al., 1986). Apart from this other proteins were also known to bind Pb in vivo (Shelton and Egle, 1982; Goering, 1993). The higher concentrations of lead in tissues following occupational or experimental exposure were associated with oxidative damage of DNA, protein and lipid which suggests that lead-induced oxidative stress play a role in lead –induced toxic effects (Monteiro et al. 1986; Patra et al., 2001).

Effects of Lead on respiratory and gastrointestinal system

Lead poisoning occurs as a result of ingestion or inhalation of inorganic lead particles or through transdermal absorption of organic alkyl lead. The respiratory tract provides the most effective route of absorption as it only depends on the size of lead particles and on the metabolic activity of the body. Airborne lead particles that are less than 0.5 to 1 microns in diameters are generally completely absorbed by the alveoli. Gastrointestinal absorption of lead is less effective and depends on a number of factors, for example, the presence of food in the stomach, the concentration of lead ingestion, the nutritional status and the age. The rate of lead absorption increases with iron, zinc and calcium

deficiencies.

Effects of lead on haematopoietic system

Lead may be rapidly absorbed and reached considerable amount in the blood (Haque et al., 2006). Once absorbed, 99% of blood lead is transported to the erythrocytes as lead diphosphate (Freeman, 1970). Increment of blood lead level -following lead acetate and lead nitrate administration was demonstrated in the experimental animals (Ferguson et al., 1998; Sharma et al., 2011d). Some reports suggested that this element is strongly bound to macromolecules in the intracellular compartment because lead binding proteins have been isolated from kidney, liver, blood and brain (Moussa et al., 2001). The half- life of lead differs for each of the compartment, ranging from 25 to 40 days in erythrocytes, 40 days in soft tissues and as many as 28 years in bone. We have also reported the protective effects of *Withania somnifera* root extract supplementation on blood profiles and serological parameters in male mice subjected to lead nitrate (Sharma et al., 2011d). We have also found the therapeutic potential of hydromethanolic root extract of *W. somnifera* on neurological parameters in Swiss albino mice subjected to lead nitrate (Sharma et al., 2011). On the basis of experiments performed in our laboratory we elucidated the protective potential of the hydromethanolic extract of *W. somnifera* in the regulation of lead nitrate induced nephrotoxicity and hepatotoxicity in Swiss albino mice (Sharma et al., 2011b, c).

Lead exposure and generation of oxidative stress

Oxidative stress occurs when generation of free radicals (i.e. substances with one or more unpaired electrons) exceed the capacity of antioxidant defense mechanisms (that is, pathways that provide protection against harmful effect of free radicals). Lead induced oxidative stress has been identified as the primary contributory agent in the pathogenesis of lead poisoning (Xu et al., 2008). Reactive oxygen species (ROS) generated as a result of lead exposure has been identified in liver, kidney, brain, lung, endothelial tissue, testes and sperm.

Lead causes oxidative stress by inducing the generation of ROS, reducing the antioxidant defense system of cells via depleting glutathione, interfering with some essential metal, inhibiting sulfhydryl dependent enzymes or antioxidant enzymes activities or increasing susceptibility of cells to oxidative attack by altering membrane integrity and fatty acid composition (Sharma et al., 2011b, c; Sharma et al., 2011).

Effect of lead on the antioxidant defense system

Although, the mechanism by which lead induce oxidative

stress is not fully understood, a large number of evidences indicate that multiple mechanism balance between reactive oxygen metabolites and antioxidant defense results in "oxidative stress" (Gibananada and Hussain, 2002). Participation of iron in fenton reaction in vivo, leading to production of more reactive hydroxyl radicals from superoxide radicals and H_2O_2 (Halliwell, 1994a) results in increased lipidperoxidation. This might be one of the reasons for significant alteration in lipidperoxidation (LPO)and significant changes in the activity of antioxidant enzymes. Usually the deleterious effects of oxidative stress are counteracted by endogenous antioxidant enzymes, mainly superoxide dismutase (SOD), Catalase (CAT) and glutathione (GSH) (Winterbourn, 1993). The binding activity of lead compounds with oxidative stress factors and the gene erythropoiesis ration of reactive oxygen species, such as hydrogen peroxide and its interaction with different metals and also toxic activity of delta-aminolevulinic acid (ALA) are reported earlier (Ariza et al., 1998; Ding et al., 2000).

SOD and catalase are considered primary enzymes since they are involved in direct elimination of ROS. SOD plays an important role in protecting the cells against the toxic effects of O_2^- by catalyzing its dismutation reactions. The enzyme requires copper and zinc for its activity. Copper ions appear to have a functional role in the reaction by undergoing alternate oxidation and reduction, where zinc ions seem to stabilize the enzyme instead of having a role in the catalytic cycle (Halliwell and Gutteridge, 1989). SOD keeps the concentration of superoxide radicals at low levels and therefore plays an important role in the defense against oxidative stress (Fridovich, 1997). Various findings demonstrated that lead has inhibitory effects on superoxide dismutase and catalase also found to inhibit antioxidant enzymes involved in the prevention of lipid peroxidation such as superoxide dismutase and catalase (Soltanianejad et al., 2003; Vaziri et al., 2003). The biological role of SOD is to dismutase superoxide ion, hydrogen peroxide (H_2O_2), produced in this reaction is eliminated by catalase, one of the most active enzymes in the human organism.

Catalase is a heme protein, which catalyzes the reduction of hydrogen peroxides (converts H_2O_2 to oxygen and water) and protects tissues from highly reactive hydroxyl radicals. Various reports regarding influence of lead on SOD and CAT activities have given divergent results. Some studies showed decreased activities of SOD and CAT (Ramstoeck et al., 1980; Chaurasia and Kar, 1997a; Chaurasia and Kar, 1997b) and others showed increased activities (Adler et al., 1993; Ahamed et al., 2006). Superoxide anions (O_2^-) itself directly affects the activity of catalase and peroxidise by affecting intracellular enzymes (Ghosh and Myers, 1998), creatine phosphokinase (Lee et al., 1998). SOD was found to be decreased in the treated animal's tissues particularly in liver, kidney and testis (Sharma et

al., 2011b, c). A decrease in SOD was explained by direct blocking action of the metal on –SH group of the enzyme (Kasperczyk et al., 2004). Decreased catalase activity observed in dead- exposed animals were attributed to the interference of lead by both processes (Sandhir et al., 1994 and 1995).

The lower activities of CAT and SOD may partly be explained by the interaction between lead and essential metals such as copper, zinc, and iron. Copper and Zinc are essential cofactors for SOD, whereas CAT also contains haem as the prosthetic group, the biosynthesis of which is inhibited by lead (Patil et al., 2006). Several studies reported alterations in antioxidant enzyme activities such as SOD, catalase and glutathione peroxidise (GPX) and changes in the concentrations of some non-enzymatic antioxidant molecules, such as glutathione (GSH) in lead exposed animals (McGowan et al., 1986) and workers (Sugawara et al., 1991; Solliway et al., 1996; Gayathri et al., 2007; Mohammad et al., 2008). These findings suggest a possible involvement of oxidative stress in the pathophysiology of lead toxicity.

One of the effects of lead exposure is on glutathione metabolism. GSH is one of the most important compounds, which helps in the detoxification and excretion of heavy metals. Glutathione is a cysteine-based molecule produced in the interior compartment of the lymphocyte. More than 90 percent of non-tissue sulphur in the human body is found in the tripeptide glutathione (Meister and Anderson, 1983). In addition to acting as an important antioxidant for quenching free radicals, glutathione is a substrate responsible for the metabolism of specific drugs and toxins through glutathione conjugation in the liver (Meister and Anderson, 1983). The sulfhydryl complex of glutathione also directly binds to toxic metals that have a high affinity for sulfhydryl groups. It binds with heavy metals. Patra and Swarup (2000) observed effect of lead on erythrocyte antioxidant defense, lipid peroxide level and thiol groups in calves. Sugawara et al. (1991) have reported a significant decrease in GSH content of erythrocytes from workers exposed to lead. Indirect depletion of GSH may occur when lead inhibits the enzyme and aminolevulinic acid dehydratase (ALAD) before it catalyzes the condensation of two molecules of d-aminolevulinic acid (δ-ALA) to porphobilinogen (Haeger-Aronsen et al., 1971). When the activity of ALAD is inhibited an effect of lead exposure which has been confirmed experimentally by several authors, the amount of δ-ALA increased (Ribarov and Bochev, 1982; Gibbs et al., 1991). Since δ-ALA itself is known to be a potent inducer of lipid peroxidation (LPO) and ROI formation both in vivo and in vitro, its accumulation may facilitate the depletion of GSH from lead- burdened cells (Monteiro et al., 1986, 1989; Hermes- Lima et al., 1991; Oteiza and Bechara, 1993). The involvement of ROS in Pb poisoning has been addressed by Schwartz et al. (2000) who found a decrease in GSH and an increase in oxidized glutathione

(GSSH) concentration in lead acetate treated rats. In addition, they also found that the effect was reduced by treatment with N- acetyl cystein, a precursor of GSH. This provided a possibility of antioxidant therapy for individuals who were exposed to lead. GSH/GSSH ratio is an important component of antioxidant defense system in mammalian cells, which was considered a sensitive indicator of oxidative stress (Wilson et al., 2000).

Mercury, arsenic, and lead effectively inactivate the glutathione molecule so it is unavailable as an antioxidant or as a substrate in liver metabolism (Christie and Costa 1984). Concentrations of glutathione in the blood have been shown to be significantly lower than control levels both in animal studies of lead exposure and in lead-exposed children and adults (Hsu, 1981; Ahamed et al., 2005). Levels of two specific sulfhydryl containing enzymes that are inhibited by lead – deltaaminolevulinic acid dehydrogenase (ALAD) and glutathione reductase (GR) – have been demonstrated to be depressed in both animal and human lead-exposure studies (Farant et al., 1982; Gurer-Orhan et al., 2004; Ahmad et al., 2005).

Lipid peroxidation, a basic cellular deteriorative change, is one of the primary effects induced by oxidative stress and occurs readily in the tissues due to presence of membrane rich in polyunsaturated highly oxidizable fatty acids (Cini et al., 1994). Lead, being a heavy metal and potent environmental pollutant in elicits variety of toxic manifestations in the living systems (Perlstein and Attala, 1966; Choice and Richter, 1972; Quinlan et al., 1988; Acharya et al., 1994). The toxic effects of lead in various tissues/ organs have hardly been believed due to some peroxidative activities, except in few tissues (Quinlan et al., 1988; Acharya et al., 1994).

On the contrary, the generation of elevated quantities of thiobarbituric acid reactive substances from the brain in lead treated mice is possibly due to the presence of high level of poly-unsaturated fatty acids and free iron. Yiin and Lin (1995) demonstrated a significant enhancement of malondialdehyde (MDA) when lead was incubated with linoic, linolenic and arachidonic acid. These initial studies for the first time and subsequent studies on lead exposed animals showed increased lipid peroxidation or decrease in antioxidant defense mechanism (Adegbesan and Adenuga, 2007; Bokara et al., 2008). A number of researchers have also shown enhanced rate of lipid peroxidation in brain of lead exposed rats (Yiin and Lin, 1995; Adegbesan and Adenuga, 2007; Bokara et al., 2008). They also showed that the level of lipid peroxidation was directly proportional to lead concentrations in brain regions (Shafiq-ur-Rehman et al., 1995; Adonaylo and Oteiza, 1999; Saxena and Flora, 2006). Similar effects were shown by Sandhir and Gill (1995) and Sharma et al. (2011b) in liver of lead exposed rats.

Lead binds to plasmic protein, where it causes alterations in high number of enzymes. Georing (1993) found that lead can also perturb protein synthesis in hepatocytes. The decrease in protein content of mice treated with Pb may be due to decreased hepatic DNA and RNA (Shalan et al., 2005). Hassanin (1994) and El-Zayat et al. (1996) reported decrease in hepatic total protein content in response to lead intoxication. They attributed that to a decreased utilization of free amino acids for protein synthesis. B-2-microglobinuria and enzymuria were reported in lead toxicity in children (Gourrier et al., 1991). According to Pachathundikandi and Varghese (2006), lead toxicity results in protein loss. Hassanin (1994) and El- Zafat et al. (1996) observed decrease in hepatic total protein content in response to lead intoxication. This may be because Pb^{2+} disturbs intracellular Ca^{2+} homeostasis (Simons, 1993) and damages the endoplasmic reticulum which in turn results in reduction of protein synthesis. In addition, lead has been shown to enter in cells through voltage dependent Ca^{2+} channels at a higher rate than Ca^{2+} as an intracellular secondary messenger. Interaction between lead and two second messenger mediators of Ca^{2+} signals (Calmodulin and protein kinase C) have been studied extensively (Goldstein, 1993). Calmodulin exhibits a higher affinity for lead than it does for Ca^{2+}, leading to an up regulation of the enzymes (Habermann et al., 1983).

Treatments for lead toxicity

Therapies to remove heavy metals from the body include chelation and supportive measures. Chelation is a chemical process that has applications in many areas, including medical treatment, environmental site rehabilitation, water purification and so forth. Several chelating compounds have been used to manage lead toxicity in the event of manage lead toxicity in the event of exposure but none are suitable in reducing lead burden in chronic lead exposure. Moreover, these chelators in turn are potentially toxic (Gilman et al., 1991) and often fail to remove Pb burden from all body tissues.

Although, lead poisoning has been studied for years, some of the toxic effects still cannot be explained (Aykir-Burns et al., 2003). The use of chelating agents and few antioxidants such as vitamin C and E (Mehta et al., 2001) can enhance lead excretion in lead poisoning but these cannot be routinely recommended as these posses many side effects (Flora and Tandon, 1990). In order to address this problem, natural therapies to promote chelation, detoxification and protection are gaining popularity because of minimal side effects. Medicinal properties of plants have also been investigated in the light of recent scientific developments throughout the world, due to their potent pharmacological activities, low toxicity and economic viability (Janmeda et al., 2011). Thus, there has been increased interest in the therapeutic potential of plant products or medicinal plants having beneficial role in reducing lead poisoning.

Lead toxicity, oxidative damage and health implications. A review

197

CONCLUSION

It was found that lead toxicity is significant but a preventable health problem. Identification of various lead sources that surround us can help towards prevention of lead toxicity. However, lead is also toxic to humans, with the most deleterious effects on the hemopoietic, nervous, hepatic and renal systems. It has now become clear that high to moderate doses of lead exposure induces generation of free radicals resulting in oxidative damage to critical biomolecules, lipids, proteins and DNA. Although, recent studies suggest that oxidative stress due to low levels lead exposure might be involved in many human diseases, the detailed mechanistic studies indicating relevance of oxidative stress markers to lead related human diseases with low exposure still warrant further investigations. Furthermore, work is needed to find the effective and safe intervention for lowering the lead exposure at the general population level.

ACKNOWLEDGEMENTS

The authors are thankful to the authorities of Banasthali University for providing support to the study.

REFERENCES

A.T.S.D.R. (Agency for Toxic Substances and Disease Registry) (1993). Toxicological profile for lead, Update. Prepared by Clement International Corporation under contact no.205-88-060 for ATSDR, U.S. Public Health Services, Atlanta, GA.

Acharya UR, Mishra P, Mishra N (1994). Effect of lead toxicity on testicular ascorbic acids and cholesterol of Swiss mice. Ad. Bios., 13: 1-10.

Adegbesan BO, Adenuga GA (2007). Effect of lead exposure on liver lipid peroxidation and antioxidant defense system of protein-undernourished rats. Biol. Trace Elem. Res., 116: 219-25.

Adeniyi TT, Ajayi GO, Akinloye OA (2008). Effect of Ascorbic acid and *Allium sativum* on tissue lead in female *Rattus navigicus*. Niger. J. Health Biomed. Sci., 7(2): 38-41.

Adler AJ, Barth RH, Berlyne GM (1993). Effect of lead on oxygen free radical metabolism: inhibition of superoxide dismutase activity. Trace Elem. Med., 10: 93-6.

Adonaylo VN, Oteiza PI (1999). Lead intoxication. Antioxidant defenses and oxidative damage in rat brain. Toxicol., 135: 77-85.

Ahamed M, Verma S, Kumar A, Siddiqui MKJ (2006). Delta-aminolevulinic acid dehydratase inhibitionand oxidative stress in relation to blood lead among urban adolescents. Human Exp. Toxicol., 25: 547-553.

Ahmad M, Saleem S, Ahmad AS, Ansari MA, Yousuf S, Hoda MN, Islam F (2005). Neuroprotective effects of *Withania somnifera* on 6-hydroxydopamine induced Parkinsonism in rats. Hum. Exp. Toxicol., 24: 137-147.

Ariza ME, Bijur GN, Williams MV (1998). Lead and Mercury mutagenesis. Role of H_2O_2, superoxide dismutase, and xanthine oxidase. Environ. Mol. Mutagen., 31: 352-361.

Aykir-Burns N, Laegeler A, Kellog G, Ercal N (2003). Oxidative effects of Lead in young and adult fisher 334 rats. Arch. Environ. Contam. Toxicol., 44: 417-420.

Barry PS (1975). A comparison of concentrations of lead in human tissues. Br. J. Ind. Med., 32: 119-139.

Bellinger DC (2008). Very low lead exposures and children's neurodevelopment current opinion in Pediat., 20: 172-177.

Bokara KK, Brown E, McCormick R, Yallapragada PR, Rajanna S,

Bettaiya R (2008). Lead- induced increase in antioxidant enzymes and lipid peroxidation products in developing rat brain. Biometals., 21: 9-16.

Bunn TL, Ladics GS, Holsapple MP (2001). Developmental immunotoxicology assessment in the rat. Age, gender and strain comparisons after exposure to Pb. Toxicol. Met., 11: 41-58.

Chaurasia SS, Kar A (1997a). Protective effects of vitamin E against lead induced deterioration of membrane associated type-1 iodothyronine 5' monodeiodinase (5' D-I) activity in male mice. Toxicol., 124: 203-209.

Chaurasia SS, Kar A (1997b). Influence of lead on type I iodothyromne 5' monadeoidinase activity in male mouse. Horm. Metab. Res., 29: 532-533.

Choice DD, Richter GW (1972). Cell proliferation in mouse kidney induced by lead. Synthesis of deoxyribonucleoic acid. Lac. Invest., 30: 647-651.

Christie NJ, Costa M (1984). In vitro assessment of the toxicity of metal compounds. IV. Disposition of metals in cells: interaction with membranes, glutathione, metallothionein, and DNA. Biol. Trace Elem. Res., 6:139–158.

Cini M, Fariello RY, Bianchettei A, Morettei A (1994). Studies on lipid peroxidation in rat brain. Neurochem. Res., 19: 283.

De Marco M, Halpern R, Barros HMT (2005). Early behavioral effects of lead perinatal exposure in rat pups. Toxicol., 211: 49- 58.

Ding Y, Gonick HC, Vaziri ND (2000). Lead promotes hydroxyl radical generation and Lipid peroxidation in cultured aortic endothelial cells. Am. J. Hypertens., 13: 552-555.

Doyle JJ, Younger RL (1984). Influence of ingested led on distribution of lead, iron, zinc, copper and manganese in bovine tissues. Vet. Human. Toxicol., 26: 201-204.

Ehle AL, McKee DC (1990). Neuropsychological effect of lead in occupational exposed workers. Crit. Rev. Toxico., 20: 237-255.

El-Zyat EM, El-Ymany NA, Kamel ZH (1996). Combined supplementation of zinc and vitamin C as protective agents against lead toxicity in growing male albino rats. 1. Liver functions. J. Egypt. Ger. Soc. Zool., 20(A): 115-139.

Farant JP, Wigfield DC (1982). Biomonitoring lead exposure with ALAD activity ratios. Int. Arch. Occup. Environ. Health, 51: 15-24.

Ferguson SA, Holson RR, Gazzara RA (1998). Minimal behavioral effects from moderate postnatal lead treatment in rats. Neurotoxicol. Teratol., 20: 637-643.

Flora SJ, Tandon SK (1990). Beneficial effects of Zinc supplementation during chelation treatment of Lead intoxication in rats. Toxicol., 64: 129-139.

Freeman R (1970). Chronic lead poisoning in children: a review of 90 children diagnosed in Sydney, 1948-67 II. Clinical features and investigations. Medical J. Australia, 1: 648-681.

Fridovich I (1997). Superoxide anion radical (O_2^-), superoxide dismutases, and related matters. J. Biol. Chem., 272: 18515-18517.

Gayathri M, Rao Beena V, Shetty, Sudha K (2007). Evaluation of lead toxicity and antioxidants in battery workers. Biomed. Res., 19(1): 1-4.

Georing PL (1993). Lead- protein interaction as a basis for toxicity. Neurotoxicol., 14: 45-60.

Ghosh J, Myers E (1998). Inhibition of arachidonate 5-lipoxyenase triggers massive apoptosis in human prostate cancer cells. Proc. Natl. Acad .Sci. USA, 95: 13–182.

Gibananada R, Hussain SA (2002). Oxidants. Ind. J. Exp. Biol., 40: 1213–1232.

Gibbs PNB, Gore MG, Jordan PM (1991). Investigation of the effect of metal ions on the reactivity of thiol groups in human 5-aminolevulinic dehydratase. Biochem. J., 225: 573-580.

Gilman AG, Rall TW, Nies AS (1991). Goodman and Gilman's. The pharmacology basis of therapeutics. Pergamon, New York.

Goldstein GW (1993). Evidence that lead acts as a calcium substitute in second messenger metabolism, Neurotoxicol., 14(3): 97-103.

Gourrier E, Lamour C, Feldmann D, Bensman A (1991). Early tubular involvements in lead poisoning in children. Arch. Fr. Pediatr., 48: 685-689.

Gurer-Orhan H, Sabir HU, Ozgunes H (2004). Correlation between clinical indicators of lead poisoning and oxidative stress parameters in controls and lead- exposed workers. Toxicol., 195: 147-154.

Habermann E, Crowell K, Janicki P (1983). Lead and other metals can substitute for Ca^{+2} in calmodulin. Arch. Toxicol., 54(1): 61-70.

Haeger-Aronsen B, Abdulla M and Fristedt BI (1971). Effect of lead on aminolevulinic acid dehydratase activity in red blood cells. Arch. Environ. Health, 23: 440–445.

Halliwell B, Gutteridge JMC (1989). Free radicals in biology and medicine- 2nd ed. Oxford.Clarendon. Press.

Hamer DH (1986). Metallothionein. Annu. Rev. Biochem., 55: 913-951.

Haque MM, Awal MA, Mostofa M, Sikder MMH, Hossain MA (2006). Effects of calcium carbonate, potassium iodine and zinc sulphate in lead induced toxicities in rat model. Bang. J. Vet. Med., 4(2): 1213-1227.

Hassanin LAM (1994). The effect of Lead pollution on the susceptibility of rats to anticoagulants rodenticides, M.sc., Thesis, Zoology Department, Faculty of Science, Cairo University, Giza, Egypt.

Hermes-Lima M, Valle GRV, Verceri AE, Bechara EJH (1991). Damage to Rat liver mitochondria promoted by and aminolevulinic acid-genrated reactive oxygen species: Connections with acute intermittent porphyria and lead poisoning. Biochem. Biophys. Acta., 1056: 57-63.

Hsu JM (1981). Lead toxicity as related to glutathione metabolism. J. Nutr., 111: 26-33.

Ikebuchi H, Teshima R, Suzuki K, Terao T, Yamane Y (1986). Simultaneous induction of Pb- metallothione like protein and Zn-thioneine in the liver of rats given lead acetate. Biochem. J., 233: 541-546.

Janmeda P, Sharma V, Singh L, Paliwal R, Sharma S, Yadav S (2011). Chemopreventive Effect of Hydro-Ethanolic Extract of *Euphorbia neriifolia* Leaves against DENA-Induced Renal carcinogenesis in Mice. Asian Pacific J. Cancer Prev., 12: 677-683.

Kasperczyk S, Birkner E, Kasperzyk A, Zalejska-Fiolka J (2004). Activity of superoxide dismutase and catalase in people protractedly exposed to lead compounds. Ann. Agric. Environ. Med., 11: 291-296.

Lee YJ, Galoforo SS, Berns CM (1998). Glucose deprivation induced cytotoxicity and alteration in mitogen activated protein kinase activation are mediated by oxidative stress in multidrug resistant human breast carcinoma cells. J. Biol. Chem., 243: 52–94.

Lockitch G (1993). Perspective on lead toxicity. Cline. Biochem., 26: 371-381.

McGowan C, Donaldson WE (1986). Changes in organ nonprotein sulfhydryl and glutathione concentrations during acute and chronic administration of inorganic lead to chicks. Biol. Trace Elem. Res., 10: 37-46.

Mehta A, Flora SJ (2001). Possible role of metal redistribution, hepatotoxicity and oxidative stress in chelating agents induced hepatic and renal metallothionein in rats. Food Chem. Toxicol., 39: 1029-1038.

Meister A, Anderson MD (1983). Glutathione. Annu. Rev. Biochem., 52: 711-760.

Mohammad IK, Mahdi AA, Raviraja A (2008). Oxidative Stress in Painters Exposed to Low Lead Levels. Arh. Hig. Rada. Toksikol. 59: 161-169.

Monteiro HP, Abdalla DSP, Faljoni-Alario A, Bechara EJH (1986). Generation of active oxygen species during coupled autooxidation of oxyhemoglobin and delta-aminolevulinic acid. Biochem. Biophys. Acta., 881: 100–106.

Moreira EG, Vassilieff I, Vassilieff VS (2001). Developmental lead exposure. Behavioral alterations in the short and long term. Neurotox. Teratol., 23: 489-495.

Mousa HM, Al- Qarawi AA, Ali BH, Abdula Rahman HA, Elmougy SA (2002). Effect of lead exposure on the erythrocytic antioxidant levels in goat. J. Vet. Med., A49: 531-534.

National Research council (1972). Lead. Airborne lead in perspective. Committee on biologic effects of atmospheric pollutants. Division of Medical Sciences. National Academy of Sciences, Washington, D.C.

Oteiza PI, Bechara EJH (1993). 5-Aminolevulinic acid induces lipid peroxidation in cardiolipin-rich lipsomes. Arch. Biochem. Biophy., 305: 282-287.

Patil AJ, Bhagwat VR, Patil JA, Dongre NN, Ambekar JG, Jailkhani R, Das KK (2006). Effect of lead (Pb) exposure on the activity of superoxide dismutase and catalase in battery manufacturing workers (BHW) of western Maharashtra (India) with reference to heme biosynthesis. Int. J. Environ. Res. Public Health, 3: 329-337.

Patra RC, Swarup D (2000). Effect of lead on erythrocyte antioxidant defence, lipid peroxide level and thiol groups in calves. Res. Vet. Sci., 68: 71.

Patra RC, Swarup D, Dwidedi SK (2001). Antioxidant effects of α-tocopherol, ascorbic acid and L-methionine on lead-induced oxidative stress of the liver, kidney and brain in rats. Toxicol., 162: 81-88.

Perlstein MA, Attala R (1966). Neurologic sequence of plumbisim in children. Clin. Pediat., 5: 292-298.

Pracheta M, Singh L (2009). Effect of lead nitrate ($Pb(NO_3)_2$ on plant nutrition, as well as physical and chemical parameters on Lobia (*Vigna unguiculata* Linn. Walp.). J. Plant Develop. Sci., 1(1 & 2): 49-56.

Quinlan GJ, Halliwell B, Moorehous CP, Gutteridge JMC (1988). Action of lead (II) and aluminium (III) ions in iron- stimulated lipid peroxidation in liposomes, erythrocytes and rat liver microsomal fractions. Biochem. Biophys. Acta., 962: 196-200.

Ramstoeck ER, Hoekstra WG, Ganther HE (1980). Trialkyl lead metabolism and lipid peroxidation inviovo in vitamin E and selenium deficient rats as measured by ethane production. Toxical. Appl. Toxicol., 54: 251-257.

Rastogi SK (2008). Renal effects of environmental and occupational lead exposure. Indian J. Occup. Environ. Med., 12: 103-106.

Razani-Boroujerdi S, Edwards B, Sopori ML (1999). Lead stimulates lymphocyte proliferation through enhanced T cell- B cell interaction. Pharma. Exp. Ther., 288: 714-719.

Ribarov SR, Bochev PG (1982). Lead-hemoglobin interaction as a possible source of reactive oxygen species—a chemiluminescent study. Arch. Biochem. Biophy., 213: 288-292.

Rosenberg CE, Fink NE, Salibian A (2007). Humoral immune alterations caused by lead. Studies on an adults lead model. Acta. Toxicol. Argent., 15(1): 16-23.

Sandhir R, Gill KD (1995). Effect of lead on lipid peroxidation in liver of rats. Biol. Trace. Elem. Res., 48: 91-97.

Sandhir R, Julka D, Gill KD (1994). Lipoperoxidative damage on lead exposure in rat brain and its implications on membrane bound enzymes. Pharmacol. Toxicol., 74: 66-71.

Saxena G, Flora SJS (2006). Changes in brain biogenic amines and heme-biosynthesis and their response to combined administration of succimer and Centella asiatica in lead poisoning rats. J. Pharm. Pharmacol., 58: 547-559.

Schwartz BS, Lee BK, Lee GS (2000). Associations of blood lead, dimercaptosuccinic acid chelatable. Lead and tibia lead with polymorphisms in the vitamin D receptor and δ- aminolevulinic acid dehydratase genes. Environ. Health Perspect., 108: 949-954.

Senapati SK, Dey S, Dwivedi SK, Swarup D (2001). Effect of garlic (*Allium sativum* L.) extract on tissue Lead level in rats . J. Ethnopharmacol., 76: 229-232.

Shafiq-ur- R, Rehman S, Chandra O, Abdulla M (1995). Evallution of malondialdehyde as an index of lead damage in rat brain homogenates. Biometals., 8: 275-279.

Shalan MG, Mostafa MS, Hassouna MM (2005). Amelioration of lead toxicity on rat liver with vitamin C and silymarin supplements. Toxicol., 206: 1-15.

Sharma S, Sharma V, Pracheta, Sharma SH (2011). Therapeutic Potential of Hydromethanolic Root Extract of Withania somnifera on Neurological Parameters in Swiss Albino Mice Subjected to Lead Nitrate. Int. J. Curr. Pharmaceu. Res., 3: 52-56.

Sharma V, Sharma S, Pracheta, Paliwal R, Sharma SH (2011c). Therapeutic efficacy of *Withania somnifera* root extract in the regulation of lead nitrate induced nephrotoxicity in Swiss albino mice. J. Pharm. Res., 4: 755-758.

Sharma V, Sharma S, Pracheta, Sharma SH (2011b). Lead Induced Hepatotoxicity in Male Swiss Albino Mice: The Protective Potential of the Hydromethanolic Extract of *Withania somnifera*. Int. J. Pharmaceu. Sci. Rev. Res., 7: 116-121.

Shelton K.R, Egle PM (1982). The proteins of lead- induced intranuclear inclusion bodies. J. Biol. Chem., 257: 11802-11807.

Singh NP, Thind IS, Vitale LF, Pawlow M (1976). Lead content of tissues of baby rats born of, and nourished by lead- poisoned mothers. J. Lab. Clin. Med., 87: 273-280.

Solliway BM, Schaffer A, Pratt H (1996). Effects of exposure to lead on selected biochemical and haematological variables. Pharmacol.

Toxicol., 78: 18-22.

Soltanianejad K, Kebriaeezadeh A, Minaiee B (2003). Biochemical and ultrastructural evidences for toxicity of lead through free radicals in rat brain. Hum. Exp. Toxicol., 22: 417-433.

Sugawara E, Nakamura K, Miyake T, Fukumura A, Seki Y (1991). Lipid peroxidation and concentration of glutathione in erythrocytes from workers exposed to lead. Br. J. Ind. Med., 48: 239-242.

Thoreux- Manlay A, Velez de la calle JF, Olivier MF (1995). Impairment of testicular endocrine function after lead intoxication in the adult rat. Toxicol., 100: 101-109.

Vargas I, Castillo C, Posadas F (2003). Acute lead exposure induces renal heme oxygenase-1 and decreases urinary Na+ excretion. Hum. Exp. Toxicol., 22: 237-244.

Vaziri ND, Lin CY, Farmand F, Sindhu RK (2003). Superoxide dismutase, catalase, glutathione peroxidase and NADPH oxidase in lead induced hypertension. Kidney Int., 63: 186-194.

Waalkes MP, Harvey MJ, Klaassen CD (1984). Relative in vitro affinity of hepatic metallothionein for metals. Toxicol. Lett., 20: 33-39.

Waalkes MP, Klaassen CD (1985). Concentrations of metallothionein in major organs of rats after administration of various metals. Fundam App. I. Toxicol., 5: 473-477.

Wilson MA, Johnston MV, Goldstein GW (2000). Neonatal lead exposure impairs development of rodent barrel field cortex. Proc. Natl. Acad. Sci. USA, 97: 5540-5545.

Winterbourn CC (1993). Superoxide as an intracellular sink. Free Rad. Biol. Med., 14: 85-90.

Xu J, Ling-jun L, Chen WU, Xiao-feng W, Wen-yu FU, Lihong X (2008). Lead induces oxidative stress, DNA damage and alteration of p53, Bax and Bcl-2 expressions in mice. Food. Chem. Toxicol., 46: 1488-1494.

Yiin SJ, Lin TH (1995). Lead-catalyzed peroxidation of essential unsaturated fatty acid. Biol Trace Elem Res., 50: 167-172.

Yucebilgic G, Bilgin R, Tamer L (2003). Effects of lead on Na^+- K^+ ATPase and Ca^{2+} ATPase activities and lipid peroxidation in blood of workers. Int. J. Toxicol., 22: 95-97.

Permissions

List of Contributors

K. I Karamba
Department of Biological Sciences, Bayero University, Kano, Nigeria

A. H. Kawo
Department of Biological Sciences, Bayero University, Kano, Nigeria

N. T Dabo
Department of Biological Sciences, Bayero University, Kano, Nigeria

M. D Mukhtar
Department of Biological Sciences, Bayero University, Kano, Nigeria

Vijaya Khader
Department of Foods and Nutrition, Post Graduate and Research Centre, Acharya N. G. Ranga Agricultural University, Rajendranagar, Hyderabad-500 030, India

K. Uma Maheswari
Department of Foods and Nutrition, Post Graduate and Research Centre, Acharya N. G. Ranga Agricultural University, Rajendranagar, Hyderabad-500 030, India

Fahriye Küçükaslan
Institute of Graduate Studies in Pure and Applied Sciences, Marmara University, 34722 Goztepe, Istanbul, Turkey

Hasibe Cingilli Vural
Department of Molecular Biology, Science Faculty, Selcuk University, Kampus Selcuklu, Konya, Turkey

Didem Berber
Institute of Graduate Studies in Pure and Applied Sciences, Marmara University, 34722 Goztepe, Istanbul, Turkey

Zeki Severoğlu
Department of Biology, Faculty of Arts and Sciences, Marmara University, 34722 Goztepe, Istanbul, Turkey

Sabri Sümer
Department of Biology, Faculty of Arts and Sciences, Marmara University, 34722 Goztepe, Istanbul, Turkey

Meltem Doykun
Department of Molecular Biology, Science Faculty, Selcuk University, Kampus Selcuklu, Konya, Turkey

Krishna Raju Patro
Regional Plant Resource Centre, Bhubaneswar, 751 015 Odisha, India

Nibha Gupta
Regional Plant Resource Centre, Bhubaneswar, 751 015 Odisha, India

Nyerhovwo J. Tonukari
Department of Biochemistry, Delta State University, P. M. B. 1, Abraka, Nigeria

AK Bhandari
Herbal Research and Development Institute (HRDI) - Mandal, Gopeshwar, Chamoli, Uttarakhand, India

M Baunthiyal
Department of Biotechnology, G. B. Pant Engineering College- Ghurdauri, Pauri-Garhwal, Uttarakhand, India

VK Bisht
Herbal Research and Development Institute (HRDI) - Mandal, Gopeshwar, Chamoli, Uttarakhand, India

Narayan Singh
Herbal Research and Development Institute (HRDI) - Mandal, Gopeshwar, Chamoli, Uttarakhand, India

JS Negi
Herbal Research and Development Institute (HRDI) - Mandal, Gopeshwar, Chamoli, Uttarakhand, India

Syed Naseer Shah
Genetics and Plant Propagation Division, Tropical Forest Research Institute, Mandla Road, Jabalpur 482 021, India

Amjad M. Husaini
Centre for Plant Biotechnology, Division of Biotechnology, SKUAST-K, Shalimar, Srinagar-191121, J&K, India

S. A. Ansari
Genetics and Plant Propagation Division, Tropical Forest Research Institute, Mandla Road, Jabalpur 482 021, India

R. Yahiaoui-Zaidi
Université de Béjaia, Faculté des Sciences de la Nature et de la Vie, Département de Biologie physico-Chimique,06000 Béjaia, Algérie. Laboratoire de Microbiologie Appliquée, Algeria

R. Ladjouzi
Université de Béjaia, Faculté des Sciences de la Nature et de la Vie, Département de Biologie physico-Chimique,06000 Béjaia, Algérie. Laboratoire de Microbiologie Appliquée, Algeria

S. Benallaoua
Université de Béjaia, Faculté des Sciences de la Nature et de la Vie, Département de Biologie physico-Chimique,06000 Béjaia, Algérie. Laboratoire de Microbiologie Appliquée, Algeria

M. O. Oniya
Department of Biology, Federal University of Technology, P. M. B. 704 Akure, Ondo State, Nigeria

O. Jeje
Department of Biology, Federal University of Technology, P. M. B. 704 Akure, Ondo State, Nigeria

Krishna Bolla
Department of Microbiology, Kakatiya University, Warangal - 506 009, A.P, India

B. V. Gopinath
Department of Microbiology, Kakatiya University, Warangal - 506 009, A.P, India

Syed Zeenat Shaheen
Department of Microbiology, Kakatiya University, Warangal - 506 009, A.P, India

M. A. Singara Charya
Department of Microbiology, Kakatiya University, Warangal - 506 009, A.P, India

Chellamani Muniandi
Pasteur Institute of India, Coonoor-643 103, The Nilgiris, Tamil Nadu, India

Kavaratty Raju Mani
Pasteur Institute of India, Coonoor-643 103, The Nilgiris, Tamil Nadu, India

Rathinasamy Subashkumar
PG and Research Department of Biotechnology, Kongunadu Arts and Science College, Coimbatore – 641 029, TamilNadu, India

Kiran Saini
Toxicology Laboratory, Department of Botany, Kakatiya University, Warangal-506009 (AP), India

S. Kalyani
Toxicology Laboratory, Department of Botany, Kakatiya University, Warangal-506009 (AP), India

M. Surekha
Toxicology Laboratory, Department of Botany, Kakatiya University, Warangal-506009 (AP), India

S. M. Reddy
Toxicology Laboratory, Department of Botany, Kakatiya University, Warangal-506009 (AP), India

V. Koteswara Rao
Department of Microbiology, Kakatiya University, Warangal-506 009, India

P. Shilpa
Department of Microbiology, Kakatiya University, Warangal-506 009, India

S. Girisham
Department of Microbiology, Kakatiya University, Warangal-506 009, India

S. M. Reddy
Department of Microbiology, Kakatiya University, Warangal-506 009, India

Kazhila C. Chinsembu
Department of Biological Sciences, Faculty of Science, University of Namibia, Windhoek, Namibia

Marius Hedimbi
Department of Biological Sciences, Faculty of Science, University of Namibia, Windhoek, Namibia
Multidisciplinary Research Centre, Science and Technology Division, University of Namibia, P/Bag 13301, Windhoek,Namibia

O. L. Shanmugasundaram
Department of Textile Technology, KSR College of Technology, Tiruchengode-637 215, Tamil Nadu, India

R. V. Mahendra Gowda
VSB Engineering College, Karur-639 111, Tamil Nadu, India

D. Saravanan
Department of Biotechnology, KSR College of Technology, Tiruchengode-637 215, Tamil Nadu, India

Shelly Goomber
Department of Biotechnology, Sector 14, Panjab University, Chandigarh 160014, India

Pushpender K. Sharma
Department of Biotechnology, Sector 14, Panjab University, Chandigarh 160014, India
Indian Institute of Sciences Education and Research, S.A.S. Nagar, Sector-81, Mohali, Punjab, 140306, India

Monika Sharma
Department of Biotechnology, Sector 14, Panjab University, Chandigarh 160014, India

Ranvir Singh
National Centre for Human Genome Studies and Research (NCHGSR), Sector 14, Panjab University, Chandigarh 160014, India

Jagdeep Kaur
Department of Biotechnology, Sector 14, Panjab University, Chandigarh 160014, India

Chinmaya Mahapatra
Dr. D. Y. Patil Biotechnology and Bioinformatics Institute, Dr. D. Y Patil University, Pune, India

Anuradha S. Tripathy
Immunology Research Group, National Institute of Virology, Indian Council of Medical Research, Pune, India

E. Y. Aderibigbe
Department of Microbiology, University of Ado-Ekiti, P. M. B. 5363, Ado-Ekiti, Nigeria

W. Visessanguan
Food Biotechnology Research Unit, National Center for Genetic Engineering and Biotechnology (BIOTEC), Pathumthani, Thailand

P. Sumpavapol
Department of Food Technology, Faculty of Agro-Industry, Prince of Songkla University, Hat Yai, Songkhla, 90112, Thailand

K. Kongtong
Food Biotechnology Research Unit, National Center for Genetic Engineering and Biotechnology (BIOTEC), Pathumthani, Thailand

O. I. OYEWOLE
Department of Biochemistry, Osun State University, Osogbo, Nigeria

F. K. AGBOOLA
Department of Biochemistry, Obafemi Awolowo University, Ile-Ife, Nigeria

Archana A. Gurav
Department of Technology, Shivaji University, Kolhapur 416004, India

Jai S. Ghosh
Department of Microbiology, Shivaji University, Kolhapur 416004, India

Girish S. Kulkarni
Department of Technology, Shivaji University, Kolhapur 416004, India

Bimrew Asmare
Department of Animal Production and Technology, Bahir Dar University, Ethiopia

R. Abbasalipourkabir
Department of Biochemistry, Faculty of Medicine, Hamedan University of medical Science, 65178-3-8736 Hamedan, Iran

A. Salehzadeh
Department of Biochemistry, Faculty of Medicine, Hamedan University of medical Science, 65178-3-8736 Hamedan, Iran

Rasedee Abdullah
Universiti Putra Malaysia (UPM), Malaysia

Nermeen Mahmoud Ahmed Abdallah
Departments of Medical Microbiology and Immunology, Ain Shams University, Cairo 11341, Egypt

Shereen Bendary Elsayed
Departments of Medical Microbiology and Immunology, Ain Shams University, Cairo 11341, Egypt

Manal Mohamed Yassin Mostafa
Departments of Medical Microbiology and Immunology, Ain Shams University, Cairo 11341, Egypt

Ghada Metwally El-gohary
Internal medicine, Faculty of Medicine, Ain Shams University, Cairo- Egypt

H. Khosravinia
Department of Animal Sciences, Faculty of Agricultural Sciences, Lorestan University, Khorrambad-68159, Lorestan, Iran

M. Manafi
Department of Animal Science, Faculty of Agricultural Sciences, Malayer University, Malayer, Iran

E. Rafiei Alavi
Department of Pathology, Faculty of Medicine, Lorestan University of Medical Sciences, Khorramabad, Iran

C. D. Mathew
Department of Biochemistry and Molecular Biology, Faculty of Medicine, University of Colombo, Sri Lanka

R. M. S. Gunathilaka
Department of Biochemistry and Molecular Biology, Faculty of Medicine, University of Colombo, Sri Lanka

Sadhana Sharma
Department of Bioscience and Biotechnology, Banasthali University, Banasthali- 304022, Rajasthan, India

Veena Sharma
Department of Bioscience and Biotechnology, Banasthali University, Banasthali- 304022, Rajasthan, India

Ritu Paliwal
Department of Bioscience and Biotechnology, Banasthali University, Banasthali- 304022, Rajasthan, India

Pracheta
Department of Bioscience and Biotechnology, Banasthali University, Banasthali- 304022, Rajasthan, India